D1716650

HEAVY ION REACTIONS
Lecture Notes

HEAVY ION REACTIONS
Lecture Notes
(In two volumes)

HEAVY ION REACTIONS
Lecture Notes

Volume I: The Elementary Processes

Part I, Elastic and Inelastic Reactions
Part II, Transfer Reactions

Ricardo A. Broglia
Dipartimento di Fisica, Università di Milano, and
INFN Sez. Milano, and
The Niels Bohr Institute, University of Copenhagen

and

Aage Winther
The Niels Bohr Institute
University of Copenhagen

ADDISON-WESLEY PUBLISHING COMPANY
The Advanced Book Program
Redwood City, California • Menlo Park, California • Reading, Massachusetts
New York • Don Mills, Ontario • Wokingham, United Kingdom • Amsterdam
Bonn • Sydney • Singapore • Tokyo • Madrid • San Juan

Publisher: *Allan M. Wylde*
Production Manager: *Jan V. Benes*
Marketing Manager: *Laura Likely*
Electronic Composition: *Peter K. Vacek*

Library of Congress Cataloging-in-Publication Data

Broglia, Ricardo A.
  Heavy Ion Reactions, Part I and
  Part II/Ricardo A. Broglia, Aage Winther.
  Includes bibliographical references.
  I. Heavy Ion Collisions. II. Winther, Aage. III. Title.
  QC794.8.H4B76   1991   91-18387
  539.7'234–dc20
  ISBN 0-201-51392-7

Part I of this book was originally published under the title *Heavy Ion Reactions, Volume I: Elastic and Inelastic Reactions* in 1981 by The Benjamin Cummings Publishing Company, Inc.
Part II consists of new material and was prepared using the TₑX typesetting language.

ABCDEFGHIJ-MA-943210

Frontiers in Physics

DAVID PINES/Editor

Volumes of the Series published from 1961 to 1973 are not officially numbered. The parenthetical numbers shown are designed to aid librarians and bibliographers to check the completeness of their holdings.

Titles published in this series prior to 1987 appear under either the W. A. Benjamin or the Benjamin/Cummings imprint; titles published since 1986 appear under the Addison-Wesley imprint.

(28)	T. Loucks	Augmented Plane Wave Method: A Guide to Performing Electronic Structure Calculations—A Lecture Note and Reprint Volume, 1967
(29)	Y. Ne'eman	Algebraic Theory of Particle Physics: Hadron Dynamics in Terms of Unitary Spin Current, 1967
(30)	S. L. Adler R. F. Dashen	Current Algebras and Applications to Particle Physics, 1968
(31)	A. B. Migdal	Nuclear Theory: The Quasiparticle Method, 1968
(32)	J. J. J. Kokkedee	The Quark Model, 1969
(33)	A. B. Migdal V. Krainov	Approximation Methods in Quantum Mechanics, 1969
(34)	R. Z. Sagdeev and A. A. Galeev	Nonlinear Plasma Theory, 1969
(35)	J. Schwinger	Quantum Kinematics and Dynamics, 1970
(36)	R ⸱. Feynman	Statistical Mechanics: A Set of Lectures, 1972
(37)	R. P. Feynman	Photo-Hadron Interactions, 1972
(38)	E. R. Caianiello	Combinatorics and Renormalization in Quantum Field Theory, 1973
(39)	G. B. Field, H. Arp, and J. N. Bahcall	The Redshift Controversy, 1973
(40)	D. Horn F. Zachariasen	Hadron Physics at Very High Energeis, 1973
(41)	S. Ichimaru	Basic Principles of Plasma Physics: A Statistical Approach, 1973 (2nd printing, with revisions, 1980)
(42)	G. E. Pake T. L. Estle	The Physical Principles of Electron Paramagnetic Resonance, 2nd Edition, completely revised, enlarged, and reset, 1973 [cf. (9)—1st edition]

Volumes published from 1974 onward are being numbered as an integral part of the bibliography.

43	R. C. Davidson	Theory of Nonneutral Plasmas, 1974
44	S. Doniach E. H. Sondheimer	Green's Functions for Solid State Physicists, 1974
45	P. H. Frampton	Dual Resonance Models, 1974
46	S. K. Ma	Modern Theory of Critical Phenomena, 1976
47	D. Forster	Hydrodynamic Fluctuations, Broken Symmetry, and Correlation Functions, 1975
48	A. B. Migdal	Qualitative Methods in Quantum Theory, 1977
49	S. W. Lovesey	Condensed Matter Physics: Dynamic Correlations, 1980
50	L. D. Faddev A. A. Slavnov	Gauge Fields: Introduction to Quantum Theory, 1980
51	P. Ramond	Field Theory: A Modern Primer, 1981 [cf. 74—2nd ed.]
52	R. A. Broglia A. Winther	Heavy Ion Reactions: Lecture Notes Vol. I, Elastic and Inelastic Reactions, 1981
53	R. A. Broglia A. Winther	Heavy Ion Reactions: Lecture Notes Vol. II, 1990
54	H. Georgi	Lie Algebras in particle Physics: From Isospin to Unified Theories, 1982
55	P. W. Anderson	Basic Notions of Condensed Matter Physics, 1983
56	C. Quigg	Gauge Theories of the Strong, Weak, and Electromagnetic Interactions, 1983
57	S. I. Pekar	Crystal Optics and Additional Light Waves, 1983

EDITOR'S FOREWORD

The problem of communicating in a coherent fashion recent developments in the most exciting and active fields of physics continues to be with us. The enormous growth in the number of physicists has tended to make the familiar channels of communication considerably less effective. It has become increasingly difficult for experts in a given field to keep up with the current literature; the novice can only be confused. What is needed is both a consistent account of a field and the presentation of a definite "point of view" concerning it. Formal monographs cannot meet such a need in a rapidly developing field, while the review article seems to have fallen into disfavor. Indeed, it would seem that the people most actively engaged in developing a given field are the people least likely to write at length about it.

FRONTIERS IN PHYSICS was conceived in 1961 in an effort to improve the situation in several ways. Leading physicists frequently give a series of lectures, a graduate seminar, or a graduate course in their special fields of interest. Such lectures serve to summarize the present status of a rapidly developing field and may well constitute the only coherent account available at the time. Often, notes on lectures exist (prepared by the lecturer himself, by graduate students, or by postdoctoral fellows) and are distributed on a limited basis. One of the principal purposes of the FRONTIERS IN PHYSICS Series is to make such notes available to a wider audience of physicists.

It should be emphasized that lecture notes are necessarily rough and informal, both in style and content; and those in the series will prove no exception. This is as it should be. The point of the series is to offer new, rapid, more informal, and, it is hoped, more effective ways for physicists to teach one another. The point is lost if only elegant notes qualify.

The informal monograph, representing an intermediate step between lecture notes and formal monographs, offers an author the opportunity to present his views of a field which has developed to the point where a summation might prove extraordinarily fruitful but a formal monograph might be feasible or desirable.

The above words, written nearly thirty years ago, continue to be applicable. During the past decade, the study of heavy ion reactions has emerged as a major subfield of nuclear physics. A wide variety of nuclear

ix

phenomena, ranging from Coulomb excitations and quasielastic transfer reactions to fusion and deep inelastic reactions come into play. In this volume, and its forthcoming companion, Professor Broglia and Winther, who have themselves made a number of significant contributions to our understanding of heavy ion reactions, provide a unified theoretical account of the field. The first four chapters of the present volume, which were first published in 1981, are here reprinted with minor corrections; they describe elastic and inelastic reactions. Chapter 5, which is published here for the first time, completes their description of elementary processes in heavy ion reactions. It deals with the fascinating topic of transfer reactions in which pairing phenomena, analogous to those encountered in superconductivity, play a central role. Special emphasis is placed throughout on the description of the reactions in terms of the elementary excitations which are involved and a considerable effort is made to make the material accessible to the beginning graduate student as well as the experienced researcher. It is a pleasure to welcome Professors Broglia and Winther once more to ranks of contributors to FRONTIERS IN PHYSICS.

DAVID PINES
Urbana, Illinois
June, 1990

PREFACE

The present volume is the first of two in which we have tried to give a survey of the field of heavy-ion reactions. The text is based on notes from lectures given in different forms since 1975 in one-semester courses in Copenhagen, Milano and Stony Brook. It aims at giving a unified description of heavy-ion reactions ranging from grazing collisions to deep-inelastic and fusion processes.

With this in mind we introduce from the outset classical approximations to describe the reaction mechanism and a description of the nuclear structure in terms of elementary modes of excitation. Although these approximations are not needed in connection with several of the subjects dealth with in the present volume, they seem to provide the natural framework for the discussion of the more complicated reactions to be dealt with in Volume II.

The description of the nuclear structure in terms of elementary modes of excitation provides a direct connection between the reactions in which single quanta are involved and those more violent collisions in which the different modes are strongly excited, leading to coherent states which behave almost classically. Using at the same time a classical treatment of the relative motion, one can establish the connection with macroscopic descriptions of strongly damped reactions.

Part I of the present volume (Chapters I–IV) which deals with elastic and inelastic reactions have been published before and is here reproduced with only minor corrections.

Part II (Chapter V) on the other hand is new, and is published here for the first time. It deals with the transfer processes which play a central role in grazing collisions.

Progress in the study of transfer reactions and on the spectroscopic information they carry have been hindered as compared to studies of inelastic scattering because of technical reasons. In fact, due to the recoil effects present in a transfer process, the nuclear and relative motion degrees of freedom are badly mixed, leading to major problems.

The discussion of these problems is one of the central subjects of the present volume. A simple, yet accurate, description of finite-range form factors, including recoil, is presented. It resembles, to a large extent, the description of the inelastic processes and associated differential cross sections discussed in Part I. In this sense, Parts I and II of the present volume form a unity, providing a simple parametrization of the building blocks of all heavy-ion reactions at moderate energy.

Finally, in Part II of the present volume, the subject of pair transfer is treated. The phenomena of pairing vibrations and rotations are amenable to a macroscopic description which parallels the well-established collective model of surface vibrations and rotations. The treatment of the pairing elementary modes of excitation in classical terms is novel and is presented here for the first time.

The results contained in the present volume provide all the elements needed to carry out a microscopic description of the more violent processes taking place in heavy-ion reactions at low energies, like fusion and deep inelastic reactions, and should eventually make it possible to relate these "macroscopic" scale phenomena to the detailed motion of single nucleons. This will be the subject of Vol. II.

Concerning the notation, we have divided each chapter, which is labeled by a Roman numeral, into sections which are labeled by Arabic numbers. Each section in turn may be broken down into subsections identified by Latin letters. Equations are identified by the number of the chapter and that of the section, aside from the sequential number. Thus (II.1.20) labels the twentieth equation of the first section of Chapter II. When referring to an equation within the same chapter, the chapter number is omitted; within the same section, both the chapter and section numbers are omitted. Figures and Tables are numbered sequentially within each chapter.

During the four years of preparation of Part I we have received invaluable help from Marcello Baldo, Pier Francesco Bortignon, Carlos Dasso, Henning Esbensen, Stephen Landowne, Roberto Liotta, Bjørn Nilsson, Giovanni Pollarolo, and Andrea Vitturi. We are indebted to many members and guests of the Niels Bohr Institute for discussions. Especially we would like to mention Aage Bohr, Jakob Bondorf, Per Rex Christensen, Jerry Garrett, Ole Hansen, Sidney Kahana, Rudi Malfliet, Ben Mottelson, Philip Siemens, and Robert Stockstad.

In the eight years which the preparation of Part II has taken, we have been helped in the task by a number of people. We thus wish to acknowledge the cooperation of M. Baldo, P. R. Christensen, C. H. Dasso, F. dos Aidos, H. Esbensen, P. Lotti, G. Pollarolo, J. M. Quesada, A. Rapisarda, J. H. Sørensen and E. Vigezzi.

The many versions of the manuscript were skillfully typed by Lise Madsen. Her support throughout the entire process has been invaluable.

The final version of the manuscript was prepared by Ms. Althea Tate, and we are indebted to her for the very professional and careful work she has done. The art work has been carried out by H. Olsen whose advise we have followed in many cases which is here gratefully acknowledged. We also want to express our gratitude to the Carlsberg Foundation for a grant, and INFN Sez. Milano for support.

RICARDO A. BROGLIA
AAGE WINTHER

CONTENTS

Contents

Contents

Part I:
Elastic and Inelastic Reactions

CHAPTER I

NUCLEAR PHENOMENA
IN HEAVY-ION COLLISIONS

Heavy-ion physics deals with the phenomena that occur when two nuclei are brought into contact such that the nuclear forces that hold the protons and neutrons together within one nucleus are felt by the other nucleus.

The couple of hundred different heavy nuclei that exist in nature practically never come in contact with each other at the temperatures that occur in our solar system, the only exception being reactions produced by cosmic rays. In recent years machines have been constructed which are able to accelerate nuclei to velocities where they can overcome the Coulomb repulsion and get to distances of the target nuclei where heavy-ion reactions can occur. The energy needed to bring two uranium ions into contact is about 1500 MeV, i.e. 6.3 MeV per nucleon, corresponding to a velocity of 12% of the velocity of light. The development in heavy-ion accelerators has been matched by a development in the detection techniques which will eventually make it possible to disentangle the information contained in the reaction products, to a high degree of accuracy. In the following we shall, unless specifically mentioned, only discuss heavy-ion reactions at bombarding energies $\lesssim 10$ MeV per nucleon.

During the last forty years a major research effort has gone into the study of nuclei by means of probes which excite them in a moderate way, mainly by bombardment with light ions such as protons, deuterons, and alpha particles. These investigations have led to a rather detailed picture of nuclei close to their ground state, and shown the rich variety of phenomena displayed by the nuclear many-body system. A special feature of the nuclear system is that the field in which protons and neutrons move in the nucleus is created by the nucleons themselves, in contrast e.g. to atoms, where the electrons move in a field which is to a large extent produced by the atomic nucleus. The richness of the nuclear phenomena manifests itself through the fact that the nucleus displays both the degrees

1

of freedom associated with the single-particle motion, as observed also for electrons in the atom, and the strong collective degrees of freedom as a droplet of quantum liquid.

When two such systems are brought into contact in a heavy-ion collision, one is in many respects entering a new field of physics. It may be tempting to compare heavy-ion physics with chemistry, where one studies the reactions between colliding molecules. In the case of "nuclear chemistry" one would not in general form new stable entities, but on a short time scale this field of research is expected to be of similar richness.

A great variety of phenomena can arise in heavy-ion collisions. In gentle encounters, when the two nuclei barely touch each other and where they essentially keep their identity, one can study the interplay between the degrees of freedom of the two nuclei, with a possibility of varying the physical situation in a practically unlimited number of ways. By appropriately selecting targets and projectiles one may be able to specifically excite different degrees of freedom and search for new ones. On the other hand, collisions which bring the two nuclei into more intimate contact lead to a combined nuclear matter system which at a late stage of the reaction may look like a normal nucleus which has been excited to a state of very high angular momentum and which is strongly deformed. The transition from the situation in which the two nuclei maintain their identity to the situation in which they become a composite system is expected to be a main theme of research in the field of heavy-ion physics.

Collisions between heavy ions have already been studied for a wide variety of target projectile combinations but in general with rather poor resolution for heavier projectiles. The experience gathered from these experiments leads to a consistent picture when the reactions are classified into categories such as grazing collisions, deep-inelastic reactions, and fusion.

1 COULOMB EXCITATION AND LOW-FREQUENCY SURFACE MODES

Some of the basic features of heavy-ion reactions can be understood in terms of an interaction potential between the centers of mass of the two colliding nuclei consisting of a Coulomb repulsion an a short-range nuclear attraction. A typical example is shown in Fig. 1.

The interaction potential (solid line) displays a maximum (the Coulomb barrier), and the two colliding nuclei must have an energy of relative motion which exceeds this barrier in order to form a composite system. The necessary bombarding energy for different projectiles is indicated in Fig. 2, as a function of the mass number A_T of the target. The numbers read from this curve differ from those in Fig. 1 because of the transformation from the center-of-mass to the laboratory system. This is also the reason why the curves in Fig. 2 turn upwards for small values of A_T.

If two heavy ions scatter with an energy of relative motion below, but close, to the Coulomb barrier, the main interaction between them is the point Coulomb interaction (displayed in Fig. 1 for $r > 10$ fm), which gives rise to a hyperbolic (Rutherford) trajectory of relative motion. Due to the finite extension of the interacting nuclei, the electric field may cause excitations in both systems. Because of the simplicity of the Coulomb interaction, the experimental results from Coulomb excitation can be analyzed in a model independent way with great accuracy, providing unambiguous information about the electromagnetic properties of nuclear states. During the last thirty years this process has been extensively studied with lighter ions and has played a crucial role in establishing the properties of collective low-lying states of nuclei throughout the mass table.

These collective states correspond to excitations of low-frequency surface modes. For nuclei which possess an equilibrium shape which is not spherically symmetric, the lowest excited states form a rotational band, by rotating the nucleus to states of higher angular

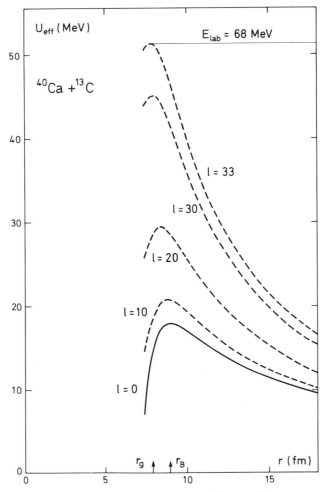

Fig. 1 Ion-ion potential for elastic scattering of ^{13}C on ^{40}Ca. The solid curve labeled $l = 0$ is the ion-ion potential obtained as a sum of the Coulomb repulsion and an exponential nuclear attraction consistent with the simplest picture of nuclear overlap. The curve is discontinued at the point where the nuclear attraction reaches the maximum value expected from nuclear surface tension. The other curves show the effective radial potential for higher values of the angular momentum l of relative motion. It is seen that for the bombarding energy of 68 MeV all impact parameters below the value $\rho = 6.6$ fm, corresponding to $l = 33$, will lead to a large overlap of the two nuclei. In these collisions the reactions will be violent and may lead to fusion of the two ions. For angular momenta above this

momentum. In the simplest picture the intrinsic structure of the deformed nucleus is independent of the rotation and the spectrum will be that of a rigid rotor with constant moment of inertia.

For both spherical and deformed nuclei the surface can vibrate around its equilibrium shape, giving rise to low-lying states of vibrational type. Depending on the number of nodes around the nuclear surface, these vibrations can be classified into quadrupole, octupole, etc. modes. In the simplest picture these vibrations give rise to a harmonic spectrum with equidistant level spacing determined by the basic frequency. Each quantum carries the angular momentum characteristic of the multipolarity of the mode (i.e. 2 for quadrupole, 3 for octupole, etc.).

Coulomb excitation below the Coulomb barrier is specifically suited to excite the rotational and low-frequency vibrational modes. Because of the slow variation of the Coulomb field in space, it can only excite modes of low multipolarity. Similarly, the slow variation of the Coulomb field in time in heavy-ion collisions makes it impossible to excite states of high frequency, since the process in this limit becomes rapidly adiabatic.

The Coulomb field at the target nucleus is proportional to the charge of the projectile, and the action (χ) responsible for the Coulomb excitation therefore increases proportionally to this charge. For heavy-ion projectiles this quantity becomes so large that the target nucleus can be excited through several steps to members of rotational and vibrational bands containing many quanta. This is illustrated in Figs. 3 and 5. In Fig. 3 we have plotted the action of the quadrupole component of the Coulomb field as a function of the charge of the projectile Z_p for the ground-state rotational band of two typical deformed nuclei. The action is given in units such that it directly measures the expected maximum transfer of angular momentum to the target nucleus. While deformed nuclei show a simple rotational spectrum at low angular momenta, the Coriolis and centrifugal forces are known to affect the intrinsic motion for high angular momenta, giving rise to major deviations from the simple behaviour (cf. Fig. 4). The

---

grazing angular momentum l_g the turning point in the classical radial motion will lie outside the distance r_g. The quasielastic reactions thus mainly take place in the region around $r = r_g$ and for l-values around $l_g = 33$. For bombarding energies above 68 MeV the maximum in the effective potential is expected to disappear, leading to a situation where a second (nuclear) rainbow angle may appear in the deflection function.

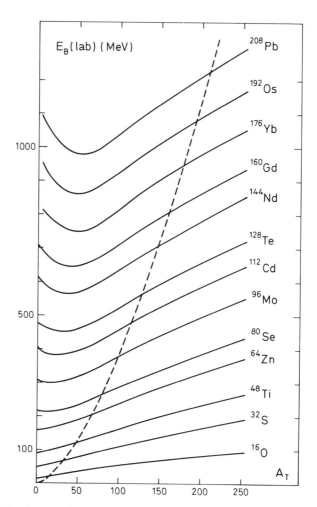

Fig. 2 The bombarding energy necessary to reach the Coulomb barrier for different projectiles labeling the curves, as a function of the target mass number. The mimima in the curves indicate that below the corresponding bombarding energy the projectile does not reach the Coulomb barrier of any target. The dashed line indicates the points in the plot where the masses of projectile and target are equal.

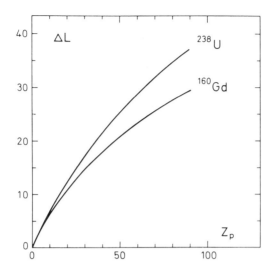

Fig. 3 Coulomb excitation of deformed nuclei. The expected maximum angular-momentum transfer ΔL to rotational motion in ^{238}U and ^{160}Gd is plotted as a function of the projectile charge at a bombarding energy below the Coulomb barrier (cf. Fig. 1), so that no interference from nuclear reactions is expected. With uranium ions on a uranium target one may thus expect to populate states of spin 36 in the ground-state rotational band. In estimating ΔL the energy of excitation has been neglected. This is expected to reduce the maximum angular momentum by a few units.

detailed study of these deviations, which can be carried out through heavy-ion induced Coulomb excitation, give important information about the phase transitions associated with the disappearance of superfluidity and with the alignment of single-particle degrees of freedom along the rotational axis which seems to take place at high angular momenta.

In Fig. 5 we have shown the square of the action of the Coulomb field for three nuclei which display low-energy quadrupole vibrations. The quantity χ^2 is the average number $\langle N \rangle$ of vibrational quanta excited in the target nucleus when bombarded by ions of charge Z_p at an energy close to the Coulomb barrier. The probability distribution on the various excited states is a Poisson distribution, and one expects therefore a sizable probability of exciting states even with higher numbers of quanta than

7

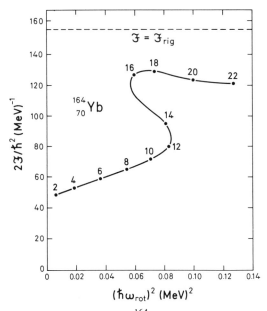

Fig. 4 Measured moment of inertia for ^{164}Yb as a function of the square of the rotational frequency. This nucleus is a typical example of a nucleus displaying an intrinsic quadrupole deformation which is prolate and axially symmetric. In the simple picture of the rigid rotation the spectrum would show the energy levels $E_I = \frac{1}{2}\hbar^2 I(I+1)/\mathscr{I}$ with spin $I = 0, 2, 4, \ldots$ and with a constant moment of inertia $\mathscr{I}$. The observed moment of inertia is here displayed as a function of the square of the rotational frequency, ω_{rot}, defined by the classical relation $\omega_{rot} = \partial E/\partial I$. The numbers labeling the points on the curve are the spin quantum numbers I of the observed levels. For low angular momenta ($I \lesssim 10$) the moment of inertia shows a slow linear increase, whereafter the rotational motion suddenly undergoes large changes; this is called "backbending". The dashed line indicates the value of the moment of inertia expected if ^{164}Yb rotated like a rigid body. The figure is from Bohr and Mottelson (1974).

$\langle N \rangle$. The estimate given in this figure is based on the harmonic picture, where e.g. the three states of spin $I = 0, 2$ and 4 corresponding to two quadrupole quanta (i.e. $N = 2$) are degenerate with twice the energy of the first 2^+ state. The actual spectra are known to deviate from this simple picture, resulting e.g. in a splitting of these levels. Through heavy-ion Coulomb excitation one may study the important questions of anhar-

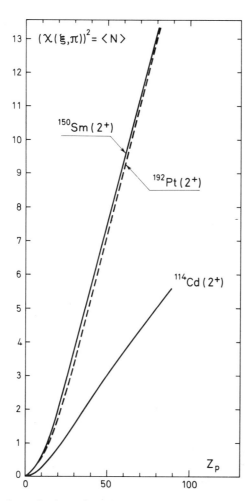

Fig. 5 Coulomb excitation of surface vibrational states in "soft" spherical nuclei. The expected average number of vibrational quanta excited in the two nuclei is plotted as a function of the projectile charge. The excitation is expected to be distributed over different numbers of quanta, $\langle N \rangle$, with a width equal to $\sqrt{\langle N \rangle}$. For each value of $\langle N \rangle$ several states of different spin ranging from 0 to $2\langle N \rangle$ are expected, the highest spin being most favored statistically.

monicities of the vibrational spectrum and of the transition that is expected to take place for the high-spin members of the many-phonon states into a rotationlike spectrum.

Special interest is attached to the excitation of vibrational states in the heaviest deformed nuclei which may undergo fission. In Fig. 6 is shown the cross-section for the excitation of the states containing five and six quadrupole vibrational quanta in ^{234}U as a function of the projectile charge for a bombarding energy close to the Coulomb barrier. In the pure vibrational picture these states will lie close to the fission barrier. The role played by the low-frequency vibrational mode in the mechanism of the fission process is an open question which can be studied by heavy-ion Coulomb excitation.

Through the Coulomb excitation processes one may also study low-lying octupole vibrational modes, though in a somewhat less efficient way than in the case of quadrupole modes. It may be possible also to study excitations caused by the magnetic field associated with the motion of the projectile, a process which has hitherto not been observed.

The well-defined experimental conditions of Coulomb excitation offer rich possibilities for studying the phenomena that take place after the excitation, when the target nucleus recoils through the target material. Heavy-ion Coulomb excitation is especially interesting in this context because of the very large recoil velocities involved. Coulomb excitation with heavy ions has a new facet as compared to the same reaction induced by light ions in that the projectile is also excited in the field of the target nucleus. This projectile excitation offers numerous new possibilities for the detection of Coulomb excitation processes.

The Coulomb excitation process is also important for collisions where the nuclear surfaces come into contact. The process has in this case to be disentangled from the nuclear reactions taking place at these distances.

The situation which arises when dealing with reactions that happen above the Coulomb barrier can be appreciated by noting that even below the Coulomb barrier the elastic scattering deviates strongly from the Rutherford cross section because of the strong Coulomb excitation of collective degrees of freedom of the two nuclei (cf. Fig. 7). This strong deviation from the elastic Rutherford cross section does not imply any significant change in the trajectory of relative motion from the classical

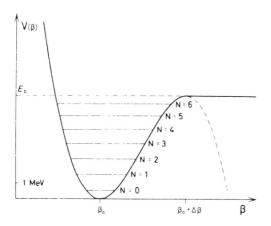

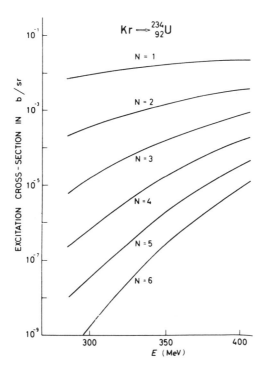

Fig. 6 Cross section for the multiple Coulomb excitation of the beta vibration in the deformed nucleus ^{234}U. In the lower part of the figure the differential cross section in barns/sr is given for backward scattering of various projectiles at the Coulomb barrier. The calculation was made in a pure vibrational model including the finite energy loss. In the upper part of the figure is given a schematic picture of the position of the multiphonon beta-vibrational states used in the calculation. The figure is from Beyer and Winther (1969).

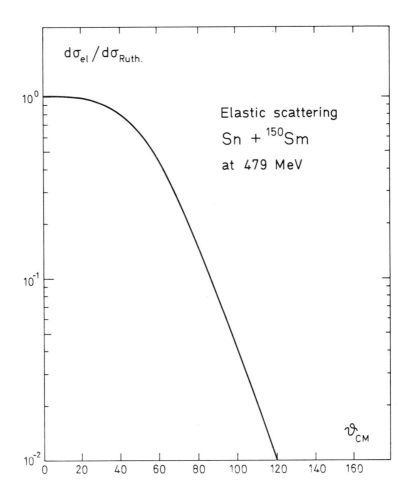

Fig. 7 Elastic scattering of 479-MeV ^{120}Sn ions on ^{150}Sm. The expected angular distribution is given in units of the Rutherford cross section at the same angle. The bombarding energy is below the Coulomb barrier, and the depopulation of the ground state (elastic channel) is entirely due to Coulomb-excitation.

hyperbolic orbit, but is solely due to the depopulation of the ground state through Coulomb excitation.

The subject of Coulomb excitation will be summarized in Chap. II.

2 QUASIELASTIC REACTIONS

In all heavy-ion experiments above the Coulomb barrier one has been able to identify a characteristic type of reaction products where the projectile has lost only a moderate amount of energy and has exchanged only a few nucleons with the target nucleus. These reactions are called quasielastic reactions and are assumed to correspond to collisions in which the surfaces of the two ions have just been in grazing contact.

Detailed analyses of these reactions have mostly been performed with relatively light projectiles where the effects associated with Coulomb excitation are of minor importance. We shall consider first such reactions where the trajectory of relative motion can be described by the elastic scattering in the combined field of Coulomb repulsion and a nuclear attraction (Fig. 1) and where the depopulation of the ground state is accounted for by an absorption.

(a) Elastic scattering

Systematic analyses of elastic scattering in terms of these ingredients have been carried out for reactions induced by a variety of projectiles. A typical example is shown in Fig. 8. The nuclear attractive potentials extracted from such (distorted-wave) analyses show systematic behavior which compares quantitatively with the potential obtained by folding the densities of the two colliding nuclei with an effective nucleon-nucleon force. The absorptive potential shows large variations from case to case, but its action outside the Coulomb barrier is always small.

The angular distribution shown in Fig. 8 can be understood in terms of the classical scattering in the potential displayed in Fig. 1, as described by the deflection function (cf. Fig. 9). This deflection function gives the relation between the scattering angle in the center-of-mass system and the angular momentum or impact parameter. The dashed curve indicates the deflection function for a pure Coulomb field. For large angular momenta, where the two nuclei pass each other at distances larger than the range of the nuclear interaction, the two curves coincide. The nuclear attraction will make the deflection function deviate toward

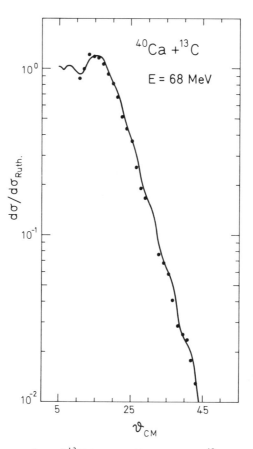

Fig. 8 Elastic scattering of ^{13}C ions at 68 MeV on a ^{40}Ca target. The points indicate the ratio of the measured cross section transformed to the center-of-mass system and the expected Rutherford cross section. In forward scattering angles (below $15°$) this ratio approaches unity. The solid curve indicates the results of a calculation of the cross-section using the potential indicated in Fig. 1. An absorptive potential of magnitude 0.6 times the nuclear attractive potential was used, but the result is rather insensitive to this choice. The associated classical deflection function is illustrated in Fig. 9. The experimental data are from Bond *et al.* (1973); the analysis was carried out by Rufino Ibarra.

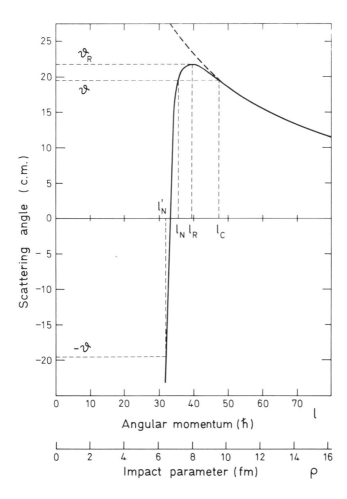

Fig. 9 Classical deflection function for the scattering of 68-MeV ^{13}C ions on ^{40}Ca. The scattering angle in the center-of-mass system is given as a function of the impact parameter ρ (in fm) or of the angular momentum l (in units of $\hbar$). The classical equations of motion were solved in the potential shown in Fig. 1. The curve is discontinued at small impact parameters where absorption is dominant. The classical differential cross section for elastic scattering is given by $d\sigma/d\Omega = \rho \, d\rho/(\sin \vartheta \, d\vartheta)$ and shows a singularity at the rainbow angle ϑ_R corresponding to the angular momentum l_R. The cross section at angles $\vartheta < \vartheta_R$ receives contributions from three angular momenta l_N, l'_N, and l_C; for the first two (of which one corresponds to a negative angle) the trajectory is strongly influenced by the nuclear attraction, while the last one corresponds to almost pure Rutherford scattering.

15

smaller angles and eventually at an angular momentum l_R give rise to a maximum deflection angle, the so-called rainbow angle ϑ_R. For small impact parameters the nuclear forces pull the projectile in to distances where a large number of nuclear reactions will occur, resulting in a strong absorption, so that no elastically scattered particles emerge from such collisions.

The sharp drop in the cross-section in Fig. 8 is associated with the transition from the lit ($\vartheta < \vartheta_R$) region to the dark region ($\vartheta > \vartheta_R$), into which no particles are scattered classically. The oscillations in the cross-section at angles $\vartheta < \vartheta_R$ are associated with the quantum mechanical interference between the three trajectories corresponding to angular momenta l_C, l_N and l'_N leading to the same angle.

The subject of elastic scattering of heavy ions is dealt with in Chapter III.

(b)   Inelastic scattering and high-frequency modes

The above simple picture of elastic scattering, which is applicable for light nuclei, can be generalized to include weak inelastic scattering, especially of vibrational states. One allows the ion-ion potential to fluctuate around the spherical equilibrium value and uses as the interaction potential, responsible for the excitation, the deviation from the equilibrium value. This is proportional to the derivative of the ion-ion potential and to the amplitude of the oscillation. The process is similar to Coulomb excitation but differs from it because the time interval over which the interaction takes place is much shorter. This means that even vibrational states of high frequency, which are inaccessible to Coulomb excitation at low bombarding energies may become accessible through nuclear excitation (cf. Fig. 10). The strength of the vibrational modes is often, as in atomic physics, expressed in terms of the oscillator strengths, and the total strength can then be estimated by the sum rule. In fact the low-frequency vibrations exhaust only a small part of the total transition strength from the nuclear ground state, and a large fraction is expected to be found in high-lying levels, often concentrated within rather narrow energy regions. These "giant resonances" play an important role in many aspects of nuclear structure, as for example in renormalizing the charge carried by

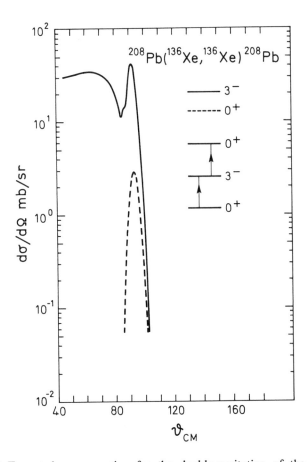

Fig. 10 Expected cross section for the double-excitation of the lowest 3^- vibrational state in ^{208}Pb by 802-MeV bombardment with ^{136}Xe ions (solid curve). The dashed line shows the expected differential cross section for the excitation of the 0^+ member of the multiplet (0^+, 2^+, 4^+ and 6^+) generated by two vibrational octupole quanta. For forward angles ($\vartheta < 80°$) the cross-section for the 3^- state is dominated by Coulomb excitation, while the peak at 90° is due to the excitation through the nuclear field. Since this field has opposite sign to the Coulomb field, there appears a minimum between the two contributions (the Coulomb nuclear interference). For the two-quantum excitations the process is dominated by the nuclear interactions because the excitation energy of the state is so high that the Coulomb excitation processes becomes almost adiabatic. The calculations are due to Stephen Landowne.

the nucleons in the single-particle motion and in determining their effective mass.

Model calculations indicate that these "giant resonances", even of high multipolarities, may be strongly excited in heavy-ion collisions when

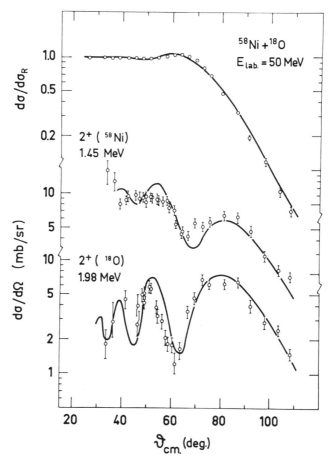

Fig. 11 Angular distribution for elastic and inelastic scattering of ^{18}O on ^{58}Ni at 50 MeV. The curves are the results of a coupled-channel calculation including, besides the ground state, the excited states of spin 2^+ in both target and projectile. The static quadrupole moment of the 2^+ state of ^{18}O was assumed in this analysis to have a large value. The experimental data and the analysis are from Videbæk *et al.* (1976).

the surfaces of the two ions have a large overlap. It is of great interest to search for these excitations in grazing collisions, where one may isolate them more easily, and for instance study the role played by the giant quadrupole modes together with the low-lying modes in the fission process.

In general a given state will be excited through both Coulomb and nuclear forces. An example is shown in Fig. 11, where states in both target and projectile are strongly excited and the associated angular distribution displays a marked interference pattern (Coulomb-nuclear interference). The analysis of this phenomenon provides information about low-lying collective states, which goes beyond the information that can be obtained from Coulomb excitation.

Inelastic scattering will be treated in Chap. IV.

(c) Transfer reactions and pairing modes

At about the same distances where nuclear interactions become important it is also possible for nucleons to move from the projectile to the target nucleus and vice versa. In the situation described above, of relatively light projectiles, transfer processes of a single nucleon can be treated in detail by a perturbation calculation. The form factors describing the transfer are constructed on the basis of the single-particle wavefunctions associated with the motion of the nucleons in the nuclei. The study of these reactions thus provides direct information about the single-particle degrees of freedom supplementing results obtained from transfer reactions induced by deuterons and other light ions. Over the years a number of single-particle transfer reactions have been studied which it has been possible to analyze in great detail (cf. e.g. Fig. 12) in terms of the distorted-wave Born approximation (DWBA).

In a single collision it is possible to transfer several nucleons from one nucleus to the other. Among these multinucleon transfer reactions special interest is attached to the simultaneous transfer of pairs of correlated neutrons or protons. The correlation between pairs of identical particles for nuclei far away from closed shells give rise to the phenomenon of nuclear superfluidity. Most heavy nuclei ($A \gtrsim 100$) are in fact many-body systems in a superfluid phase, displaying the typical features of this phase

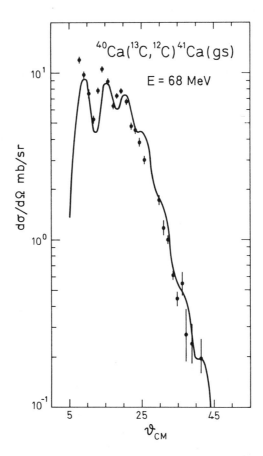

Fig. 12 Angular distribution for the single-neutron-transfer reaction ^{40}Ca $+ ^{13}$C $\rightarrow ^{41}$Ca $+ ^{12}$C at 68 MeV. The points indicate the measured cross-section transformed to the center-of-mass system, and the curve shows the result of a distorted-wave Born-approximation calculation with the potential used in Fig. 8. The formfactor for the transfer was calculated from the single-particle wavefunctions appropriate for the motion of the transferred neutron in the ground states of ^{13}C and ^{41}Ca. Because the reaction takes place at shorter distances (i.e. angular momenta of order l_N in Fig. 9) the result is more sensitive to the absorptive potential than the elastic scattering given in Fig. 8. The experimental data are from Bond *et al.* (1973); the DWBA calculation was carried out by Rufino Ibarra.

such as an energy gap, a low moment of inertia (cf. Fig. 4), etc. This condensate can be described by a BCS wavefunction which does not have a fixed number of particles, in much the same way as the surface deformed nuclei are described in terms of an intrinsic wavefunction which does not have a definite angular momentum. Thus, the nuclear condensate can be viewed as a superposition of the ground state of even neighboring isotopes in a similar way in which the intrinsic state of a quadrupole deformed nucleus is a superposition of the different states which are members of the rotational band, the number of particles playing the role of the angular momentum and the energy gap that of the quadrupole deformation.

In some nuclei the density of levels around the Fermi surface is so low that the superfluid condensate cannot be established and the energy gap vanishes. Fluctuations in the gap give then rise to a vibrational mode, called pairing vibration. The quantum of excitation associated with this mode changes the number of particles by two, just as a quadrupole surface mode changes the angular momentum by two units.

From the above discussion it follows that the two-particle transfer reaction is a powerful tool to study pair correlation phenomena. Detailed studies of these phenomena have been carried out with the help of light-ion reactions. With heavy ions one may extend the study of the collective nature of the pairing phenomena by, for instance, inducing reactions between two superfluid systems. One may thus hope to observe a transfer of several pairs of nucleons in a single collision both in superfluid and in normal systems, in the same way as one is able to transfer many quanta of surface rotations and vibrations in heavy-ion collisions. Another transfer process which is commonly observed in grazing collisions is the transfer of an alpha particle between target and projectile. It is expected that this reaction contains important information relevant to nuclear dynamics, and one of the intriguing questions is the extent to which the alpha particle is transferred as a whole, rather than as a combination of a transfer of a pair of neutrons and a pair of protons.

Transfer reactions between heavy ions will be one of the main subjects of the present notes, and will be discussed mainly in Chap. V.

(d) Combined reactions and yrast spectroscopy

The above discussion has been based on experience gathered in studying collisions with projectiles of mass number $\lesssim$ 30. With the availability

21

of much heavier ions as projectiles at energies above the Coulomb barrier, one must envisage that the two nuclei at the time where they come into contact have already been Coulomb excited to a state of high rotational or vibrational motion (cf. Figs. 3, 5, and 7). The transfer reactions would therefore take place mainly between excited states. This gives a new dimension to the study of nuclear structure. A few examples of obvious interest can already be identified at the present time. Thus, as mentioned in connection with Coulomb excitation, the rotational motion is expected to influence the superfluidity of the intrinsic motion in a major way (cf. Fig. 4). This effect can be studied by transferring pairs of neutrons or protons into a high rotational state excited prior to the transfer by the Coulomb field of the projectile.

Another possibility of combining Coulomb and nuclear inelastic scattering with transfer processes lies in the study of the spectroscopy of the so-called yrast levels. If one classifies the spectrum of a nucleus into subspectra consisting of levels of a given spin and parity, one finds a situation like the one indicated in Fig. 13. The lowest state of each subspectrum, the yrast level, has an energy which on the average increases with angular momentum. Even for high excitation energies the states of correspondingly high angular momentum are in a region of very low level density for that spin and parity. Some of these levels can be thought of as members of a rotational band built on an intrinsic state of low angular momentum. Thus, the lowest states of spin and parity 2^+, 4^+, 6^+, etc. in Fig. 13 are states formed by setting the intrinsic state corresponding to the ground state into rotational motion. Similarly, the lowest 3^-, 5^-, 7^-, etc. states form the rotational band built on the lowest 1^- state. In some cases, the high-angular-momentum states appearing close to the yrast level are due to intrinsic motion of few particles.

The study of the subspectra of Fig. 13 for high angular momenta is in many aspects similar to the study of nuclear levels close to the ground state in that the nuclear level density near the yrast line is low. One may view the study of the spectroscopy of the high-angular-momentum states close to the yrast levels as the study of cold nuclear matter under conditions of violent rotation. The dominating macroscopic effect of the rotation is the influence of the centrifugal force on the equilibrium shape of the nucleus, which tends to produce an oblate deformation which increases wtih increasing angular momentum. One expects that even

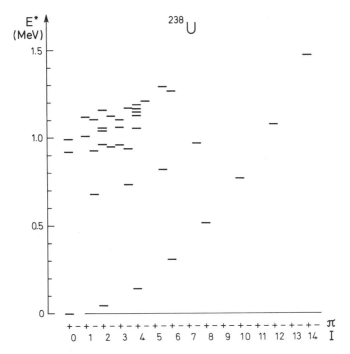

Fig. 13 Observed energy levels of ^{238}U as a function of angular momentum I (and parity). The low-lying energy levels of ^{238}U, taken from *Nuclear Data B4* (1970), p. 641, are divided into subsets characterized by the spin and parity. The lowest member of each group is called the yrast level. The general average behavior of the yrast levels is displayed in Fig. 14.

nuclei for which the shell structure produces a prolate deformation close to the ground state, may be forced into an oblate shape at large angular momenta.

In the oblate shape the angular momentum will be pointing along the symmetry axis, and one cannot talk about a collective rotational motion. In fact the total angular momentum in this situation is equal to the sum of the angular momenta of the individual nucleons. In this regime the yrast spectrum is essentially different from the rotational spectra known from the lower part of the yrast region where the rotational motion takes place around an axis perpendicular to the symmetry axis. The states along the

23

yrast line in the oblate region do not form collective rotational bands, but fluctuate in a manner determined by the intrinsic single-particle motion. One might in this region find states that have an exceptionally long lifetime.

The transition from prolate to oblate shapes. caused by the centrifugal forces, is expected to take place gradually, so that the nucleus passes through ellipsoidal forms with three different axes. At the same time the single-particle configurations characterizing the yrast line will change, since the sequence of single-particle levels varies with the changes in the deformation of the average potential.

The above features are illustrated in Fig. 14, which schematically displays the different phases occurring along the yrast line. At the very highest values of the angular momentum I, the nuclear forces are not able to withstand the centrifugal forces and the nucleus is expected to take a shape which is not symmetric around the rotational axis and finally becomes unstable against fission. This critical angular momentum has been estimated on the basis of the liquid-drop model and is displayed in Fig. 15 as a function of the mass number of the nucleus in question.

Reactions induced by heavy ions are ideal for investigating the high-angular-momentum states of nuclei. In grazing collisions one may provide the nucleus with high angular momentum both through Coulomb excitation and other inelastic processes and by transferring particles to intrinsic states of high angular momentum. The study of high-angular-momentum states in this way may lead to rather detailed information about states of angular momentum up to 30–40 $\hbar$, which is considerably beyond the region hitherto accessible to detailed spectroscopy. To reach states of higher angular momenta the two nuclei must undergo a more violent reaction, and in particular the fusion reactions are expected to provide important information about the nucleus up to its angular-momentum limit of stability.

The fission mode has hitherto been studied only by giving the nucleus enough energy at a low angular momentum to overcome the fission barrier, so that in some cases, in statistical competition with neutron emission and gamma decay, it may eventually undergo fission. Special interest is attached to the study of the fission of highly rotating systems where the corresponding fission barrier, as measured from the yrast level, is lowered (cf. Fig. 14). Through heavy-ion collisions one may excite the

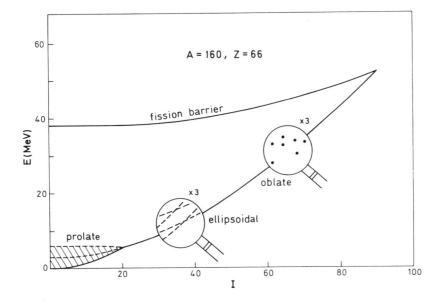

Fig. 14 Expected level structure of region close to the yrast level. The figure gives a schematic picture of the different phases that are expected to be found close to the yrast line in different angular-momentum intervals. The nucleus under consideraton has an intrinsic prolate deformation close to the ground state due to the nuclear shell structure. The hatched area indicates the region of nuclear superfluidity which is destroyed both by rotation and by thermal excitation. The magnified regions display the possible spectra close to the yrast line. The upper line indicates the energy needed for the nucleus to overcome the fission barrier for different angular momenta. At the point where this line crosses the yrast line the nucleus in unstable towards fission even at zero temperature, and above the corresponding critical angular momentum ($I \approx 90$) the nuclear forces cannot keep the system together (cf. Fig. 15). The figure is taken from Bohr and Mottelson (1974a).

heaviest nuclei in the specific fission degrees of freedom, as mentioned earlier, and at the same time provide it with a high rotational angular momentum.

The multistep reactions discussed above are specially dealt with in Chap. VI.

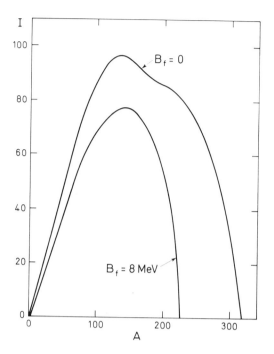

Fig. 15 Stability against fission for a rotating nucleus. The critical angular momentum I for which the nucleus becomes unstable against fission has been calculated in the liquid-drop model as a function of the mass number of the nucleus, and the corresponding curve is labelled $B_f = 0$. The curve labeled $B_f = 8$ MeV shows the angular momentum for which the fission barrier is found at an energy of 8 MeV above the ground state corresponding to the average neutron separation energy. The figure is based on Cohen *et al.* (1974) (cf. also Bohr and Mottelson (1974)).

3   FUSION REACTIONS AND COMPOUND STATES

The heavy-ion reactions discussed in the previous sections are character-istic of collisions with impact parameters corresponding approximately to the grazing angular momentum l_N (cf. Fig. 9). For smaller impact parameters, where the two nuclei have a sizable overlap during the collision, nuclear reactions become very prolific, populating a large number of states in regions of high level density. In such a situation it becomes appropriate to use statistical concepts to describe the processes which take place. As seen from the simple quasielastic reactions, these processes act like an effective absorption.

As a result of these interactions, a large amount of energy and angular momentum is lost from the relative motion to the internal degrees of freedom. In many cases this loss is so large that the two nuclei cannot overcome the nuclear attraction and fuse. One then expects that the composite system will approach a statistical equilibrium which may be described through thermodynamical concepts like temperature and entropy. Such systems have been studied since the early days of nuclear physics under the name of compound nuclei. These nuclei are known to exist over periods of time typically of order $10^{-17} - 10^{-19}$ sec, which is several orders of magnitude longer than the time it takes the projectile to go through the interaction region (the collision time $\approx 10^{-21}$ sec).

To illustrate the range of possibilities offered by heavy-ion compound formation we display in Fig. 16 the excitation energy and maximum angular momentum of three different compound nuclei, formed by different combinations of target and projectile leading to the nucleus in question. The numbers labeling the open circles along the different lines indicate the bombarding energy in the laboratory system needed to populate states in the compound nucleus at the excitation energy E^*. The angular momentum of the states formed will lie between zero and the maximum value indicated by the abcissa of the point. The dashed line indicates the expected average position of the yrast levels of the compound nucleus. The steep part of this line has been drawn at about the value of the angular momentum where one expects the nucleus to become unstable towards fission. The compound nucleus which will be formed in heavy-ion reactions at the excitation energy E^* will, at formation, have a temperature which depends on the angular momentum. The ones formed with angular momentum close to zero have the largest temperature, while those formed closed to the maximum angular momentum will have a smaller temperature, determined by the distance between E^* and the yrast line.

The compound nucleus will cool down mainly by evaporation of neutrons, but also by evaporation of protons and alpha particles and by gamma decay. However, for high angular momenta where the fission barrier is below 8 MeV (cf. Fig. 15) the compound system is more likely to undergo fission than to cool down by particle or gamma emission.

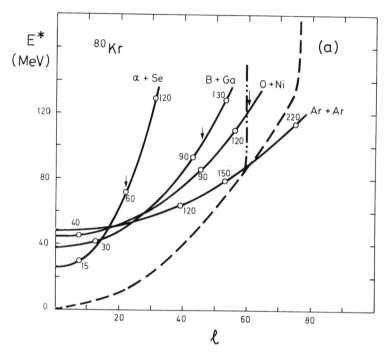

Fig. 16 Maximum angular momentum in the formation of three compound nuclei by diferent combinations of target and projectile. For each of the three compound systems ^{80}Kr, ^{164}Er and ^{248}Fm the solid curves indicate, for different bombarding energies (in MeV), the connection between the angular momentum of relative motion for a grazing collision and the excitation in the compound nucleus which would result if the projectile and target were to fuse in this situation. In the simplest picture fusion occurs in the angular momentum range $0 < l < l_g$, l_g being the grazing angular momentum. The arrows indicate the point at which the maximum in the effective potential is expected to disappear. The dashed line indicates the expected limit of the region where states in the compound system exists. The vertical part of this curve shows the expected critical angular momentum at which the compound system can no longer sustain the centrifugal forces of rotation (cf. Fig. 15). The lower part of the curve indicates the approximate average position of the yrast line (cf. Fig. 14). In the region between the dashed and the dash-dot curves the compound system is expected to decay predominantly by fission and not by neutron emission. The Q-values for the reactions are estimated on the basis of a simple empirical mass formula.

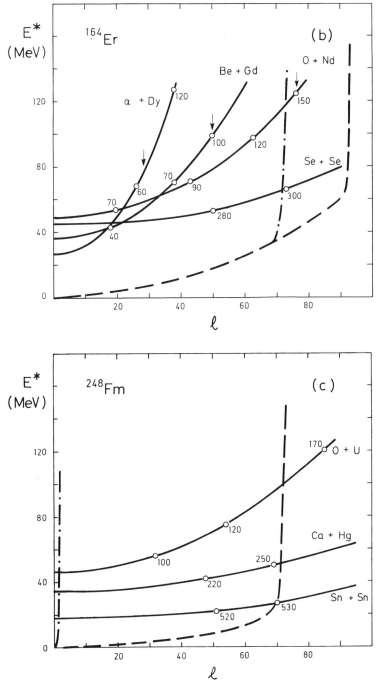

In the beginning of the cooling process the emission of gamma quanta cannot compete with particle evaporation. However, towards the end of the process, the nucleus mainly cools down by gamma emission and a large number of gamma quanta are emitted in cascades until the ground state of the final evaporation residue is reached. This will most often be a nucleus of mass and charge number several units less than the original compound nucleus.

The evaporated neutrons are expected to carry only a few units of angular momentum away from the compound system, and the evaporation residue therefore possesses an angular momentum which is not very different from that of the original compound system. By the evaporation the nucleus is therefore brought down to a region relatively close to the yrast region, and the gamma quanta, each carrying an angular momentum of one or two units of $\hbar$, will therefore contain information about the nuclear states close to the yrast level. By sophisticated experimental techniques one hopes to be able to measure not only the energy of the individual gamma quanta but also to determine which gamma quanta belong together in a cascade. Great expectations are attached to the prospect of learning in this way details about the yrast spectra even up to the region approaching the critical angular momentum. One of the exciting problems in this field is the possibility of identifying the yrast traps mentioned in the previous section.

It is seen from Fig. 16 that with alpha particle bombardment one may not populate states of very high angular momentum and the states populated have a high excitation energy. With heavy ions one may, even at relatively low bombarding energies, populate high-angular-momentum states at relatively low excitation energy.

In Fig. 16 we have completely disregarded the fact that the configuration associated with the bombarding condition of the two heavy nuclei getting into contact is very far removed from the situation of a compound nucleus in thermal equilibrium. It is to be expected that the transition from the one situation to the other, which involves a major rearrangement, may require a relatively long time during which particles can be emitted from a system which is not in thermal equilibrium. It may also happen that the dynamics of the process will prevent the formation of the compound nuclei, e.g. for small impact parameters, modifying the qualitative picture indicated in Fig. 16.

Since the decay of the compound nucleus is expected to be independent of the way in which it was formed, one can get information about the abovementioned effects by studying the decay of a given compound nucleus formed through different combinations of targets and projectiles. It is in this context an interesting observation that the curves in Fig. 16 cross each other and that one should be able to thus form e.g. ^{164}Er at the same excitation energy and with the same spin distribution with different combinations of target and projectile.

The statistical aspects of heavy-ion reactions and the subject of fusion and compound-nucleus formation are dealt with in Chap. VII.

4 DEEP-INELASTIC REACTIONS

Besides the two types of reactions discussed in the previous sections which were classified under the labels of quasielastic collisions and fusion, a third category of processes has been identified. The basic feature of these processes, which are known under the name of deep-inelastic reactions, is that one observes scattered particles, which, in contrast to the quasielastic reactions, have lost a large fraction of their kinetic energy, but still have mass and charge rather close to those of the projectile. The study of such reactions have already revealed rather spectacular features.

A particularly striking observation is illustrated in Fig. 17. Xenon ions of 1130 MeV from the Berkeley Heavy Ion Linear Accelerator were scattered on a ^{209}Bi target. The observed angular distribution of the scattered ions, which have a mass distribution around Xe, and which have undergone some energy loss, is peaked at an angle that corresponds to the expected rainbow angle (cf. Fig. 9), within a half width of 10°. At first sight this looks like the angular distribution of a quasielastic reaction. However, the peak contains, within experimental error, the full reaction cross section, from all impact parameters below the one corresponding to the grazing collision. The detected projectile-like particles have a charge distribution centered around the charge of Xe and have a broad energy distribution ranging from a small energy loss down to values about 400 MeV below the incident energy, that is, down to values which are approximately 100 MeV below the Coulomb barrier of the (Xe, Bi) system. The deep-inelastic collision cross section is roughly 70% of the reaction cross section.

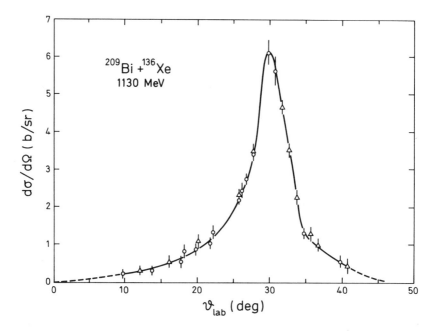

Fig. 17 Differential reaction cross section as a function of laboratory angle for the scattering of 1130-MeV ^{136}Xe ions on ^{209}Bi. The cross-section is strongly peaked at an angle of $\approx 30°$, and the peak contains a total cross section of 2.8 barns. The scattered particles have a charge and mass distribution centered around those of ^{136}Xe. The energy loss of these particles ranges from 10 MeV to 400 MeV, the latter energy being well below the kinetic energy that two spherical ions would achieve due to the Coulomb repulsion at the grazing distance. No indication of fragments from a fusion reaction is observed. The data are from Schröder *et al.* (1978).

In a number of other experiments similar large energy losses have been observed, but with an angular distribution which does not show the sharp maximum (focusing) but a gradual increase in cross-section towards forward angles. The example of Fig. 18 shows a tendency in this direction.

Some features of the deep-inelastic reactions may be connected with the fact that no state of the compound nucleus exists below the yrast line or above the critical angular momentum. The part of the full drawn curves

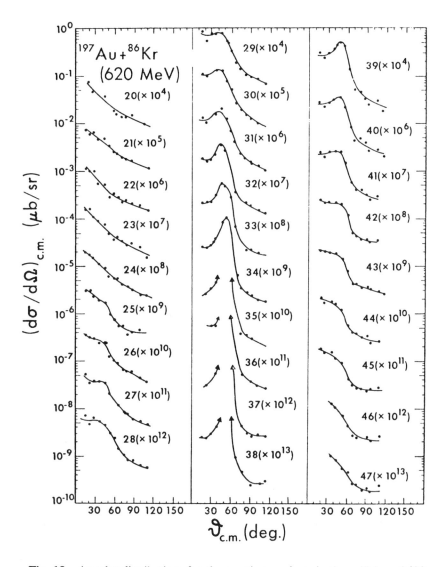

Fig. 18 Angular distributions for the reaction products in the collision of 620-MeV $^{86}_{36}$Kr on $^{197}_{79}$Au. The differential cross sections for different outgoing particles are shown as a function of the center-of-mass scattering angle for the different charge numbers indicated. The curves have been displaced by multiplying the cross-sections by different powers of ten as indicated. The peak around charge number 36 contains the reaction products for grazing collisions. The energy associated with the different detected particles reaches down to values below the Coulomb energy of two touching spheres. The bombarding energy is 1.5 times the Coulomb barrier. The data are from Moretto and Schmitt (1976).

in Fig. 16 that lies outside the region of states of the compound systems, delimited by the dashed curve, indicates collisions in which target and projectile come to distances of large overlap, but still with angular momenta beyond those that can be sustained by the compound nucleus. In particular for the system Xe + Bi no states of the compound nucleus exists for any angular momentum. For such collisions with large overlap, but where compound-nucleus formation is not possible, the nuclear inter-action seems to be so strong that a major loss of energy from the relative motion will take place.

Although a detailed understanding of the deep-inelastic processes does not exist at present, it appears that the exchange of mass between the two ions provides a mechanism for the loss of kinetic energy, whether this exchange takes place as a coherent flow or more like a diffusion process. It is to be expected also that the excitation of surface modes in both target and projectile may play an important role in providing a mechanism for the fast energy loss. The importance of these modes is strongly supported by the fact that the two ions often emerge with kinetic energies well below the Coulomb barrier of two touching spheres. This implies large deformations at least towards the final stages of the collision.

It will be a major challenge in the field of heavy-ion reactions to understand in detail the deep-inelastic collisions. One way to approach this problem will be to study the boundary between these reactions and the reactions taking pace in grazing collisions. For the latter we have a rather solid knowledge of both the nuclear-structure and the reaction-mechanism aspects, and one may, on this basis, direct one's efforts towards the study of multistep reactions of large energy losses. Another way is to study the boundary between the fusion and the deep-inelastic processes. There are indications that the simple models in terms of a critical angular momentum are insufficient and that specific dynamic effects play an important role.

A discussion of these and related questions will be the subject of Chap. VIII.

5 NUCLEAR PHENOMENA IN HIGHER-ENERGY COLLISIONS

The above discussion and the main text of these lecture notes are concerned with heavy-ion collisions at bombarding energies below 10 MeV per nucleon. This is partly a practical limit governed by the fact that it is the energy region covered by several heavy-ion accelerators. It is also the region where the relative velocity of the nuclei is notably smaller than the Fermi velocity of the nucleons within each of the two nuclei.

Very little is known experimentally about heavy-ion collisions at higher but still nonrelativistic energies. Still, several extrapolations can be made from the lower energies, especially concerning Coulomb excitation and reactions in grazing collisions. Thus for Coulomb excitation the decrease in collision time implied by the high velocities leads to an increase in the adiabatic cutoff, so that nuclear states at increasingly higher energies can be excited even in distant collisions. The dramatic increase in the cross section for Coulomb excitation will make high-energy heavy-ion beams a powerful probe for the study of the electromagnetic response of the nucleus in the energy region of the giant resonances above 10 MeV where most of the oscillator strength is concentrated.

The Coulomb-excitation cross-section is concentrated at angles below the rainbow angle ϑ_R, which approximately satisfies the relation $\vartheta_R \, \varepsilon \approx 1$, where ε is the bombarding energy in the center-of-mass system measured in units of the height of the Coulomb barrier. The reaction products from the grazing encounters are expected to be found in the same angular region. Because of the large centrifugal forces, no capture (fusion) will happen close to grazing impacts, and the cross section for quasielastic processes is expected to increase. To the extent that these processes can be experimentally disentangled from the background, they contain information e.g. about the nuclear compression degrees of freedom which are not excited by the Coulomb field.

In grazing collisions compression modes are expected to be excited by the attractive (surface-surface) force, which pulls the surfaces and tends to increase the nuclear volumes. The result of more central high-energy collisions will depend on the much more violent forces which come into play when two pieces of nuclear matter are forced together. It is a main challenge in high-energy heavy-ion collisions to explore nuclear matter under these conditions.

Heavy-ion beams at relativistic energies have recently become available. The nuclear phenomena that can be studied in such collisions seem rather detached from the low-energy phenomena to the extent that they are dominated by particle production. Still, at distant collisions Coulomb excitation will provide a connection to nuclear spectroscopy. In the relativistic region the Lorentz contraction of the Coulomb field will be responsible for a further decrease in the collision time and a corresponding increase in the maximum impact parameter where Coulomb excitation can take place. The total Coulomb-excitation cross section may in this way eventually become comparable to the full geometric cross section. Peripheral collisions at very high energies may also be able to take snapshots of the zero-point fluctuations induced in the ground state by different collective modes, for example the isovector mode. When in a fast collision a number of nucleons are suddenly removed from the target nucleus, the charge-to-mass ratio of the remaining product will reflect the local imbalance of protons and neutrons associated with this mode.

CHAPTER II

COULOMB EXCITATION

Heavy-ion reactions are characterized by the strong Coulomb interaction between the two ions, which is a dominating feature for the relative motion. The field is also responsible for a strong excitation of low-lying nuclear states in both target and projectile. Since most nuclei possess such low-lying collective states, it is rather the exception when two nuclei can reach the region of nuclear interaction without being excited during the approach.

The subject of Coulomb excitation with heavy ions has been dealt with in detail by Alder and Winther (1975). In the present chapter we shall summarize some basic results, as the subject constitutes an important background for all heavy-ion experiments.

1 CLASSICAL DESCRIPTION

A very accurate description of Coulomb excitation is obtained by assuming that the relative motion of the centers of mass of the two ions can be described classically. For large impact parameters the trajectory is such that the two nuclei do not get into the region of nuclear interaction and the reaction can be described as a pure Coulomb excitation. For smaller impact parameters pure Coulomb excitation will take place in the first part of the trajectory, leading to a coherently excited state of the two nuclei which is the doorway state to the nuclear reactions taking place thereafter.

The impact parameter which divides these two regions can be estimated by demanding that at the distance of closest approach, the nuclear attractive force is e.g. $\approx$ 2% of the Coulomb repulsive force. From the expression (III.1.36) for the nuclear potential and the estimate (III.2.6) of the Coulomb barrier, one finds this distance to be given by

$$r_c = [1.07(A_a^{1/3} + A_A^{1/3}) + 5.2] \text{ fm,} \tag{1}$$

where A_a and A_A are the mass numbers of the two nuclei. This distance is the distance of closest approach for a hyperbola with eccentricity ε_C given by

$$\varepsilon_C = \frac{r_c}{a_o} - 1, \tag{2}$$

where

$$a_o = \frac{Z_a Z_A e^2}{2E} \tag{3}$$

is half the distance of closest approach in a head-on collision, Z_a and Z_A being the charge numbers of projectile and target, while E is the kinetic energy in the center-of-mass system.

The angular momentum of relative motion in this trajectory is

$$L_C = \hbar \, \eta (\varepsilon_C^2 - 1)^{1/2}, \tag{4}$$

where η is the Sommerfeld parameter

$$\eta = \frac{Z_a Z_A e^2}{\hbar v}, \tag{5}$$

v being the relative velocity at large distances.

All trajectories with angular momenta larger than L_C where no other reaction than Coulomb excitation occurs emerge with center-of-mass scattering angles less than

$$\vartheta_C = 2 \, \arcsin \frac{1}{\varepsilon_C}. \tag{5a}$$

Trajectories with $L < L_C$ may also emerge at these scattering angles, leading to interference with the Coulomb trajectory.

In the classical description the electromagnetic interaction between target and projectile is a known function of time, and the nuclear wavefunction $\psi(t)$ can thus be found from the equation

$$i\hbar \frac{\partial}{\partial t} \mid \psi(t) \rangle = H(t) \mid \psi(t) \rangle, \qquad (6)$$

with the time dependent Hamiltonian

$$H(t) = H_a + H_A + V_E(a, -\mathbf{r}_{aA}(t)) + V_E(A, \mathbf{r}_{aA}(t)). \qquad (7)$$

The quantities H_a and H_A are the Hamiltonians of nuclei a and A respectively, while $V_E(A, \mathbf{r})$, given by

$$V_E(A, \mathbf{r}(t)) = \sum_{\substack{\lambda \geq 1 \\ \mu}} V_{\lambda\mu}(\mathbf{r}(t)),$$

with

$$V_{\lambda\mu}(\mathbf{r}) = \frac{4\pi Z_a e}{2\lambda + 1} \mathscr{M}_A(E\,\lambda - \mu)(-1)^\mu Y_{\lambda\mu}(\hat{r})r^{-\lambda-1}, \qquad (8)$$

is the electric interaction arising from nucleus a. The interaction $V_E(a, \mathbf{r})$ arising from the monopole moment of nucleus A is given by an analogous expression. The quantity $\mathscr{M}_A(E\lambda\mu)$ is the electric multipole moment of order λ, μ of nucleus A, i.e.

$$\mathscr{M}_A(E\lambda\mu) = \int \rho^A(\mathbf{r}')(r')^\lambda Y_{\lambda\mu}(\hat{r}') \, d^3r', \qquad (9)$$

ρ^A being the charge density of this nucleus. In (7) we have neglected the magnetic interactions as well as multipole-multipole interactions.

The equation (6) should be solved with the initial condition that the two colliding nuclei are in the ground state, i.e.

$$\mid \psi(-\infty) \rangle = \mid \psi_0^a \rangle \mid \psi_0^A \rangle. \qquad (10)$$

II   Coulomb Excitation

The quantity $\mathbf{r}_{aA}(t)$ indicates the position of the center of mass of a with respect to A. Insofar as the trajectory is not affected by the excitation of the two nuclei, the wavefunction $\psi(t)$ can be written as a product

$$| \psi(t) \rangle = | \psi^a(t) \rangle \ | \psi^A(t) \rangle. \tag{11}$$

The wavefunction ψ^A thus satisfies the equation

$$i\hbar \, \frac{\partial | \psi^A(t) \rangle}{\partial t} = [H_A + V_E(A, \mathbf{r}_{aA}(t))] | \psi^A(t) \rangle, \tag{12}$$

with the initial condition $| \psi^A(-\infty) \rangle = | \psi_0^A \rangle$.

It is convenient to solve (12) by expanding $| \psi^A(t) \rangle$ on the eigenstates $| \psi_n^A \rangle$ of H_A defined by

$$H_A | \psi_n^A \rangle = E_n^A | \psi_n^A \rangle, \tag{13}$$

i.e.

$$| \psi^A(t) \rangle = \sum_n a_n^A(t) \, e^{-iE_n^A t/\hbar} \, | \psi_n^A \rangle. \tag{14}$$

The excitation amplitudes $a_n^A(t)$ then satisfy the coupled equations

$$i\hbar \, \dot{a}_n^A(t) = \sum_m \langle \psi_n^A | V_E(A, \mathbf{r}_{aA}(t)) | \psi_m^A \rangle$$

$$\times \, e^{i(E_n^A - E_m^A)t/\hbar} \, a_m^A(t). \tag{15}$$

The matrix elements of the interaction (8) are essentially the product of a nuclear electromagnetic matrix element and a function of time.

The differential cross section for Coulomb excitation of the nucleus A to the state n and of the nucleus a to the state n' is given by

$$\left(\frac{d\sigma}{d\Omega} \right)_{nn'} = P_n^A P_{n'}^a \left(\frac{d\sigma}{d\Omega} \right)_{\text{Ruth}}, \tag{16}$$

where $(d\sigma/d\Omega)_{\text{Ruth}}$ is the Rutherford cross section, while

$$P_n^A = |\, a_n^A(\infty)|\,^2, \tag{17}$$

is the excitation probability of nucleus A. Similarly $P_{n'}^a$ is the excitation probability of nucleus a, which is determined by equations similar to the ones given above for A.

   In order for the classical approximation utilized above to be accurate, it is necessary not only that the Sommerfeld parameter be large, i.e.

$$\eta \gg 1, \tag{18}$$

but also that the energy loss $\Delta E = (E_n^A - E_0^A) + (E_{n'}^a - E_0^a)$ be small compared to the bombarding energy E, i.e.

$$\frac{\Delta E}{E} \ll 1, \tag{19}$$

and that the angular momentum loss ΔL be small compared to the total angular momentum of relative motion, i.e.

$$\frac{\Delta L}{L} \ll 1. \tag{20}$$

If these conditions are only marginally fulfilled, it is a major improvement to use trajectories in (15) which have an energy and an angular momentum corrected by the amount transferred to the states n and n' (cf. Sec. 2 below).

   Explicit expressions for the excitation probabilities P_n, relevant for large impact parameters, are given in Appendix A.

2   ADIABATICITY

Since the electric field varies slowly in space, the multipole expansion (1.8) converges rapidly and in practice one need only consider multipolarities λ less than 4.

   Similarly the slow variation in time of the Coulomb field, for the nonrelativistic collisions considered here, limits the states that are excited in Coulomb excitation to states of low excitation energy. Thus the states that can be excited must have an excitation energy ΔE for which the corresponding frequency $\Delta E/\hbar$ is of the order or smaller than $1/\tau_{coll}$, where τ_{coll} is the collision time. A convenient estimate is given by

$$\tau_{coll} = \frac{b(\vartheta)}{v} . \tag{1}$$

The quantity $b(\vartheta)$ is the distance of closest approach which is related to the scattering angle ϑ in the hyperbolic motion by

$$b(\vartheta) = a_0(1 + \varepsilon), \tag{2}$$

where a_0 is given in (1.3) and ε is the eccentricity of the trajectory i.e.

$$\varepsilon = \frac{1}{\sin \frac{1}{2}\vartheta} . \tag{3}$$

(a) Adiabaticity parameter

The degree of adiabaticity for the excitation from the state i to the state f is measured by the parameter

$$\xi_{fi}(\vartheta) = \frac{E_f - E_i}{2\hbar}\tau_{coll}. \tag{4}$$

If the quantity (4) is large compared to unity, the state f cannot be excited. This feature is evident from the coupled equations (1.15), where ξ_{nm} measures the number of oscillations of the exponential function during the collision. Insofar as the matrix element in (1.15) is a smooth function of time (with the maximum at $t = 0$), the oscillations of the exponential function tend to cancel the effective transition rate from the state m to the state n.

In practice the matrix element in (1.15) may also display oscillatory behavior due to the angular-momentum transfer. This is most easily seen if V_E [cf. (1.8)] is evaluated in a coordinate system with z-axis perpendicular to the plane of the orbit (cf. Fig. 1). In this system

$$\langle I_n M_n \mid V_E(A,\mathbf{r}(t)) \mid I_m M_m \rangle \sim e^{i\mu\phi(t)}, \tag{5}$$

where $\phi(t)$ is the azimuthal angle in the scattering plane. We shall assume that $t = 0$ for $\phi = 0$. Combining (5) with the exponential factor in (1.15), the total phase is

$$\varphi(t) = \mu\phi(t) + \Delta E t/\hbar \tag{6}$$

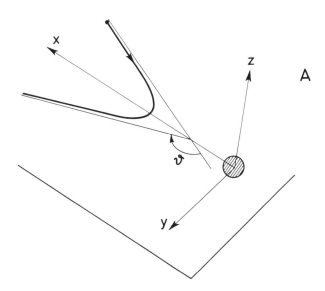

Fig. 1 The coordinate system A which is often used for the description of Coulomb excitation. The origin is chosen to be the center of mass of the target nucleus. The z-axis is perpendicular to the plane of the orbit, the x-axis is along the apex line towards the projectile, and the y-axis is chosen such that the y-component of the projectile velocity is positive. The scattering angle of the projectile is indicated.

Because the matrix element is largest at $t = 0$, the number of oscillations during the collision is measured by

$$\tilde{\xi} = \tfrac{1}{2} \left(\mu \dot{\phi}(0) + \frac{\Delta E}{\hbar} \right) \tau_{\text{coll}}. \tag{7}$$

For a given (positive) excitation energy ΔE the adiabaticity parameter $\tilde{\xi}$ is minimal, and thus the transition rate maximal, if the quantity μ satisfies the equation

$$\mu \dot{\phi}(0) = -\frac{\Delta E}{\hbar}. \tag{8}$$

As is seen from (5), $-\mu$ is the difference between the magnetic quantum numbers M_n and M_m of the nuclear states n and m, i.e.,

$$-\mu = M_n - M_m \tag{9}$$

is the angular momentum transferred to the nucleus in the transition. The quantity $\dot{\phi}(0)$ can be expressed in terms of the orbital angular momentum L and the distance of closest approach, b, as

$$\dot{\phi}(0) = \frac{L}{m_{aA} b^2}, \tag{10}$$

where m_{aA} is the reduced mass. Utilizing (10) and the definitions above, we can write the adiabaticity paramenter (7) in the form

$$\tilde{\xi} = \tilde{\xi}_\mu(\vartheta) = \eta \left(\frac{\Delta E}{2E} \frac{\sin \tfrac{1}{2}\vartheta + 1}{2 \sin \tfrac{1}{2}\vartheta} + \frac{\mu \hbar}{L} \frac{1 - \sin \tfrac{1}{2}\vartheta}{2 \sin \tfrac{1}{2}\vartheta} \right). \tag{11}$$

Note that since the only states which can be excited in Coulomb excitation are those for which $\tilde{\xi} \lesssim 1$, and since $\eta \gg 1$, the two relations (1.19) and (1.20) are automatically satisfied for these states.

The relation (8) has a simple classical interpretation. Representing the reaction of the excitation process on the relative motion by the force $\mathbf{F}$, the rate of change of the energy of relative motion, $\dot{E}$, is at time $t = 0$ given by

$$\dot{E} = \mathbf{F} \cdot \mathbf{v}(0). \tag{12}$$

Similarly, the rate of change of the component of the angular momentum along the angular momentum is

$$\begin{aligned}
\mathbf{L} \cdot \dot{\mathbf{L}} &= [m_{aA}\mathbf{r}(0) \times \mathbf{v}(0)] \cdot [\mathbf{r}(0) \times \mathbf{F}] \\
&= m_{aA}b^2\mathbf{v}(0) \cdot \mathbf{F} \\
&= m_{aA}b^2\dot{E}.
\end{aligned} \tag{13}$$

Utilizing (10) and noting that ΔE in (8) is the energy taken from the relative motion, it is seen that Eq. (8) is equivalent to the classical relation (13).

In a strongly coupled situation where the matrix elements in (1.15) are "large", the states which will be mainly populated as a function of time will tend to be those which satisfy the relation $\mu\dot{\phi}(t) = -\Delta E/\hbar$ at each instant of time. The optimal trajectory is thus one in which the energy and angular momentum of relative motion are continuously adjusted according to this rule, corresponding to the solution of the equation of motion in the presence of a frictional force.

The coupled equations (1.15) become especially simple if all states n considered have excitation energies smaller than $\hbar/\tau_{\text{coll}}$, i.e.

$$\xi_{n0} \ll 1. \tag{14}$$

In this limit the nuclear degrees of freedom do not change during the collision and one may use the sudden approximation, which corresponds to neglecting the exponential factors $\exp\{i(E_n^A - E_m^A)t/\hbar\}$ in the coupled equations. These equations have then the explicit solution (cf. Appendix B)

$$a_n^A(t) = \langle \psi_n^A| \exp\{-\frac{i}{\hbar} \int_{-\infty}^{t} V_E(A,\mathbf{r}(t'))\, dt'\} \, | \psi_0^A \rangle. \tag{15}$$

This approximation gives a rather good description of the excitation of the ground-state rotational band in strongly deformed nuclei.

(b)   Polarization by Coulomb field

High-lying nuclear states z for which the parameter

$$\xi_{z0} \gg 1 \tag{16}$$

cannot be excited. However, through virtual excitation, they can still contribute, giving rise to a polarization of the nucleus. One may include the effect of high-lying states satisfying (16) in the coupled equations by introducting a polarization potential

$$V_{pol}(A,\mathbf{r}(t)) = - \sum_z \frac{V_E(A,\mathbf{r}(t))| z \rangle \langle z| V_E(A,\mathbf{r}(t))}{E_z - E_0}. \tag{18}$$

The solution of the problem is then reduced to the solution of the equations

$$i\hbar \, \dot{a}_n^A(t) = \sum_m \langle \psi_n^A | V_E(A,\mathbf{r}(t)) + V_{pol}(A,\mathbf{r}(t))| \psi_m^A \rangle$$

$$\times \, e^{i(E_n^A - E_m^A)t/\hbar} \, a_m^A(t), \tag{19}$$

where the states z have been left out of the summation. It is noted that the polarization potential has both diagonal and nondiagonal matrix elements. The diagonal matrix elements in the ground state give rise to a contribution which should be added to the point Coulomb field to determine the trajectory of relative motion.

The most important polarization effects are expected to arise from the virtual excitation of the giant dipole and giant quadrupole modes. The giant dipole resonance gives rise to the following potential:

$$V_{pol}(A,\mathbf{r}) = - \tfrac{1}{2}P_1 \frac{Z_a^2 e^2}{r^4}, \tag{20}$$

where the dipole polarizability is given by [cf. Bohr and Mottelson (1975) p. 460]*

$$P_1 = \tfrac{8}{9}\pi \sum_z \frac{B(E1;0 \to z)}{(E_z - E_0)}$$

$$\approx 2.2 \times 10^{-3} A_A^{5/3} \text{ fm}^3. \tag{21}$$

To get a feeling for the importance of the polarizability, we may write V_{pol} as

$$V_{pol} = -\tfrac{1}{2}DE, \tag{22}$$

where E is the electric field $-Z_a e/r^2$, and D is the induced electric dipole moment

$$D = -P_1 \frac{Z_a e}{r^2}. \tag{23}$$

At the relative distance corresponding to the radius of the Coulomb barrier [cf. (III.2.5)] one finds for ^{208}Pb on ^{208}Pb

$$D = -5.6 \, e \text{ fm}, \tag{24}$$

which corresponds to a shift of the 82 charges with respect to the center of mass of ≈ 0.07 fm.

The polarization potential arising from the giant quadrupole resonances is estimated to be

$$V_{pol}(A,\mathbf{r}) = -\tfrac{1}{2}\frac{Z_a^2 e^2}{r^6} P_2 \tag{25}$$

---

*In Alder and Winther (1975), Appendix J) the induced dipole moment is in error by a factor of 2 [cf. (22) and (23) below], which is compensated by a factor $\tfrac{1}{2}$ in the definition of the polarizability.

with

$$P_2 = \tfrac{8}{25}\pi \; \sum_z \frac{B(E2;0 \rightarrow z)}{E_z - E_0}$$

$$\approx 0.6 \times 10^{-2} A^{7/3} \; \text{fm}^5, \tag{26}$$

where we have distributed the total electric oscillator strength [cf. (4.16) below] with 40% on the isoscalar mode at $\sqrt{2} \times 41 \times A^{-1/3}$ MeV and 50% at the isovector mode at $\sqrt{11} \times 41 \times A^{-1/3}$ MeV. Writing the polarization potential in terms of the induced quadrupole moment Q_0, i.e.

$$V_{\text{pol}}(A, \mathbf{r}) = \tfrac{1}{4} \frac{Z_a e Q_0}{r^3}, \tag{27}$$

we may infer that Q_0 is given by

$$Q_0 = \frac{-2 Z_a e}{r^3} P_2. \tag{28}$$

At the relative distance corresponding to the radius of the Coulomb barrier one finds for ^{208}Pb on ^{208}Pb

$$Q_0 = -69 \; e \; \text{fm}^2. \tag{29}$$

This corresponds to a deformation parameter given by

$$\beta = \frac{\sqrt{5\pi}}{3} \frac{Q_0}{Z_A e (1.2 A_A^{1/3})^2} \approx -0.02. \tag{30}$$

We may thus conclude that the polarization effects induced by the Coulomb field are rather small.

3   STRENGTH OF COULOMB EXCITATION

The strength of the transition from the state m to the state n induced by the Coulomb field is measured by the action integral of this field measured in units of $\hbar$. This is true insofar as the exponential factor in (1.15) can be neglected, i.e. the transition rate is not influenced by the adiabatic cutoff.

Using the multipole expansion of the field (1.8), the action integral for a given multipole order, is given by

$$\chi_{m \to n}^{(\lambda \, \mu)}(\vartheta) = \frac{1}{\hbar} \int_{-\infty}^{\infty} \langle \psi_{I_n M_n}^A \mid V_{\lambda\mu}(\mathbf{r}(t)) \mid \psi_{I_m M_m}^A \rangle \, dt$$

$$= \frac{\sqrt{16\pi}(\lambda - 1)!}{(2\lambda + 1)!!} \frac{Z_a e}{\hbar \, v} \frac{\langle \psi_m^A \mid\mid \mathcal{M}(E\lambda)\mid\mid \psi_n^A \rangle^*}{a_0^\lambda \sqrt{2I_m + 1}}$$

$$\times \sqrt{2\lambda + 1} \langle I_n M_n \lambda \mu \mid I_m M_m \rangle R_{\lambda\mu}(\vartheta,0), \tag{1}$$

where I and M specify the spin and magnetic quantum numbers of the nuclear states, while a_0 is defined in (1.3).

Table 1. The orbital integrals $R_{\lambda\mu}(\theta,0)$ [cf. (3.1)] in coordinate system B for $\lambda = 1$, 2 and 3. The angle ϑ is the scattering angle in the center-of-mass system for Rutherford trajectories. For more details see Alder and Winther (1975), Appendix H.

$R_{10}^B(\theta,0) = \sin \frac{1}{2}\theta$

$R_{1\,\pm1}^B(\theta,0) = 0$

$R_{20}^B(\theta,0) = \dfrac{3}{4} \dfrac{\sin^2 \frac{1}{2}\theta}{\cos^2 \frac{1}{2}\theta}(\cos^2 \frac{1}{2}\theta + 1 - \frac{1}{2}(\pi - \theta) \tan \frac{1}{2}\theta)$

$R_{2\,\pm1}^B(\theta,0) = 0$

$R_{2\,\pm2}^B(\theta,0) = \dfrac{3}{4}\sqrt{\dfrac{3}{2}} \dfrac{\sin^2 \frac{1}{2}\theta}{\cos^2 \frac{1}{2}\theta}(\frac{1}{3}\cos^2 \frac{1}{2}\theta - 1 + \frac{1}{2}(\pi - \theta) \tan \frac{1}{2}\theta)$

$R_{30}^B(\theta,0) = \dfrac{45}{16} \dfrac{\sin^3 \frac{1}{2}\theta}{\cos^4 \frac{1}{2}\theta}(\frac{5}{9}\cos^4 \frac{1}{2}\theta - \frac{1}{3}\cos^2 \frac{1}{2}\theta + 1 - \frac{1}{2}(\pi - \theta) \tan \frac{1}{2}\theta)$

$R_{3\,\pm1}^B(\theta,0) = 0$

$R_{3\,\pm2}^B(\theta,0) = \dfrac{45}{16}\sqrt{\dfrac{5}{6}} \dfrac{\sin^3 \frac{1}{2}\theta}{\cos^4 \frac{1}{2}\theta}(\frac{2}{15}\cos^4 \frac{1}{2}\theta + \frac{1}{3}\cos^2 \frac{1}{2}\theta - 1 + \frac{1}{2}(\pi - \theta) \tan \frac{1}{2}\theta)$

$R_{3\,\pm3}^B(\theta,0) = 0$

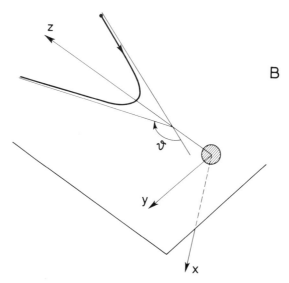

Fig. 2 The coordinate system B which is often used in Coulomb excitation. The origin is chosen at the center of mass of the target nucleus. The x-axis is perpendicular to the plane of the orbit, the z-axis is along the apex line towards the projectile, and the y-axis is chosen such that the y-component of the projectile velocity is positive. The scattering angle ϑ of the projectile is indicated.

Apart form the geometrical factor, the dependence on the direction of the transferred angular momentum as well as on the scattering angle is contained in the function R. This function can be evaluated in terms of elementary functions, and the result is given in Table 1 for $\lambda = 1, 2$ and 3, in the coordinate system B in which the z-axis is chosen along the symmetry axis of the hyperbola [cf. Fig. 2].

The transition amplitude (1) including the adiabatic cutoff is given by a similar expression:

$$\chi_{m \to n}^{(\lambda \mu)}(\vartheta, \xi) = \frac{1}{\hbar} \int_{-\infty}^{\infty} \langle \psi_{I_n M_n}^A \mid V_{\lambda \mu}(\mathbf{r}(t)) \mid \psi_{I_m M_m}^A \rangle e^{(i/\hbar)(E_n - E_m)t} \, dt$$

$$= \frac{\sqrt{16\pi}(\lambda - 1)!}{(2\lambda + 1)!!} \frac{Z_a e}{\hbar v} \frac{\langle \psi_m^A \parallel \mathcal{M}(E\lambda) \parallel \psi_n^A \rangle^*}{a_0^\lambda \sqrt{2I_m + 1}}$$

$$\times \sqrt{2\lambda + 1} \langle I_n M_n \lambda \mu \mid I_m \ M_m \rangle R_{\lambda \mu}(\vartheta, \xi), \qquad (2)$$

where

$$\xi = \frac{\Delta E}{\hbar} \frac{a_0}{v}. \tag{3}$$

The functions $R_{\lambda\mu}(\vartheta,\xi)$ are tabulated in Alder and Winther (1975).

For qualitative estimates it may be useful to write (2) in the coordinate system A in a form suggested in the previous section, i.e.

$$\chi_{m \to n}^{(\lambda\mu)}(\vartheta,\xi) = \frac{1}{2} \left(\frac{4\pi}{\lambda + \frac{1}{4}} \right)^{3/2} \frac{Z_a e}{\hbar v} Y_{\lambda\mu}(\tfrac{1}{2}\pi,0)$$

$$\times \frac{\langle \psi_{I_n M_n}^A | \mathcal{M}(E\lambda\mu) | \psi_{I_m M_m}^A \rangle}{[b(\vartheta)]^\lambda} J_\lambda(\tilde{\xi}_\mu(\vartheta)), \tag{4}$$

where the function J_λ is connected to the functions $R_{\lambda\mu}^A$ in the coordinate system A by the relation

$$\sqrt{\frac{4\pi}{2\lambda + 1}} \ Y_{\lambda\mu}(\tfrac{1}{2}\pi,0) J_\lambda = \left(\frac{1+\varepsilon}{2} \right)^\lambda R_{\lambda\mu}^A(\vartheta,\xi). \tag{5}$$

The function J_λ is mainly a function of the one variable $\tilde{\xi}_\mu(\vartheta)$ defined in (2.11). The expression

$$J_\lambda(\tilde{\xi}) \approx \begin{cases} \dfrac{4.2}{1 + \exp\left(\dfrac{|\tilde{\xi}| + 0.42}{0.36}\right)} & \text{for} \quad \lambda = 1, \\[3ex] \dfrac{1.8}{1 + \exp\left(\dfrac{|\tilde{\xi}| - 0.12}{0.43}\right)} & \text{for} \quad \lambda = 2 \end{cases} \tag{6}$$

is accurate to about $\pm$ 50%.

Note that the functions $R_{\lambda\mu}(\vartheta,0)$ in coordinate system B are very small for $\mu \neq 0$ as compared to the value for $\mu = 0$ (cf. Table 1). This means that in the sudden approximation, where χ^2 measures the probability for Coulomb excitation, only states with magnetic quantum number $M_f = M_i$ will be excited. This is quite different from the case of finite excitation energy, where the adiabatic cutoff, as discussed in Sec. 2, will favor the population of states with a positive magnetic quantum number in coordinate system A. The state of polarization of a nucleus that has been Coulomb excited thus changes as a function of the excitation energy from being aligned in a plane perpendicular to the z-axis in coordinate system B to being polarized along the z-axis in coordinate system A.

The total transition rate between the nuclear states m and n is measured by the quantiy

$$\chi^{(\lambda)}_{m \to n}(\vartheta,\xi) = \left[\frac{1}{2I_m + 1} \sum_{M_n M_m \mu} |\chi^{(\lambda\mu)}_{m \to n}|^2 \right]^{1/2}$$

$$= \frac{\sqrt{16\pi}(\lambda - 1)!}{(2\lambda + 1)!!} \frac{Z_a e}{\hbar v} \frac{\langle \psi^A_m \| \mathscr{M}(E\lambda) \| \psi^A_n \rangle}{a_0^\lambda \sqrt{2I_m + 1}} R_\lambda(\vartheta,\xi), \qquad (7)$$

where

$$R_\lambda(\vartheta,\xi) = [\sum_\mu |R_{\lambda\mu}(\vartheta,\xi)|^2]^{1/2}. \qquad (8)$$

The functions $R_\lambda(\vartheta,\xi)$ are tabulated in Alder and Winther (1975). For $\xi = 0$ they are illustrated in Fig. 3 for $\lambda = 1, 2$ and 3.

For qualitative estimates the transition rate can be written

$$|\chi^{(\lambda)}_{m \to n}(\vartheta,\xi)|^2 = \frac{4\pi^2}{(\lambda + \frac{1}{4})^3} \left(\frac{Z_a e}{\hbar v}\right)^2 \frac{B(E\lambda, m \to n)}{[b(\vartheta)]^{2\lambda}} f^C_\lambda(\xi(\vartheta)), \qquad (9)$$

where the adiabatic cutoff function is defined by

$$f^C_\lambda(\xi(\vartheta)) = \left(\frac{1 + \varepsilon}{2}\right)^{2\lambda} R^2_\lambda(\vartheta,\xi)$$

$$\approx \sum_\mu \frac{4\pi}{2\lambda + 1} |Y_{\lambda\mu}(\tfrac{1}{2}\pi, 0)|^2 |J_\lambda(\tilde{\xi}_\mu(\vartheta))|^2. \qquad (10)$$

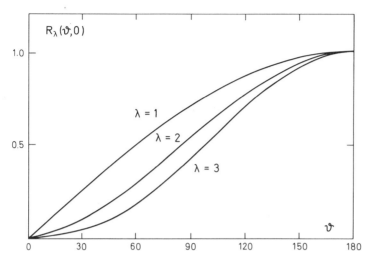

Fig. 3 The quantity $R_\lambda(\vartheta,0)$ defined in Eq. (3.8) as a function of ϑ for $\lambda = 1$. 2, and 3. It is the strength parameter associated with a trajectory of scattering angle ϑ normalized with respect to the strength parameter for the same bombarding energy in a pure Coulomb field for backward scattering.

This function is approximately unity for $\tilde{\xi} = 0$ and essentially a function of the quantity

$$\xi(\vartheta) = \frac{1 + \varepsilon}{2}\xi = \tilde{\xi}_0(\vartheta) \tag{11}$$

only. To within a factor of 3 it is given by

$$f_\lambda^C(\xi(\vartheta)) = \begin{cases} \exp(-|\xi(\vartheta)|/0.20) & \text{for } \lambda = 1, \\ \exp(-|\xi(\vartheta)|/0.29) & \text{for } \lambda = 2, \\ \exp(-|\xi(\vartheta)|/0.39) & \text{for } \lambda = 3. \end{cases} \tag{12}$$

In a heavy-ion collision the transitions from the ground state which carry the largest strengths are the giant dipole and giant quadrupole resonances and the transitions to the low-lying quadrupole and octupole modes. As discussed in the previous section, the giant dipole and quadrupole modes, at low bombarding energies, only give rise to a polarization effect, which is actually small.

The low-lying quadrupole mode, which systematically appears in the low-lying spectra of even nuclei, is strongly coupled to the ground state, giving rise to strength parameters of the order of unity or larger for most projectile-target combinations. In Fig. 4 we have indicated the value of $\chi^{(2)}_{0 \to 2}$ for the lowest 2^+ state in even-even nuclei. It was obtained from the approximate expression (9) with $b(\vartheta) = r_B$, where r_B is the radius of the Coulomb barrier given in (III.2.5). The corresponding $B(E2; 0 \to 2)$ were taken from experiment. The ordinate indicates the mass number of the nucleus excited in the collision, while the abscissa indicates the mass number of the nucleus which creates the exciting field. The contours of equal $\chi^{(2)}_{0 \to 2}$-values are shown, labeled by the corresponding values.

The large values of $\chi^{(2)}_{0 \to 2}$ encountered in the mass region $60 \lesssim A_A \lesssim 140$ correspond to the excitation of low-lying surface vibrations. The even larger values of $\chi^{(2)}_{0 \to 2}$ in the regions $150 \lesssim A_A \lesssim 190$ and $A \gtrsim 220$ correspond to the excitation of the ground-state rotational band in strongly deformed nuclei.

Since the condition for the applicability of first-order perturbation theory is that

$$\chi_{0 \to n} \ll 1, \tag{13}$$

it is seen that for most target-projectile combinations, both target and projectile will be excited in a way which goes beyond a perturbative description. In fact, in most cases the coupled equations (1.15) must be solved with a large number of nuclear states, from time $t = -\infty$ up to the time of nuclear contact, to determine the Coulomb-excitation doorway state to the nuclear reactions.

In the mass region where the $\chi^{(2)}$-values become very large, such calculations become difficult because of the very large number of nuclear matrix elements which enter as parameters in the description. One may however in such a situation take advantage of the fact that the nuclear models describing the nuclear modes as surface vibrations and rotations may give an adequate estimate of these matrix elements. In situations where very many quanta are involved, the excitation process becomes classical and one may treat the surface degrees of freedom as classical

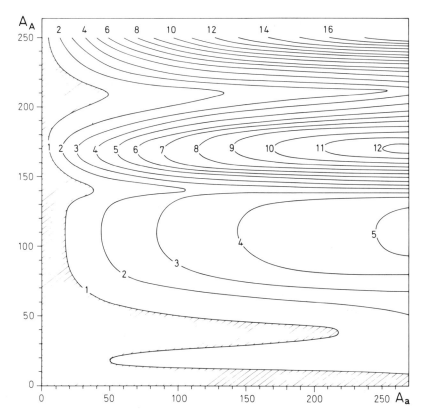

Fig. 4   The quadrupole strength parameter $\chi_{0\to 2}^{(2)}$ for Coulomb excitation at the Coulomb barrier. The parameter (3.7) [cf. also Eq. (3.9)] was evaluated for $\xi = 0$ ($f_2^C = 1$) and for $b(\vartheta)$ equal to the radius of the Coulomb barrier for different projectiles (mass number A_a) exciting the lowest 2^+ state in even target nuclei (mass number A_A) using the experimental $B(E2)$ values cf. e.g. Bohr and Mottelson (1975), Vol. II, Fig. 4.5). The curves are contour lines of equal values of $\chi_{0\to 2}^{(2)}$ as indicated by the numbers by which they are labeled. The hatched area indicates the region where perturbation theory may be applied.

variables together with the relative coordinate $\mathbf{r}(t)$. Such descriptions, where at least the orientation angles of the deformed nucleus are taken as classical variables, become especially useful in the treatment of nuclear reactions involving heavy nuclei.

Coulomb excitation as it proceeds within the framework of the rotational and vibrational models is discussed in the following sections.

4   EXCITATION OF LOW-LYING VIBRATIONAL STATES

(a)   Surface modes

The spectra of most spherical nuclei show low-lying states which can be interpeted as collective vibrations of the surface around the equilibrium shape. A great simplification in the description is achieved if one can consider these vibrations to be harmonic, i.e. described by a Hamiltonian of the form*

$$H_A = \sum_{n,\lambda\mu} \left\{ \frac{D_\lambda(n)}{2} \mid \dot{\alpha}_{\lambda\mu}(n) \mid^2 + \frac{C_\lambda(n)}{2} \mid \alpha_{\lambda\mu}(n) \mid^2 \right\}. \quad (1)$$

Here, $\alpha_{\lambda\mu}(n)$ describes the vibrational amplitude of multipolarity λ, while $D_\lambda(n)$ and $C_\lambda(n)$ are parameters describing the inertia and the restoring force respectively.

The energy spectrum corresponding to each term in the Hamiltonian (1) is given by

$$E_{N_{\lambda\mu}(n)} = N_{\lambda\mu}(n)\hbar\ \omega_\lambda(n) = N_{\lambda\mu}(n)\hbar\ \left(\frac{C_\lambda(n)}{D_\lambda(n)}\right)^{1/2}, \quad (2)$$

where $N_{\lambda\mu}(n)$ is the number of phonons of multiplority $\lambda\mu$. We have introduced the index n to indicate that there usually are more than one surface mode of a given multipolarity. Thus besides the low-lying quadrupole vibration there exists high-lying, e.g. giant modes.

It is convenient to describe the vibrational spectrum in terms of bosons each carrying an angular momentum (λ, μ). The corresponding creation and annihilation operators c and $c^\dagger$, respectively, which fulfill the usual commutation relations, are related to $\alpha_{\lambda\mu}$ by the equation

---

*Since we consider only the excitation of nucleus A, we shall leave out the index A on the various intrinsic variables.

$$\alpha_{\lambda\mu}(n) = \left(\frac{\hbar\,\omega_\lambda(n)}{2C_\lambda(n)}\right)^{1/2} [c_{\lambda\mu}^\dagger(n) + (-1)^\mu c_{\lambda-\mu}(n)]. \qquad (3)$$

The Hamiltonian (1) then takes the form

$$H_A = \sum_{n,\lambda\mu} \hbar\,\omega_\lambda(n)[c_{\lambda\mu}^\dagger(n)c_{\lambda\mu}(n) + \tfrac{1}{2}]. \qquad (4)$$

An eigenstate of the vibrator (λ,n) is given by

$$| N_{\lambda-\lambda}(n),\cdots,N_{\lambda\mu}(n),\cdots,N_{\lambda\lambda}(n) \rangle$$
$$= \frac{1}{\sqrt{N_{\lambda-\lambda}(n)!\cdots N_{\lambda\mu}(n)!\cdots N_{\lambda\lambda}(n)!}} \qquad (5)$$
$$\times [c_{\lambda-\lambda}^\dagger(n)]^{N_{\lambda-\lambda}(n)}\cdots[c_{\lambda\mu}^\dagger(n)]^{N_{\lambda\mu}(n)}\cdots[c_{\lambda\lambda}^\dagger(n)]^{N_{\lambda\lambda}(n)}| 0 \rangle,$$

where $| 0 \rangle$ is the ground state of the system. The energy of the states only depends on the quantity

$$N_\lambda(n) = \sum_\mu N_{\lambda\mu}(n). \qquad (6)$$

The state (5) has a well-defined component of the angular-momentum quantum number along the z-axis which is given by

$$M = \sum_\mu N_{\lambda\mu}(n)\mu. \qquad (7)$$

The total angular momentum of this state is however not well defined.

The states of good total angular momentum I_N can be constructed utilizing the techniques of fractional-parentage coefficients [see e.g. Bayman and Landé (1966)]. We shall work out in detail only the case where all quantum numbers $N_{\lambda\mu}(n) = 0$ except those for $\mu = 0$. These are the only states excited in backward scattering. The amplitude of the state (5) on the state $| N\,\alpha_N\,I_N\,M_N = 0 \rangle$, where α_N are the additional quantum numbers (like seniority) which are needed to characterize the many-phonon states, is given by

$$\langle 0, \ldots, N_{\lambda 0} = N, \ldots, 0 \mid N \, \alpha_N \, I_N \, M_N = 0 \rangle$$

$$= \frac{1}{\sqrt{N!}} \sum_{\substack{I_{N-1}.I_{N-2}.\cdots.I_2 \\ \alpha_{N-1}.\alpha_{N-2}.\cdots.\alpha_2}}$$

$$\begin{pmatrix} I_N & \lambda & I_{N-1} \\ 0 & 0 & 0 \end{pmatrix} \langle N\alpha_N I_N \parallel c_\lambda^\dagger \parallel (N-1) \, \alpha_{N-1} \, I_{N-1} \rangle$$

$$\times \begin{pmatrix} I_{N-1} & \lambda & I_{N-2} \\ 0 & 0 & 0 \end{pmatrix} \langle (N-1)\alpha_{N-1} I_{N-1} \parallel c_\lambda^\dagger \parallel (N-2) \, \alpha_{n-2} \, I_{N-2} \rangle$$

$$\times \cdots$$

$$\times \begin{pmatrix} I_2 & \lambda & \lambda \\ 0 & 0 & 0 \end{pmatrix} \langle 2\alpha_2 I_2 \parallel c_\lambda^\dagger \parallel 1 \, \alpha_1 \, I_1 = \lambda \rangle. \tag{8}$$

For $\lambda = 2$ and $N = 1, 2, 3$ and 4 the amplitudes (8) are given in Table 2.

We shall assume that the electric multipole moments (1.9) are linear in $\alpha_{\lambda\mu}$, i.e.

$$\mathcal{M}(E\lambda\mu) = \sum_n (2\lambda + 1)^{-1/2} F_\lambda \alpha_{\lambda\mu}(n). \tag{9}$$

The quantity F_λ is connected with the reduced matrix element of $\mathcal{M}(E\lambda\mu)$ between the ground state and the state containing one phonon of type (λ, n), by the relation

$$\langle N_\lambda(n) = 1, I = \lambda \parallel \mathcal{M}_A(E\lambda) \parallel \{N_\lambda(n)\} = 0, I = 0 \rangle$$

$$= \sqrt{\frac{\hbar \, \omega_\lambda(n)}{2C_\lambda(n)}} \, F_\lambda. \tag{10}$$

To the extent that the vibrations can be interpreted as oscillations of the nuclear shape, the nuclear radius in the direction $\hat{r}$ is given by

$$R(\hat{r}) = R^{(0)}\left(1 + \sum_{n\lambda\mu} \alpha_{\lambda\mu}(n) Y_{\lambda\mu}^*(\hat{r})\right), \tag{11}$$

Table 2. The amplitude of the state with N quadrupole phonons of $\mu = 0$ on the eigenstate of total angular momentum I (and $M = 0$). For four and more phonons, several states of given I exist which can be distinguished by the seniority quantum number v. The table has been calculated utilizing the expression (4.8) and the table of fractional-parantage coefficients of Bayman and Landė (1966). The numbers in that reference should be multiplied by $[N(2I_N + 1)]^{1/2}$ to comply with the notation used in (4.8). Note that because $\mu = 0$, all amplitudes for odd values of I are zero, and that the sum of the squares of the amplitudes is unity.

I	N	1	2	3	4
0			0.447	−0.239	0.293
2		1	−0.535	0.655	−0.436 ($v = 2$)
					−0.263 ($v = 4$)
3				0.0	
4			0.717	−0.530	−0.586 ($v = 2$)
					−0.155 ($v = 4$)
5				0.0	
6				0.483	0.432
8					0.317

where $R^{(0)}$ is the equilibrium radius of the nucleus,

$$R^{(0)} = 1.2A^{1/3} \text{ fm}. \qquad (12)$$

Assuming furthermore that the charge distribution follows the vibration together with the mass distribution, one can express the electric multipole moment (1.9) of nucleus A in terms of (9) with

$$F_\lambda = \frac{3}{4\pi} Ze(R^{(0)})^\lambda \sqrt{2\lambda + 1}. \qquad (13)$$

For more details see Sec. IV.3 below.

A convenient measure of the collectivity of the state (λ, n) is given by the associated oscillator strength. We define

$$s_n(E\lambda) \equiv \hbar w_\lambda(n) B(E\lambda; 0 \to (\lambda, n)) = \frac{\hbar^2 F_\lambda^2}{2D_\lambda(n)}$$

$$= (2\lambda + 1)\frac{9}{16\pi^2} Z^2 e^2 (R^{(0)})^{2\lambda} \frac{\hbar^2}{2D_\lambda(n)}, \qquad (14)$$

where the reduced transition probability is defined by

$$B(E\lambda; I_i \to I_f) = \frac{1}{2I_i + 1} |\langle I_f \parallel \mathcal{M}(E\lambda) \parallel I_i \rangle|^2. \qquad (15)$$

The total electric oscillator strength for a given multipolarity satisfies the sum rule [Bohr and Mottelson (1975)] (cf. also Sec. IV.6)

$$S(E\lambda) = \sum_n s_n(E\lambda)$$

$$= \frac{3(2\lambda + 1)\lambda}{4\pi} \frac{\hbar^2 Z e^2}{2M} (R^{(0)})^{2\lambda - 2}, \qquad (16)$$

where M is the nuclear mass unit. The quantity $D_\lambda(n)$ can thus be expressed in terms of the fraction of the sum rule (9) which is associated with the state n, i.e. the oscillator strength

$$f_n(E\lambda) = \frac{s_n(E\lambda)}{S(E\lambda)} = \frac{3}{4\pi\lambda} Z(R^{(0)})^2 \frac{M}{D_\lambda(n)}. \qquad (17)$$

For quadrupole vibrations the lowest 2^+ state carries $\lesssim 10\%$ of the energy weighted sum rule, the rest being associated with higher-lying states, especially with the quadrupole giant resonance. The lowest octupole vibration often carries a larger percentage of the corresponding energy weighted sum rule than the lowest quadrupole state. The above estimate for the giant quadrupole resonance was used in calculating the polarization effect in Sec. 2.

(b)   Coulomb excitation

For a Hamiltonian of type (1) or (4), the time dependent Schrödinger equation (1.12) can be solved explicitly (cf. Appendix B). One finds the result

$$| \psi(t) \rangle = \exp\{-\tfrac{1}{2} \sum_{n\lambda} B(E\lambda;0 \to (\lambda,n)) q_\lambda^2(\omega_\lambda(n),t)\} \, e^{-iH_A t/\hbar}$$

$$\times \exp\{-i \sum_{n\lambda\mu} \langle 1_{\lambda n} \| \mathscr{M}(E\lambda)_{-} \| 0 \rangle$$

$$\times (-1)^\mu q_{\lambda\mu}(\omega_\lambda(n),t) c_{\lambda-\mu}^\dagger(n)\} \, | 0 \rangle, \tag{18}$$

where we have left out an overall time dependent phase. The reduced matrix element and transition probability are defined in (10) and (15). The orbital integral $q_{\lambda\mu}$ is defined by

$$q_{\lambda\mu}(\omega,t) = \frac{4\pi Z_a e}{(2\lambda+1)^{3/2}\hbar} \int_{-\infty}^{t} Y_{\lambda\mu}(\hat{r}(t')) r^{-\lambda-1}(t') e^{i\omega t'} \, dt', \tag{19}$$

while

$$q_\lambda^2(\omega,t) = \sum_\mu | q_{\lambda\mu}(\omega,t) |^2. \tag{20}$$

One may expand (18) in the eigenstates (5) of the unperturbed Hamiltonian (4). The corresponding excitation amplitude [cf. (1.14)] is given by the product of the excitation amplitudes for each mode (λ,μ,n), i.e.

$$a_{\{N_{\lambda\mu}(n)\}}(t) = \prod_{\lambda\mu n} a_{N_{\lambda\mu}(n)}(t), \tag{21}$$

where

$$a_N(t) = e^{-\frac{1}{2}B(E\lambda)| q_{\lambda\mu}(\omega,t)|^2} \frac{[-i\langle 1_{\lambda n} \| \mathscr{M}(E\lambda) \| 0 \rangle (-1)^\mu q_{\lambda-\mu}(\omega,t)]^N}{(N!)^{1/2}}. \tag{22}$$

61

The reduced matrix element, the electromagnetic transition probability and the frequency ω are those corresponding to the excitation of the mode (λ,n), while N stands for the number $N_{\lambda\mu}(n)$.

For a pure Coulomb excitation the amplitudes at $t = +\infty$ can be written in terms of the strength parameter defined in (3.2), i.e.

$$a_N(\infty) = \frac{[-i\chi_{0-1}^{\{\lambda\}}(-1)^{\lambda+\mu}R_{\lambda-\mu}(\vartheta,\xi)]^N}{(N!)^{1/2}}e^{-\frac{1}{2}|\chi_{0-1}^{\{\lambda\}}|R_{\lambda-\mu}(\vartheta,\xi)|^2}, \quad (23)$$

where $\chi_{0-1}^{\{\lambda\}} \equiv \chi_{0-1}^{\{\lambda\}}(\pi,0)$. The quantity $R_{\lambda\mu}(\vartheta,\xi)$ is related to $q_{\lambda\mu}(\omega,\infty)$ by

$$q_{\lambda\mu}(\omega,\infty) = \frac{\sqrt{16\pi(\lambda-1)!}}{(2\lambda+1)!!}\frac{Z_a e}{\hbar\, v a_0^\lambda}R_{\lambda\mu}(\vartheta,\xi), \quad (24)$$

where $\xi = a_0\omega/v$ and ϑ is the scattering angle.

The corresponding excitation probabilities are the product of Poisson distributions. The total probability of exciting N phonons in the mode (λ,n) irrespective of the magnetic quantum number and of the excitation of all other modes is thus given by

$$P_N = \frac{1}{N!}[|\chi_{0-1}^{\{\lambda\}}|^2 R_\lambda^2(\vartheta,\xi)]^N\, e^{-|\chi_{0-1}^{(\lambda)}|^2 R_\lambda^2(\vartheta,\xi)}, \quad (25)$$

where $R_\lambda(\vartheta,\xi)$ is defined in (3.8). The average number of phonons in this Poisson distribution is

$$\langle N \rangle = |\chi_{0-1}^{\{\lambda\}}|^2 R_\lambda^2(\vartheta,\xi). \quad (26)$$

The expected average number of phonons which can be excited in the bombardment of ^{114}Cd, ^{150}Sm and ^{192}Pt with various projectiles at energies close to the Coulomb barrier is illustrated in Fig. I.5.

The validity of the harmonic description of the low-frequency surface modes is limited because the coupling between the phonons and the coupling to other degrees of freedom give rise to anharmonicity effects and to a spreading of the oscillator strength. The detailed study of these effects is a major challenge in the field of nuclear structure as probed by heavy-ion reactions. The effects are expected to be rather different for the different members of the multiphonon multiplets, depending on the total angular momentum. Thus the member where all the phonons are aligned, i.e. the state of angular momentum $I_N = N\lambda$, is likely to be the yrast level in some energy interval.

One may evaluate the excitation probability of the various spin members of the multiplet from Eq. (21) utilizing fractional-parentage techniques. Explicit results are given in Alder and Winther (1975), p. 202. For the case of backward scattering the results are especially simple, since in the coordinate system B only states of spin $\mu = 0$ are excited (cf. Fig. 2). The squares of the amplitudes (8) thus directly measure the relative excitation probabilities of the various spin members in a given multiplet. It is seen from Table 2 that for $N \gtrsim 3$ the maximum excitation probability should occur for some intermediate spin value. Only the high-spin members have been experimentally identified (cf. Fig. 5).

Note that as the scattering angle is diminished and the excitation becomes more adiabatic, there is a tendency to excite predominantly phonons with $\mu = -\lambda$ in coordinate system A (cf. Sec. 2). In the limit of forward scattering the aligned state of maximum spin is the only one that will be excited.

Equation (25) can be used to calculate the depopulation of the target ground state due to the excitation of surface vibrations. Thus, the probability that the target stays in the ground state after a scattering through the angle ϑ is given by

$$P_0 = \exp \{-\mid \chi_{0 \to 1}^{(\lambda)} \mid^2 R_\lambda^2(\vartheta, \xi)\}. \tag{27}$$

The cross-section for elastic scattering is given by (1.16), where the probability of projectile and target excitation should both be included. The calculated cross-section for the scattering of Sn on ^{150}Sm at an energy below the Coulomb barrier is given in Fig. I.7. For spherical nuclei the result (27) is expected to give a rather good estimate of the effect of Coulomb excitation on elastic scattering, since the anharmonicities and spreading of the multiphonon states are not going to influence very much the depopulation of the ground state.

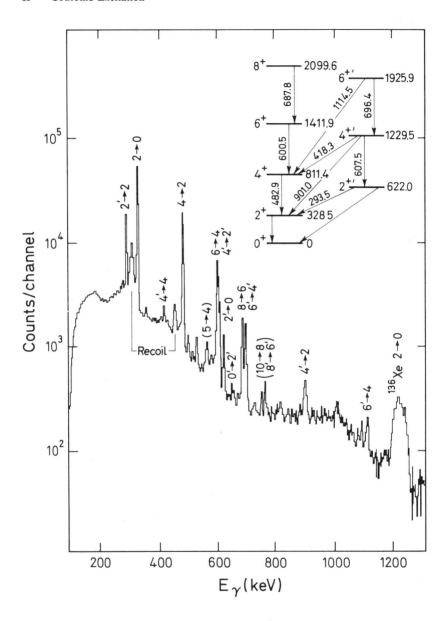

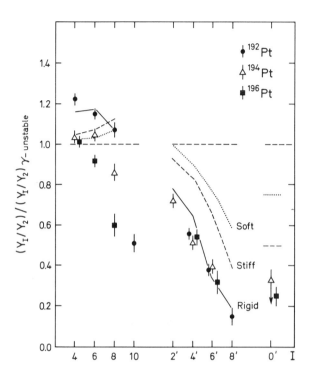

Fig. 5 Coulomb excitation of Pt isotopes by 620-MeV Xe ions as measured by Lee *et al.* (1977). Part (a) shows the gamma spectrum from a thin ^{194}Pt target as measured in a Ge(Li) detector in coincidence with the scattered Xe ions at about 90°. The inferred level diagram is also shown. This spectrum is typical of transitional nuclei in that it shows both rotational and vibrational features. While in the pure vibrational model one would expect an average of more than eight vibrational quanta to be excited [cf. Fig. I.5], an estimate in terms of the rotational model predicts a maximum angular-momentum transfer of about $10\hbar$ [cf. Eq. (7.11)], which seems to be a more realistic estimate. In (b) the resulting relative gamma yields for the three Pt isotopes are shown in comparison with predictions of the collective model obtained utilizing potential-energy surfaces which cover the range from γ-deformed (rigid) to completely γ-unstable nuclei (horizontal dashed line).

The expectation value of the energy loss from relative motion to excitation of the target is, according to (26), given by

$$\langle E \rangle = \sum_{\lambda n} s_n(E\lambda) q_\lambda^2(\omega_\lambda(n), \infty)$$

$$= \sum_{\lambda n} \hbar\, \omega_\lambda(n) \mid \chi_{0 \to \lambda, n}^{(\lambda)} \mid^2 R_\lambda^2(\vartheta, \xi). \qquad (28)$$

The quantity $R_\lambda^2(\vartheta, \xi)$ vanishes exponentially for $\xi \gtrsim 1$. For $\xi \ll 1$ we may use (3.4) and we find

$$\langle E \rangle \approx r\, \frac{\eta^2}{Z_A} \left(\frac{R^{(0)}}{b(\vartheta)} \right)^{2\lambda} \frac{100}{[(R^{(0)})_{\text{fm}}]^2}\ \text{MeV}, \qquad (29)$$

for quadrupole and octupole excitations, where r is the fraction of the energy weighted sum rule which is below the adiabatic cutoff.

The energy loss (29) is always small compared to the bombarding energy, implying that the elastic Rutherford trajectory is essentially not affected by the Coulomb excitation.

5   COHERENT EXCITATION OF SURFACE MODES

It is a characteristic result of the previous section that the Coulomb excitation of vibrational states can be thought of as the independent excitation of all the vibrational modes characterized by the quantum numbers λ, μ and n, leading to Poisson distributions. The effect of Coulomb excitation on nuclear reactions above the Coulomb barrier is however determined by the nuclear state vector (4.18) at time $t \approx 0$, which is the doorway state for the nuclear reactions to take place. In this wavefunction the coherence between the different vibrational amplitudes is essential to reproduce the effect which in the classical limit corresponds to the macroscopic deformation of the surface.

On the basis of the wavefunction (4.18) we can calculate the expectation value of the amplitude $\alpha_{\lambda\mu}$ at time t. One finds

$$\langle \alpha_{\lambda\mu}(n) \rangle = \langle \psi(t) \mid \alpha_{\lambda\mu}(n) \mid \psi(t) \rangle$$

$$= -i \left(\frac{\hbar \, \omega_\lambda(n)}{2C_\lambda(n)} \right) F_\lambda [q_{\lambda\mu}(\omega_\lambda(n),t)e^{-i\omega\lambda(n)t}$$

$$- (-1)^\mu q^*_{\lambda-\mu}(\omega_\lambda(n),t)e^{i\omega\lambda(n)t}]. \tag{1}$$

This average excitation amplitude satisfies the classical equation of motion corresponding to the Hamiltonian [cf. (1.7)]

$$H(t) = H_A + V_E(A, \mathbf{r}(t)). \tag{2}$$

In fact we obtain from (1)

$$\frac{d\langle \alpha_{\lambda\mu}(n) \rangle}{dt} = -\left(\frac{\hbar \, \omega_\lambda(n)}{2C_\lambda(n)} \right) F_\lambda \, \omega_\lambda(n)$$

$$\times [q_{\lambda\mu}(\omega_\lambda(n),t)e^{-i\omega\lambda(n)t}$$

$$+ (-1)^\mu q^*_{\lambda-\mu}(\omega_\lambda(n),t)e^{i\omega\lambda(n)t}], \tag{3}$$

and

$$\frac{d^2\langle \alpha_{\lambda\mu}(n) \rangle}{dt^2} = i \left(\frac{\hbar \, \omega_\lambda(n)}{2C_\lambda(n)} \right) F_\lambda$$

$$\times \Bigg(\omega_\lambda^2(n)[q_{\lambda\mu}(\omega_\lambda(n),t)e^{-i\omega\lambda(n)t}$$

$$- (-1)^\mu q^*_{\lambda-\mu}(\omega_\lambda(n),t)e^{i\omega\lambda(n)t}]$$

$$+ i\omega_\lambda(n) \frac{8\pi Z_a e}{(2\lambda + 1)^{3/2}\hbar} Y_{\lambda\mu}(\hat{r}(t))r^{-\lambda-1}(t) \Bigg). \tag{4}$$

Utilizing the definition (4.9), we see that $\langle \alpha_{\lambda\mu} \rangle$ satisfies the equation

$$\frac{d^2}{dt^2} \langle \alpha_{\lambda\mu}(n) \rangle + \omega_\lambda^2(n) \langle \alpha_{\lambda\mu}(n) \rangle$$

$$= - \frac{F_\lambda}{D_\lambda(n)} \frac{4\pi Z_a e}{(2\lambda + 1)^{3/2}} Y_{\lambda\mu}(\hat{r}(t)) r^{-\lambda-1}(t)$$

$$= - \frac{(-1)^\mu}{D_\lambda(n)} \frac{\partial V_E(\mathbf{r}(t))}{\partial \alpha_{\lambda-\mu}}, \tag{5}$$

which is the classical equation for a forced harmonic oscillator.

Another way of writing the solution (1) of Eq. (5) is

$$\langle \alpha_{\lambda\mu}(n) \rangle = \frac{\hbar F_\lambda}{2\pi D_\lambda(n)} \int_{-\infty}^{\infty} d\omega \frac{q_{\lambda\mu}(\omega,\infty) e^{-i\omega t}}{\omega^2 - [\omega_\lambda(n)]^2 + 2i\varepsilon\omega}, \tag{6}$$

where ε is a small positive quantity. By contour integration utilizing the pole structure of the denominator one finds

$$\langle \alpha_{\lambda\mu}(n) \rangle = - \frac{4\pi Z_a e}{(2\lambda + 1)^{3/2}} \frac{F_\lambda}{\omega_\lambda(n) D_\lambda(n)}$$

$$\times \int_{-\infty}^{t} dt' \; Y_{\lambda\mu}(\hat{r}(t')) r^{-\lambda-1}(t') \sin[\omega_\lambda(n)(t - t')], \tag{7}$$

which is identical to (1).

By means of this formula we may estimate the deformation which a nucleus will suffer through the Coulomb excitation of the surface vibrational modes before the projectile reaches the nuclear surface. The largest effect is expected for backward scattering at a bombarding energy equal to that of the Coulomb barrier. At time $t = 0$ one finds from (7)

$$\langle \alpha_{\lambda\mu}(n) \rangle_{t=0} = - \delta(\mu,0) \frac{\sqrt{4\pi}}{2\lambda + 1} \frac{Z_a e F_\lambda}{b^{\lambda+1} \omega_\lambda^2(n) D_\lambda(n)} g_\lambda(\xi_n), \tag{8}$$

where b is the distance of closest approach and

$$g_\lambda(\xi) = 2^{\lambda+1} \xi \int_0^\infty \frac{\sin[\xi(\sinh w + w)]}{(\cosh w + 1)^\lambda} dw. \tag{9}$$

In obtaining (9) we used the parametric representation of the hyperbolic orbit (cf. Appendix III.C) and the adiabaticity parameter

$$\xi_n = \omega_\lambda(n) \frac{a_0}{v} = \omega_\lambda(n) \frac{b}{2v}, \tag{10}$$

defined in Sec. 2. The function $g_\lambda(\xi)$ is illustrated in Fig. 6 for $\lambda = 1$ to 4. For $\xi \gg 1$ it approaches the asymptotic value $g(\xi) = 1$. This property can be proved by partial integration of (9). For small values of ξ we find

$$g_\lambda(\xi) = \gamma_\lambda \xi^2, \tag{11}$$

with

$$\gamma_\lambda = 2^{\lambda+1} \int_0^\infty \frac{\sinh w + w}{(\cosh w + 1)^\lambda} \, dw$$

$$= \begin{cases} 6.363, & \lambda = 2, \\ 3.491, & \lambda = 3, \\ 2.421, & \lambda = 4. \end{cases} \tag{12}$$

Utilizing Eqs. (4.13) and (4.14), one may write (8) in the form

$$\langle \alpha_{\lambda 0} \rangle_{t=0} \equiv \beta = - \frac{6\sqrt{4\pi}}{(2\lambda + 1)^{3/2}(\lambda + 3)^2} \left(\frac{R_A^{(0)}}{b} \right)^{\lambda+1}$$

$$\times \frac{Z_a}{Z_A} \frac{e^2}{R_A^{(0)} \Delta E} (B(E\lambda))_{sp} \, g_\lambda(\xi), \tag{13}$$

where $\Delta E = \hbar \, \omega_\lambda(n)$ and the reduced transition probability is measured in single-particle units

$$B_w(E\lambda) = \frac{1}{4\pi} \left(\frac{3}{3 + \lambda} \right)^2 (R^{(0)})^{2\lambda} e^2. \tag{14}$$

69

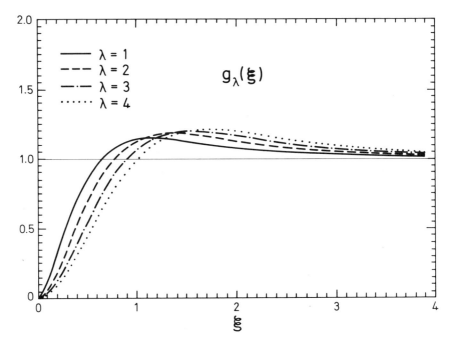

Fig. 6 The function $g_\lambda(\xi)$ which describes the vibrational amplitude at the distance of closest approach. It is defined in Eq. (5.9); it approaches unity for $\xi \gg 1$ and depends quadratically on ξ for $\xi \ll 1$.

For the 558-keV 2^+ state in ^{114}Cd, which carries 36 single-particle units, one finds

$$\beta = -0.022 \qquad (15\,a)$$

for Sn projectiles, and

$$\beta = -0.040 \qquad (15\,b)$$

for Pb projectiles. For the 334-keV 2^+ state in ^{150}Sm, which carries 55 single-particle units, we obtain

$$\beta = -0.036 \tag{16}$$

for Pb projectiles. For the 1.63-MeV 2^+ state in ^{144}Sm, which carries 11 single-particle units, we find

$$\beta = -0.011, \tag{17}$$

again with Pb projectiles.

For $\xi \gtrsim 1$ Eq. (13) becomes identical to (2.30) describing the adiabatic polarization effects of the giant resonances. The effects are of similar magnitude and may add up to a total deformation parameter $\beta \approx 0.08$ in very-heavy-ion collisions.

The velocity given to the surface can be estimated from Eq. (3), which for backward scattering at time $t = 0$ leads to the expression

$$\langle \dot{\alpha}_{\lambda 0} \rangle_{t=0} = - \frac{\sqrt{4\pi}(\lambda - 1)!}{(2\lambda + 1)!!} \frac{F_\lambda}{D_\lambda(n)} \frac{Z_a e}{a_{0}^{\lambda} v} R_{\lambda 0}(\pi, \xi). \tag{18}$$

We have here introduced the orbital integral

$$R_{\lambda 0}(\pi, \xi) = \frac{(2\lambda - 1)!! a_{0}^{\lambda} v}{2(\lambda - 1)!} \int_{-\infty}^{\infty} e^{i\omega t} r^{-\lambda-1} \, dt. \tag{19}$$

The velocity of the nuclear surface in units of the velocity of light can then be written

$$\begin{aligned}
\frac{v_{\text{surf}}}{c} &= \left(\frac{2\lambda + 1}{4\pi} \right)^{1/2} \langle \dot{\alpha}_{\lambda 0} \rangle_{t=0} \frac{R_A^{(0)}}{c} \\
&= \frac{6(\lambda - 1)!}{(2\lambda + 1)!!(\lambda + 3)^2} \left(\frac{2R_A^{(0)}}{b} \right)^{\lambda+1} \frac{Z_a}{Z_A} \frac{e^2}{\hbar c} \\
&\quad \times (B(E\lambda))_{\text{sp}} \xi R_{\lambda 0}(0, \xi).
\end{aligned} \tag{20}$$

For the examples mentioned above v_{surf} is of the order of $10^{-3}c$, which is small compared to the typical relative velocities of the ions.

6   EXCITATION OF ROTATIONAL STATES

Many nuclei have a nonspherical equilibrium shape. The Hamiltonian describing the surface degrees of freedom then differs from (4.1) in that the equation $\partial H/\partial \alpha_{\lambda\mu} = 0$ leads to a set of deformation parameters not all zero. One defines an intrinsic coordinate system through the principal axes of the corresponding inertial tensor and assumes, in lowest order, harmonic vibrations around the equilibrium shape in this system.

The orientation of the intrinsic system with respect to the laboratory system is defined by the three Eulerian angles θ, ϕ and ψ. Assuming the deformation to be of quadrupole type, the equilibrium shape of the nucleus is completely determined by the two remaining quadrupole variables

$$\alpha_{20}^{(0)} = \beta_0 \cos \gamma_0,$$

$$\alpha_{22}^{(0)} = \alpha_{2-2}^{(0)} = \sqrt{\tfrac{1}{2}}\, \beta_0 \sin \gamma_0. \tag{1}$$

The nucleus can thus rotate and perform oscillations of quadrupole type and of higher multipolarities around this equilibrium shape. In the present section we discuss only the rotational motion of a nucleus with axially symmetric shape ($\gamma_0 = 0$).

The Hamiltonian may then be written in the form

$$H = \frac{\hbar^2}{2\mathscr{I}} \mathbf{R}^2 + H', \tag{2}$$

where $\mathbf{R}^2$ is the square of the rotational angular momentum and $\mathscr{I}$ is the moment of inertia. In terms of the Eulerian angles we have

$$\mathbf{R}^2 = -\left\{ \frac{\partial^2}{\partial \theta^2} + \cot \theta \frac{\partial}{\partial \theta} + \frac{1}{\sin^2 \theta} \frac{\partial^2}{\partial \phi^2} \right\}. \tag{3}$$

The function H' is the Hamiltonian associated with the remaining degrees of freedom. The eigenstates of (2) can be classified according to the quantum number K, which denotes the components of the intrinsic angular momentum along the symmetry axis. The eigenvectors of (2) may thus be written in terms of the product

$$| IKM \rangle = \left(\frac{2I + 1}{4\pi} \right)^{1/2} D_{MK}^{I}(\phi,\theta,0) | NK \rangle, \tag{4}$$

where I and M are the total-angular-momentum and magnetic quantum numbers, respectively. The wavefunction (4) describes the rotation of the intrinsic state $| NK \rangle$ which is a function of the intrinsic coordinates only and which is specified by the quantum numbers N and K.

The eigenvalues of H_A corresponding to the wavefunction (4) are given by

$$E_{INK} = \frac{\hbar^2}{2\mathscr{I}} I(I + 1) + E_{NK}, \tag{5}$$

where $I = K, K+1, \ldots$ $(K>0)$ and where E_{NK} is the eigenvalue of the intrinsic Hamiltonian. The proper eigenstates of the Hamiltonian are linear combinations of the product states (4) with $K = \pm| K |$.

The electric multipole moments are conveniently expressed in terms of the moments $\mathscr{M}'(E\lambda\mu)$ in the intrinsic system, i.e.

$$\mathscr{M}(E\lambda\mu) = \sum_{\nu} D_{\mu\nu}^{\lambda}(\phi,\theta,0) \mathscr{M}'(E\lambda\nu). \tag{6}$$

The expectation value of $\mathscr{M}'$ defines the intrinsic multipole moment, i.e.

$$\langle N,K| \mathscr{M}'(E2,0) | N,K \rangle = \sqrt{\frac{5}{16\pi}} Q_{NK}, \tag{7}$$

where Q_{NK} is the intrinsic quadrupole moment characteristic for the rotational band specified by the intrinsic state $| N,K \rangle$.

We shall only be interested in the excitation of the ground-state rotational band ($Q_{NK} = Q_0$), where

$$\langle NK | \mathscr{M}'(E2\mu) | NK \rangle = \tfrac{1}{2} Q_0 Y_{2\mu}(\theta,\phi). \tag{8}$$

The reduced matrix element between two states in the ground state band is

$$\langle I_m \| \mathscr{M}(E2) \| I_n \rangle =$$

$$\sqrt{\frac{5}{16\pi}} \sqrt{2I_n + 1} \langle I_n K 20 | I_m K \rangle Q_0. \tag{9}$$

The Coulomb excitation of the ground-state rotational band can be worked out analytically only in the sudden approximation (2.15). For the rotational bands in heavy nuclei this is a good approximation, since the adiabaticity parameter associated with the transition between the ground state and the first rotational state turns out to be $\xi \lesssim 0.5(E_{2+})_{\mathrm{MeV}} \approx 0.05$.

For the excitation of the members of the ground-state rotational band the interaction energy (1.8) can be written in terms of the Eulerian angles specifying the orientation of the rotor, since the intrinsic state is not changed in this process. We shall only consider the excitation of even nuclei. For backward scattering in coordinate system B where only states with $M = 0$ are excited, one finds from (2.15) using (1.8), (4) and (8)

$$a_{I0}(t) = \sqrt{2I + 1}\, A_{I0}(\pi, q(t)), \tag{10}$$

where

$$A_{I0}(\pi, q) = \tfrac{1}{2} e^{i\, 2/3q} \int_{-1}^{1} dx\, P_I(x)\, \exp(-2iqx^2)$$

$$= \frac{\Gamma((I+1)/2)}{2\Gamma((2I+3)/2)} e^{-i\, 4/3q} (-2iq)^{I/2}$$

$$\times {}_1F_1(\tfrac{1}{2}(I+2), \tfrac{1}{2}(2I+3); 2iq). \tag{11}$$

The confluent hypergeometric function appearing in (11) can be written in terms of Fresnel integrals. The quantities A_{I0} are tabulated by Alder and Winther (1960). The quantity $q(t)$ is defined by

$$q(t) = \frac{15}{16\sqrt{\pi}} Q_0 q_{20}(0,t) = \frac{Z_a e Q_0}{4\hbar\, v a_0^2} \frac{q_{20}(0,t)}{q_{20}(0,\infty)}, \tag{12}$$

in terms of the intrinsic quadrupole moment and the orbital integrals (4.19) for backward scattering.

For pure Coulomb excitation, the excitation probability at $t = +\infty$ for the ground-state rotational band in even nuclei (at $\vartheta = \pi$) is given by

$$P_I = (2I + 1)|A_{I0}(\pi, q(\infty))|^2, \tag{13}$$

where

$$q(\infty) = \frac{Z_a e Q_0}{\hbar \, v b^2}$$

$$= \sqrt{\frac{45}{16}} \chi_{0 \rightarrow 2} . \tag{14}$$

These excitation probabilities are illustrated in Fig. 7 as a function of q. It is seen that for a given value of q one excites all states in the rotational band up to the state of spin

$$I_{\max} \approx 2q(\infty). \tag{15}$$

The excitation probability of the different states is an oscillating function of the spin, and the maximum excitation probability is found for a spin value a few units smaller than $I_{\max}$. The maximum angular momentum transfer $\Delta L = \hbar \, I_{\max}$ for the excitation of the ground-state rotational band in ^{238}U and ^{160}Gd with various projectiles at a bombarding energy well below the barrier is shown in Fig. I.3. The values of $q(\infty)$ which will be inferred from (14) utilizing $\chi_{0 \rightarrow 2}$ from Fig. 4 would be somewhat larger because of the smaller distance of closest approach used in constructing this figure.

Figure I.3 was only meant for qualitative estimates. The actual excitation of a rotational band is influenced by several effects, as e.g. the deviation from the pure rotational model and the influence of the finite excitation energy on the excitation probability. The latter effect has been estimated by Alder and Winther (1975), where an improved expression for the excitation probability is found to be

$$P_I = P_I(\xi = 0) + \Lambda_I(q)\xi, \tag{16}$$

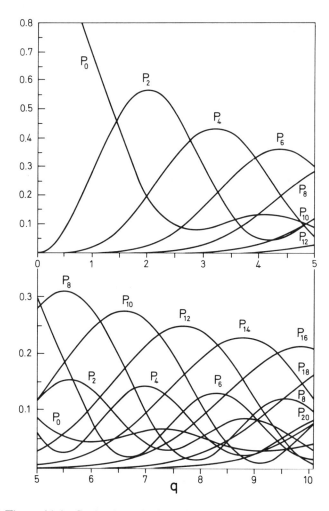

Fig. 7 The multiple Coulomb excitation of a pure rotational band in an even-even nucleus in the sudden approximation. The excitation probability P_I of the state with spin I is given as a function of the parameter q for backward scattering. Note the change of scale between the values of P_I for $q < 5$ and $q > 5$. This figure is from Alder and Winther (1975).

with

$$\xi = \frac{\Delta E_{0\rightarrow 2}}{\hbar}\frac{a_0}{v} = \frac{3\hbar}{\mathcal{I}}\frac{a_0}{v} \tag{17}$$

The quantity $\Lambda_f(q)$, which is tabulated in Alder and Winther (1960) is of the order of magnitude unity.

The expression (13) can to a good approximation be generalized to other scattering angles by replacing q with

$$q_{\text{eff}} = q(\infty)R_2(\vartheta,0), \tag{18}$$

where $R_2(\vartheta,0)$ is defined in (3.8) and illustrated in Fig. 3. Utilizing (17), one can calculate the depopulation of the ground state as a function of the scattering angle and therefore estimate the elastic cross section. One finds

$$\frac{d\sigma_{\text{el}}}{d\sigma_{\text{Ruth}}} = |A_{00}(\pi,q_{\text{eff}})|^2. \tag{19}$$

For the range of angles where the ratio $d\sigma_{\text{el}}/d\sigma_{\text{Ruth}}$ is $\gtrsim 0.1$, this angular distribution coincides with the corresponding expression for the excitation of vibrational states, (4.27), within 10%. For larger scattering angles (19) oscillates, and the two expressions are widely different.

As was done for the excitation of vibrational states, we may estimate the expectation value of the excitation energy. One finds [cf. Alder and Winther (1975)]

$$\langle E \rangle = E_2 |\chi_{0\rightarrow 2}(\vartheta)|^2, \tag{20}$$

where $\chi_{0\rightarrow 2}(\vartheta)$, in analogy to (14), is defined by

$$\chi_{0\rightarrow 2}(\vartheta) = \sqrt{\tfrac{16}{45}}\, q_{\text{eff}}$$

$$= \sqrt{\frac{16}{45}}\frac{Z_a e Q_0}{\hbar v b^2}R_2(\vartheta,0). \tag{21}$$

The expression (20) is identical to the expression (4.28) for $\xi = 0$. Again in this case the energy loss is very small as compared to the bombarding energy.

7   COHERENT EXCITATION OF ROTATIONAL STATES

For bombarding energies above the Coulomb barrier the Coulomb excitation of a rotational state enters through the wavefunction

$$| \psi^A(t) \rangle = \sum_{IM} a_{IM}^A(t)\dot{e}^{-iE_I^A t/\hbar} | IKM \rangle, \qquad (1)$$

which depends on the excitation amplitudes (6.10) in a coherent fashion.

Most properties of this state can be obtained by a classical description of Coulomb excitation in the same way as was done in the case of vibrational states in Sec. 5.

The classical Hamiltonian for the excitation of rotational motion is

$$H = \frac{v\xi}{6a_0\hbar}\left[p_\theta^2 + \frac{1}{\sin^2\theta}\, p_\phi^2 \right] + \sqrt{\frac{\pi}{5}}\, Z_a e Q_0 r^{-3}(t) Y_{20}(\theta,\phi), \qquad (2)$$

where we have specialized to the case of backward scattering. We have here used (6.2), (6.8), (6.17) and (1.8), and have introduced the generalized momenta p_ϕ and p_θ conjugate to the Eulerian angles ϕ and θ describing the orientation of the axis of the symmetric top.

We shall consider the situation in which the top is at rest before the collision. Since ϕ is a cyclic coordinate, this means $p_\phi = 0$ and we may choose $p_\phi = \phi = 0$. The classical equations of motion are

$$\dot{\theta} = \frac{\partial H}{\partial p_\theta} = \frac{v\xi}{3a_0\hbar}\, p_\theta \qquad (3)$$

and

$$\dot{p}_\theta = -\frac{\partial H}{\partial \theta} = \tfrac{3}{4} Z_a e Q_0 r^{-3}(t) \sin 2\theta. \qquad (4)$$

Since $\xi \ll 1$, the angle of rotation does not change appreciably during the collision from its initial value θ_0. We may thus estimate p_θ at time t by

$$p_\theta(t) = \tfrac{3}{4} Z_a e Q_0 \sin 2\theta_0 \int_{-\infty}^{t} [r(t')]^{-3} \, dt'. \tag{5}$$

The angle of rotation up to time $t = 0$ is then

$$\Delta\theta(0) = \frac{v\xi}{3a_0\hbar} \int_{-\infty}^{0} p_\theta(t') \, dt'$$

$$= a_0 v^2 \xi \int_{0}^{\infty} t[r(t)]^{-3} \, dt \, \sin 2\theta_0. \tag{6}$$

The integral in (6) may be expressed in terms of the coefficient γ_2 in (5.12), i.e.

$$\Delta\theta(0) = \frac{A_a A_A}{A_a + A_A} \frac{(Q_0/e)_{\mathrm{fm2}}}{160 Z_A} (\Delta E)_{\mathrm{MeV}} (\sin 2\theta_0) \tfrac{1}{8}\gamma_2. \tag{7}$$

For an initial orientation angle $\theta_0 = 0$ or $\theta_0 = \pi$, the torque vanishes and thus $\Delta\theta(0) = 0$. The maximum value of $\Delta\theta(0)$ is obtained for $\theta_0 = \tfrac{1}{4}\pi$ and $\theta_0 = \tfrac{3}{4}\pi$, where for heavy nuclei one finds

$$\Delta\theta(0) \lesssim 25°. \tag{8}$$

This rotation is similar to the vibrational deformation calculated in Eq. (5.13) in giving rise to a shift in the position of the nuclear surface at the time of contact. The effect, however, may be much larger than the estimate (5.15)–(5.17).

The angular momentum transferred [cf. (5)] until $t = 0$,

$$\mathscr{I}\dot\theta(0) = p_\theta(0) = \hbar \, q(\infty) \sin 2\theta_0, \tag{9}$$

is half of the total angular momentum ΔL transferred during the whole collision,

$$\Delta L = 2\hbar\, q(\infty)\, \sin 2\theta_0. \tag{10}$$

Note that the maximum value of this quantity,

$$\Delta L_{max} = 2q(\infty)\hbar\ , \tag{11}$$

is identical to the value estimated in (6.15).

The fact that the rotational angular momentum transferred in heavy-ion collision depends on the initial orientation angle of the deformed nucleus makes it possible to determine this quantity experimentally. If, in a grazing collision, one considers those reactions which lead to states in the target or in neighboring nuclei with rotational angular momenta close to the maximum value (11), one may conclude that the initial orientation was $\theta_0 \approx \frac{1}{4}\pi$. On the other hand, a reaction which leads to final states with no rotational angular momentum must be reactions in which the initial orientation was $\theta_0 = 0$ or $\theta_0 = \frac{1}{2}\pi$. Although considerable angular momentum may be transferred through particle transfer, the low-rotational-angular-momentum states will still be distinguishable as the states close to the band heads of the corresponding rotational bands. Since the probability for particle transfer depends sensitively on the position of the nuclear surface and therefore on the initial orientation of the deformed nucleus, the above effect should be readily discernable in transfer reactions between heavy deformed nuclei.

There is a striking difference between the Coulomb excitation of rotational and vibrational states in that the excitation probabilities for the vibrational excitation are smooth functions of χ [cf. (4.25)], while for the rotational case they are oscillating functions of χ [cf. (6.13)]. The oscillatory nature of the excitation probability may be put even more in relief by comparing the excitation probability with the classical probability of transferring a given angular momentum ΔL to the rotor.

Considering all initial orientations of the top to be equally probable, this classical angular-momentum distribution after the collision is given by

$$\begin{aligned}
P(\Delta L) &= \frac{1}{2\Delta L_{max}} \left(\frac{\sin(\frac{1}{2}\arcsin f)}{\sqrt{1-f^2}} + \frac{\cos(\frac{1}{2}\arcsin f)}{\sqrt{1-f^2}} \right) \\
&= \frac{1}{2\Delta L_{max}} \frac{1}{\sqrt{1-f}},
\end{aligned} \tag{12}$$

where

$$f = \frac{\Delta L}{\Delta L_{max}}. \qquad (13)$$

The two contributions to (12) reflect the fact that for a fixed value of ΔL, Eq. (10) has two solutions in the interval $0 \leq \theta_0 \leq \frac{1}{2}\pi$. The quantal result of the previous section [cf. (6.13)] for $q = 10$ are illustrated in Fig. 8 in comparison with the classical result (12). The quantal result shows a pronounced oscillatory behaviour around the classical result, and extends beyond the classical allowed region. This oscillation can be understood as an interference between the two classical final states with the same angular momentum but corresponding to different initial orientations. The semiclassical expression for the probability can be written as

$$P(\Delta L) = |\sum_i \sqrt{P_i}\, e^{i\phi_i}|^2, \qquad (14)$$

where ϕ_i is essentially the action integral in units of $\hbar$. The index i labels the various initial conditions leading to the final angular momentum-transfer ΔL. In the present case there are two possible values of i corresponding to the two solutions of (10). The corresponding quantities P_i are the first and second terms in (12), respectively. We thus find

$$P(\Delta L) = P_1 + P_2 + 2\sqrt{P_1 P_2}\cos(\phi_1 - \phi_2)$$

$$= \frac{1}{2\Delta L_{max}}\left(\frac{1}{\sqrt{1-f}} + \sqrt{\frac{2f}{1-f^2}}\cos(\phi_1 - \phi_2)\right). \qquad (15)$$

As shown in Appendix C, the phase difference in the limit of small values of $q(\infty)/\eta$ is given by

$$\phi_2 - \phi_1 = 2q(\infty)(\sqrt{1-f^2} - f\arccos f) - \tfrac{1}{2}\pi. \qquad (16)$$

The resulting probabilities are indicated in Fig. 8 by the solid curve. For $\Delta L < \Delta L_{max}$ they follow the quantal results quite closely.

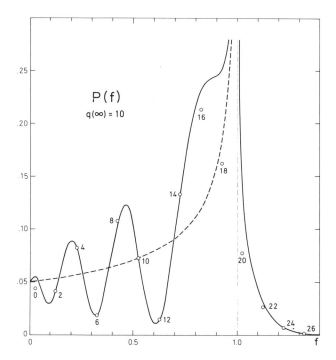

Fig. 8 Comparison between classical and quantal excitation probabilities of Coulomb excitation of the ground-state rotational band in deformed nuclei. The abscissa is the ratio of the transferred angular momentum in units of the maximum classical value ΔL_{max} [cf. (7.11)]. The classical probability distribution (7.12) is indicated by a dashed curve, while the quantal results (6.13) are indicated by open circles for the case of $q(\infty) = 10$. The connection between the spin I and the quantity f was assumed to be $f = (I + \frac{1}{2})/[2q(\infty)]$. The solid curve indicates the semiclassical results (7.15) and (7.18) times $2(=\Delta I)$. The discrepancy between this curve and the quantal result for $I = 0$ may be associated with the somewhat smaller angular momentum interval available for this spin.

The expression (14) can also be used in the classically forbidden region for $f > 1$ by an analytic continuation. One may thus solve the equation (10) by means of the complex angles

$$\theta_0 = \tfrac{1}{2}\pi \pm \tfrac{1}{2}i \operatorname{arccosh} f, \tag{17}$$

and use only the one solution which leads to an action integral with a positive imaginary part. The result is (cf. Appendix C)

$$P(\Delta L) = \frac{\sqrt{f}}{2\,\Delta L_{max}\sqrt{2(f^2-1)}}$$

$$\times \exp[2q(\infty)(\sqrt{f^2-1} - f \operatorname{arccosh} f)]. \qquad (18)$$

The probability of transferring angular momenta above the classically allowed value thus decreases exponentially in accurate agreement with the calculations of the previous section [see Fig. 8].

The expression (14) is quite reminiscent of the one-dimensional WKB approximation, where a turning point in the motion divides the axis into a classically allowed and a classically forbidden region. In the first region the particle can travel in both directions and there will appear interferences between the corresponding waves. In the classical forbidden region the wavefunction and the corresponding probability decrease exponentially, while at the turning point the probability diverges like $1/v$. Improved wavefunctions can be obtained by solving the wave equation in the neighborhood of the turning point in terms of an Airy function. By modifying the argument of the Airy function it is possible to construct a wavefunction which not only is correct at the singularity but joins smoothly with the WKB wavefunction in the classically allowed as well as in the classically forbidden regions. Such a uniform approximation can be introduced to improve (14) [cf. Levit et al (1974) and Massmann and Rasmussen (1975); also Appendix III.E].

8   EXCITATION OF VIBRATIONAL STATES IN DEFORMED NUCLEI

In deformed nuclei with axially symmetric equilibrium shape the vibrational amplitudes are specified in an intrinsic system with z-axis along the symmetry axis. As mentioned in Sec. 6, there are two intrinsic quadrupole degrees of freedom, which can be described in terms of the parameters β and γ [cf. (6.1)]. For higher multipole orders the vibrational amplitudes

$\alpha'_{\lambda\nu}$ in the intrinsic frame are expected to occur with components $\nu = 0,1,2,\ldots,\lambda$ along the symmetry axis.

For small vibrations around the spheroidal equilibrium shape the nuclear Hamiltonian is assumed to be [cf. Eq. (6.2)]

$$H_A = H_{\text{vib}} + \frac{\hbar^2}{2\mathscr{I}} \mathbf{R}^2 , \tag{1}$$

with

$$
\begin{aligned}
H_{\text{vib}} = {} & \tfrac{1}{2}D_\beta\dot{\beta}^2 + \tfrac{1}{2}C_\beta(\beta - \beta_0)^2 \\
& + \tfrac{1}{2}D_\gamma\dot{\gamma}^2 + \tfrac{1}{2}C_\gamma\gamma^2 \\
& + \sum_{\substack{\lambda>2 \\ 0\leq\nu\leq\lambda}} \{\tfrac{1}{2}D_{\lambda\nu}|\,\dot{\alpha}'_{\lambda\nu}|^2 + \tfrac{1}{2}C_{\lambda\nu}|\,\alpha'_{\lambda\nu}|^2\}.
\end{aligned}
\tag{2}
$$

The vibrational amplitudes are conveniently expressed in terms of the creation and annihilation operators (cf. (4.3)) as

$$\alpha'_{20} = \beta = \sqrt{\frac{\hbar\,\omega_\beta}{2C_\beta}}\,(c_\beta'^\dagger + c_\beta') + \beta_0 , \tag{3}$$

$$\alpha'_{2\,\pm2} = \frac{\beta_0}{\sqrt{2}}\,\gamma = \frac{\beta_0}{\sqrt{2}}\sqrt{\frac{\hbar\,\omega_\gamma}{2C_\gamma}}(c_\gamma'^\dagger + c_\gamma'), \tag{4}$$

$$\alpha'_{\lambda\nu} = \sqrt{\frac{\hbar\,\omega_{\lambda\mu}}{2C_{\lambda\nu}}}\,(c_{\lambda\nu}'^\dagger + c_{\lambda\nu}'). \tag{5}$$

The electric multipole moments in the intrinsic frame are assumed to be proportional to these amplitudes, i.e.

$$\mathscr{M}'(E\lambda\nu) = (2\lambda + 1)^{-1/2}F_\lambda\alpha'_{\lambda\nu} , \tag{6}$$

with F_λ defined in (4.13). In the laboratory system the electric multipole moments are then

$$\mathscr{M}(E\lambda\mu) = \sum_{\nu} \mathscr{M}'(E\lambda\nu) D^{\lambda}_{\mu\nu}(\phi,\theta,0) \, , \tag{7}$$

where θ and ϕ are the polar angles specifying the symmetry axis of the rotor.

Since the nucleus often rotates little during the collision time (as was shown in Sec. 7), we can calculate the excitation of the vibrational modes as if the nuclear axis were fixed in space. We thus solve the time dependent Schrödinger equation (1.12) with H_A given by (1), assuming θ and ϕ to be constants. The probability of exciting a state with $N_{\lambda\nu}$ phonons in the mode λ, ν is then given by a Poisson distribution

$$P_{N_{\lambda\nu}}(\theta_0,\phi_0) = \frac{1}{N_{\lambda\nu}!} |\, t^{\lambda\nu}(\theta_0,\phi_0)|^{\,2N_{\lambda\nu}} \exp\{ - |\, t^{\lambda\nu}(\theta_0,\phi_0)|^{\,2}\} \, , \tag{8}$$

where

$$t^{\lambda\nu}(\theta_0,\phi_0) = \sqrt{2\lambda + 1} \; \chi_{0\to\lambda\nu} \sum_{\mu} (-1)^{\mu} D^{\lambda}_{\mu\nu}(\phi_0,\theta_0,0) R_{\lambda-\mu}(\vartheta,\xi), \tag{9}$$

and

$$\chi_{0\to\lambda\nu} = \frac{\sqrt{16\pi} \; Z_a e(\lambda - 1)!}{\hbar \, va_0^{\lambda}(2\lambda + 1)!!} F_{\lambda} \sqrt{\frac{\hbar \, \omega_{\lambda\nu}}{2C_{\lambda\nu}}} \, , \tag{10}$$

is the strength parameter for the excitation of the vibrational mode λ, ν. Thus we find in perturbation theory that the total excitation probability of this mode for backward scattering is

$$P_{\lambda\nu} = \int |\, t^{\lambda\nu}(\theta,\phi)|^2 \frac{1}{4\pi} \sin\theta \, d\theta \, d\phi$$

$$= |\, \chi_{0\to\lambda\nu}|^2 \, R_{\lambda}^2(\pi,\xi) \, . \tag{11}$$

Inserting the $B(E\lambda, 0 \to \lambda)$ value for the transition from the ground state to the rotational state of spin λ in the λ, ν band, one finds that the strength parameters are at the maximum, of the order of 0.5, for quadrupole as well as octupole excitations.

The small excitation probabilities associated with higher members of the vibrational bands may still be interesting for the study of fission induced by Coulomb excitation. To the extent that the fission mode in a deformed nucleus is associated with the β degree of freedom, one may relate the fission probability to the probability of multiple excitation of the β-vibrational mode to energies above the fission barrier. A schematic model for ^{234}U was illustrated in Fig. I.6(a), where a cubic potential was used which was adjusted so that the equilibrium deformation β_0 as well as the fission barrier E_b and the corresponding deformation $\beta_0 + \Delta\beta$ are realistic. Utilizing also the experimental value for the transition probability $B(E2, 0 \rightarrow 2^+(\beta))$, one can calculate the total excitation probability of the states with N β-vibrational phonons as

$$P_N = \int P_N(\theta,\phi)\frac{1}{4\pi}\sin\theta \, d\theta \, d\phi \ . \tag{12}$$

The result in Fig. I.6(b) shows the Coulomb excitation of ^{234}U by Kr ions for different bombarding energies below the Coulomb barrier. The results are given as a cross section

$$\left(\frac{d\sigma}{d\Omega}\right)_N = P_N \left(\frac{d\sigma}{d\Omega}\right)_{\text{Ruth}} \tag{13}$$

for backward scattering. The deviations of the matrix elements from the simple model due to the cubic potential were included.

In the reaction, angular momentum will be transferred to rotational motion. The amount of angular momentum is classically given by

$$\Delta L = 2q\hbar \ \sin 2\theta_0 \ , \tag{14}$$

with

$$q = \frac{Z_a e Q_0}{4\hbar \, v a_0^2} , \tag{15}$$

Q_0 being the intrinsic quadrupole moment $\approx 11 \ e$ b. A fast rotational motion may lower the fission barrier because of the associated centrifugal potential. The lowering ΔE of the barrier is estimated by

$$\Delta E = \tfrac{1}{2}(\Delta L)^2 \left(\frac{1}{\mathscr{I}(\beta_0)} - \frac{1}{\mathscr{I}(\beta_0 + \Delta\beta)} \right)$$

$$= \frac{(\Delta L)^2}{2\,\mathscr{I}(\beta_0)} \left(1 - \frac{1}{1 + (\Delta\beta/\beta_0)\alpha} \right) , \tag{16}$$

where $\mathscr{I}(\beta)$ is the moment of inertia for deformation β, and α is between 0.4 and 1.

The amount of angular momentum transferred depends strongly on the vibrational state which is excited. Thus we find for backward scattering from (9)

$$t^{20}(\theta_0\phi_0) = \sqrt{5}\ \chi_{0\to\beta}P_2(\cos\theta_0)R_{20}(\pi,\xi) , \tag{17}$$

and the probability distribution (8) is therefore strongly peaked at $\theta_0 = 0$ for high quantum numbers $N_{\lambda\nu}$. The integral in (12) receives its main contribution from a narrow region around $\theta_0 = 0$, and the angular momentum transfer is correspondingly small according to (14).

A more quantitative estimate [cf. Beyer *et al.* (1970)] leads to a rotational energy $(\Delta L)^2/[2\,\mathscr{I}(\beta_0)]$ of less than 0.5 MeV and therefore to a lowering of the barrier not larger than this value.

9   COULOMB EXCITATION AS AN ABSORPTION

Through the Coulomb excitation process the ground state is depopulated as a function of time during the collision. In nuclear reaction theory one often describes such a depopulation in terms of an absorptive potential $W(r)$, which in the semiclassical description is equivalent to a position dependent mean free path. In terms of this potential the probability P_0 of being in the entrance channel is given by [cf. (III.2.11) and (III.2.13)]

$$P_0(t) = e^{-(2/\hbar)\int_{-\infty}^{t} W(t')\,dt'} . \tag{1}$$

For the case of the excitation of a vibrational band, the expression for the probability that the target remains in the ground state has the same structure as this expression. One thus finds from (4.22)

$$P_0(t) = |a_0(t)|^2 = e^{-B(E\lambda)q_\lambda^2(\omega,t)} . \tag{2}$$

For the case where several vibrational degrees of freedom can be excited the probability P_0 will be

$$P_0(t) = \prod_{\lambda,\,n} P_0^{(\lambda,n)}(t) = e^{-\Sigma_{\lambda,n}B(E\lambda;0\to(\lambda n))q_\lambda^2(\omega_n,t)} . \tag{3}$$

For the excitation of other degrees of freedom the probability P_0 may be quite different, as was discussed e.g. in connection with (6.19). Only for small values of the exponent is the equation (2) correct in all cases, since it then coincides with the result in lowest-order perturbation theory for the depopulation of the ground state. Identifying the two expressions (1) and (3), we obtain the following expression for $W(r(t))$:

$$W(r(t)) = \frac{\hbar}{2} \sum_{\lambda,n} B(E\lambda;0 \to (\lambda,n)) \frac{d}{dt} q_\lambda^2(\omega_n,t). \tag{4}$$

An analytic expression for (4) can be found in the sudden approximation where $\omega_n = 0$. For $\lambda = 2$ and backward scattering we find from (4.19)

$$W(r(t)) = \frac{4\pi}{25} \frac{Z_a^2 e^2}{\hbar} \frac{1}{[r(t)]^3} \int_{-\infty}^{t} \frac{dt'}{[r(t')]^3} \sum_n B(E2;0 \to n) . \tag{5}$$

The result may be written in the form

$$W(r(t)) = \frac{8\pi}{25} \frac{Z_a^2 e^2}{\hbar\, v} \frac{\sum_n B(E2;0 \to n)}{b^5} f(r/b) , \tag{6}$$

where

$$
f(r/b) = \frac{b^3}{3r^3}
\begin{cases}
2 - \dfrac{v(r)}{v}\left[3 - \left(\dfrac{v(r)}{v}\right)^2\right], & t < 0, \\[4mm]
2 + \dfrac{v(r)}{v}\left[3 - \left(\dfrac{v(r)}{v}\right)^2\right], & t > 0.
\end{cases}
\tag{7}
$$

The quantity $v(r)$ is the local velocity:

$$
\frac{v(r)}{v} = \frac{1}{v}\sqrt{\left(E - \frac{Z_a Z_A e^2}{r}\right)\frac{2}{m_{aA}}} = \sqrt{1 - \frac{b}{r}},
\tag{8}
$$

while $v = \sqrt{2E/m_{aA}}$ is the velocity at infinity.

The function $f(r(t))$ is quite unsymmetric about $t = 0$, as can be seen from Fig. 9, meaning that the depopulation mainly takes place after the two nuclei have reached the distance of closest approach. The absorptive potential $W(r)$ can be calculated numerically also for nonvanishing excitation energies $\hbar\,\omega_n$. In this case the potential may become negative for $t > 0$, indicating that the entrance channel is repopulated. The general expression (4) thus includes a depopulation through polarization effects which in fact only give rise to a virtual depopulation.

In the above discussion we determined W from the condition that (1) describes the population in the ground state at any time. If instead we ask only that (1) reproduce the probability at time $t = +\infty$, there is a considerable ambiguity in the determination of $W(r)$. One may thus delete the terms proportional to $v(r)$ in (7), since they cancel out in the integral from $-\infty$ to $+\infty$. Such potential, generalized to other impact parameters, has been calculated by Baltz et al. (1978) and Love et al. (1977).

More systematically, one may impose additional constraints to determine $W(r)$. Thus one may require that $W(r)$ be independent of the bombarding energy, which is not the case for the potential (6). Again in

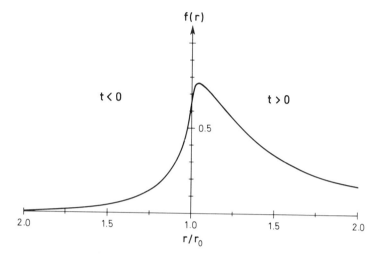

Fig. 9 The function of $f(r)$ [cf. (9.7)] associated with the depopulation of the ground state in Coulomb excitation. The abcissa is r measured in units of the distance of closest approach, $b = Z_a Z_A e^2 / E$.

the sudden approximation and for $\lambda = 2$ and backward scattering we find, utilizing (4.24), the following condition for W:

$$\frac{2}{\hbar} \int_{-\infty}^{\infty} W(r(t)) \, dt = \frac{16\pi Z_a^2 e^2}{225 \hbar^2 v^2 a_0^4} \sum_n B(E2; 0 \to n), \qquad (9)$$

which should be satisfied as an identity in the bombarding energy. Writing the equation instead as an identity in the distance of closest approach $r_0 = 2a_0$, we find

$$\int_1^{\infty} \frac{dx \, W(r_0 x)}{\sqrt{1 - 1/x}} = \frac{32\pi}{225} \left(\frac{2 Z_a^3 m_{aA} e^2}{r_0^9 Z_A \hbar^2} \right)^{1/2} \sum_n B(E2; 0 \to n), \qquad (10)$$

where we have utilized

$$dt = \frac{dr}{\sqrt{\dfrac{2}{m_{aA}}\left(E - \dfrac{Z_a Z_A e^2}{r}\right)}}. \tag{11}$$

The solution of this integral equation is seen to be

$$W(r) = \beta r^{-9/2} , \tag{12}$$

where β is a constant. Inserting this in (10), we obtain

$$W(r) = \frac{512}{1125} \sqrt{\frac{2m_{aA} Z_a^3 e^2}{Z_A \hbar^2}} \;\; \frac{\sum\limits_{n} B(E2; 0 \to n)}{r^{9/2}} . \tag{13}$$

The absorptive potentials derived above, when generalized to other scattering angles, can be used to analyze elastic scattering for bombarding energies below the Coulomb barrier. At higher bombarding energies, however, depopulation of the entrance channel takes place also through transfer reactions and inelastic processes caused by the nuclear field. Since the latter interfers with Coulomb excitation, the total depopulation due to inelastic processes is not simply the sum of the two contributions but contains also an interference term. In fact this term is negative, since the Coulomb and nuclear forces have opposite signs. The corresponding term in the absorptive potential will thus act as a source. The use of a Coulomb absorptive potential together with an absorptive potential due to nuclear interaction is therefore expected to be correct only to the extent that depopulation due to transfer reactions dominates at distances where nuclear inelastic processes start to become important.

The technique used in deriving (13) is however rather general and will be used in Chaps. IV and V to calculate the absorptive potentials due to nuclear inelastic scattering and to transfer reactions.

APPENDIX A   COULOMB EXCITATION FOR
SMALL-ANGLE SCATTERING

For bombarding energies above the Coulomb barrier, Coulomb excitation is mixed with nuclear excitation except for large impact parameters leading to forward scattering angles. At such forward angles one is usually able to use perturbation theory for calculating the differential cross section, which is then given by the formula

$$\left(\frac{d\sigma}{d\Omega}\right)_n = P_n \left(\frac{d\sigma}{d\Omega}\right)_{\text{Ruth}} , \tag{1}$$

where

$$P_n = |\chi_{0 \to n}(\vartheta, \xi)|^2. \tag{2}$$

For the population of a definite magnetic substate the quantity χ is defined in Eq. (3.2) in terms of the functions $R_{\lambda\mu}(\vartheta, \xi)$. These functions are tabulated in Alder and Winther (1975), but for the forward scattering angles of interest one may use the simpler expression

$$R_{\lambda\mu}^A(\vartheta, \xi) = (-1)^{(\lambda+\mu)/2} \frac{(2\lambda - 1)!!}{(\lambda - 1)!} e^{-\pi\xi/2} \xi^\lambda$$

$$\times \sum_{v=-\lambda}^{\lambda} d_{\mu v}^\lambda(\tfrac{1}{2}\pi) \frac{K_v(\varepsilon\xi)}{[(\lambda + v)!(\lambda - v)!]^{1/2}} \tag{3}$$

[cf. Winther and Alder (1979)]. In (3), $K_v(x)$ indicates the modified Bessel function and $d_{\mu v}^\lambda$ a rotation matrix. Explicit expressions for $\lambda = 1, 2$ and 3 are given in Table 3. The result (3) is valid for eccentricity $\varepsilon = (\sin \tfrac{1}{2}\vartheta)^{-1} \gg 1$ and for all values of ξ. In particular one finds for $\xi = 0$

$$R_{\lambda\mu}^A(\vartheta, 0) = (-1)^{(\lambda+\mu)/2} \frac{(2\lambda - 1)!!}{[(\lambda + \mu)!(\lambda - \mu)!]^{1/2}} \varepsilon^{-\lambda} . \tag{4}$$

For the total transition probability the quantity χ is given by the expression (3.7), where $R_\lambda^2(\vartheta, \xi)$ in the same approximation is given by

Table 3. Approximate expressions for orbital integrals $R_{\lambda\mu}(\vartheta,\xi)$ in coordinate system A *for* $\lambda = 1, 2$ and 3. The expressions are valid for small-angle scattering [cf. Eq. (A.3)].

λ	μ	$R_{\lambda\mu}^{A}(\vartheta,\xi)$
1	± 1	$\mp\frac{1}{2}\sqrt{2}\,\xi e^{-\pi\xi/2}[K_1(\xi\varepsilon) \mp K_0(\xi\varepsilon)]$
2	± 2	$\frac{1}{8}\sqrt{6}\,\xi^2 e^{-\pi\xi/2}\,[K_2(\xi\varepsilon) \mp 4K_1(\xi\varepsilon) + 3K_0(\xi\varepsilon)]$
2	0	$-\frac{3}{4}\xi^2 e^{-\pi\xi/2}\,[K_2(\xi\varepsilon) - K_0(\xi\varepsilon)]$
3	± 3	$\mp\frac{1}{32}\sqrt{5}\,\xi^3 e^{-\pi\xi/2}\,[K_3(\xi\varepsilon) \mp 6K_2(\xi\varepsilon) + 15K_1(\xi\varepsilon) \mp 10K_0(\xi\varepsilon)]$
3	± 1	$\pm\frac{5}{32}\sqrt{3}\,\xi^3 e^{-\pi\xi/2}\,[K_3(\xi\varepsilon) \mp 2K_2(\xi\varepsilon) - K_1(\xi\varepsilon) \pm 2K_0(\xi\varepsilon)]$

$$R_{\lambda}^{2}(\vartheta,\xi) = \left[\frac{(2\lambda - 1)!!}{(\lambda - 1)!}\right]^2 \xi^{2\lambda}e^{-\pi\xi}$$

$$\times \sum_{\mu}\frac{[K_{\mu}(\xi\varepsilon)]^2}{(\lambda + \mu)!(\lambda - \mu)!}. \tag{5}$$

Insofar as the quantity $\chi(\vartheta,\xi)$ is smaller than unity, perturbation theory applies and the cross-section is given by (1).

APPENDIX B   SEMICLASSICAL S-MATRIX

It is often convenient to write the solution of the time dependent Schrödinger equation (1.6),

$$i\hbar\,\frac{\partial}{\partial t}|\,\psi(t)\,\rangle = [H_0 + V(t)]\,|\,\psi(t)\,\rangle\,, \tag{1}$$

in terms of the S-matrix, i.e.

$$|\,\psi(\infty)\,\rangle = S|\,\psi(-\infty)\,\rangle\,. \tag{2}$$

93

II   Coulomb Excitation

The standard expression for S is

$$S = \mathcal{T} \exp\{-\tfrac{i}{\hbar} \int_{-\infty}^{\infty} \tilde{V}(t)dt\},$$
(3)

where $\tilde{V}$ is the interaction $V(t)$ in the interaction representation,

$$\tilde{V}(t) = e^{(i/\hbar)H_0 t} V(t) e^{-(i/\hbar)H_0 t},$$
(4)

and $\mathcal{T}$ indicates the time-ordering operator.

The series expansion of (3) leads to the usual perturbation series

$$S = 1 - \frac{i}{\hbar} \int_{-\infty}^{\infty} \tilde{V}(t')\, dt'$$

$$+ \left(\frac{-i}{\hbar}\right)^2 \int_{-\infty}^{\infty} dt' \int_{-\infty}^{t'} dt''\ \tilde{V}(t')\tilde{V}(t'')$$

$$+ \cdots .$$
(5)

If the interaction $V(t)$ is only different from zero for a short time τ compared to the times characteristic for the system H_0, i.e. if

$$\xi = \frac{\Delta E}{\hbar} \tau \ll 1,$$
(6)

then the quantity $\tilde{V}(t)$ can be replaced by $V(t)$ and the time ordering in (5) is unimportant. The series can then be summed to give the result

$$S = \exp\{-\tfrac{i}{\hbar} \int_{-\infty}^{\infty} V(t)dt\}.$$
(7)

In general it may be convenient to write the S-matrix in a form where the expression (7) appears as a natural first approximation. An expansion of this type is

$$S = e^{i(R+G+\cdots)},$$
(8)

where

$$R = -\frac{1}{\hbar} \int_{-\infty}^{\infty} \tilde{V}(t)\, dt \qquad (9)$$

and

$$G = \frac{i}{2\hbar^2} \int_{-\infty}^{\infty} dt' \int_{-\infty}^{t'} dt'' [\tilde{V}(t'), \tilde{V}(t'')] . \qquad (10)$$

The remaining terms contain higher-order commutators [cf. e.g. Alder and Winther (1975)].

This expansion is especially convenient for harmonic oscillators if the interaction $V(t)$ is linear in the amplitude. For the one-dimensional case this means

$$V(t) = f(t)(c^\dagger + c), \qquad (11)$$

where c is the annihilation operator. The quantity $\tilde{V}$ is then

$$\tilde{V}(t) = f(t)(c^\dagger e^{i\omega t} + c e^{-i\omega t}), \qquad (12)$$

as can be seen for instance by using the expansion

$$e^{iA} B e^{-iA} = B + i[A,B]$$
$$+ \frac{1}{2!} i^2 [A,[A,B]] + \cdots \qquad (13)$$

for (4). The quantities R and G are then

$$R = -\frac{1}{\hbar} \int_{-\infty}^{\infty} f(t) e^{i\omega t}\, dt\, c^\dagger + \text{H.c.} \qquad (14)$$

and

$$G = \frac{1}{\hbar^2} \int_{-\infty}^{\infty} dt' \int_{-\infty}^{t'} dt'' \, f(t')f(t'') \sin \omega(t' - t'') . \qquad (15)$$

Since the quantity G is a c-number, all higher-order terms in (8) vanish and we find

$$S = e^{-i(\chi c^{\dagger} + \chi^* c) + iG}, \qquad (16)$$

where

$$\chi = \frac{1}{\hbar} \int_{-\infty}^{\infty} f(t)e^{i\omega t} \, dt . \qquad (17)$$

In order to evaluate the matrix elements of S we use the expression

$$e^{A+B} = e^A e^B e^{-[A,B]/2}, \qquad (18)$$

which is correct if $[A,B]$ is a c-number. We thus find

$$S = e^{-|\chi|^2/2 + iG} \, e^{-i\chi c^{\dagger}} e^{-i\chi^* c} . \qquad (19)$$

When this operator acts on the ground state of the harmonic oscillator, only the first term in a series expansion of the last exponential contributes and we find

$$S|0\rangle = e^{-|\chi|^2/2 + iG} \, e^{-i\chi c^{\dagger}} |0\rangle. \qquad (20)$$

This result is essentially equivalent to (4.18) except that there we used the upper limit t on all integrals instead of $+\infty$.

APPENDIX C   COULOMB EXCITATION OF ROTATIONAL STATES BY CLASSICAL-LIMIT QUANTUM MECHANICS

In recent years there has been an important development in the use of semiclassical methods for solving scattering and reaction problems, especially within the field of physical chemistry [Miller (1974)]. These methods are particularly useful in those cases in which all the relevant degrees of freedom have a simple classical interpretation, like rotations and vibrations as they are coupled to the relative motion of two colliding systems.

In this appendix we apply these methods to the Coulomb excitation of rotational states as carried out by Massmann and Rasmussen (1975)[cf. also Levit *et al.* (1974)].

For zero impact parameter, the classical Hamiltonian of a charged particle impinging on a quadrupole deformed nucleus can be written as

$$H = \frac{p_r^2}{2m_{aA}} + p_\chi^2 \left(\frac{1}{2m_{aA}r^2} + \frac{1}{2\mathcal{I}} \right)$$

$$+ \frac{Z_a Z_A e^2}{r} + \frac{Z_a e Q_0}{2r^3} P_2(\cos\chi) . \tag{1}$$

The angle χ is the difference between the polar angle of the symmetry axis of the top and the polar angle of the scattered particle measured in a coordinate system with z-axis along the beam and pointing towards the incoming particle. It is further assumed that the rotor is initially at rest.

The equations of motion are

$$\dot{r} = \frac{p_r}{m_{aA}} , \tag{2}$$

$$\dot{p}_r = \frac{p_\chi^2}{m_{aA}r^3} + \frac{Z_a Z_A e^2}{r^2} + \frac{3Z_a e Q_0}{4r^4} (3\cos^2\chi - 1) , \tag{3}$$

$$\dot{\chi} = \left(\frac{1}{m_{aA}r^2} + \frac{1}{\mathcal{I}} \right) p_\chi, \tag{4}$$

$$\dot{p}_\chi = \frac{Z_a e Q_0}{4r^3} 3\sin 2\chi . \tag{5}$$

The change in the angle χ can be estimated by

$$\Delta\chi \approx \frac{Z_a e Q_0}{2b^3}\left(\frac{1}{m_{aA}b^2} + \frac{1}{\mathscr{I}}\right)\tau_{\text{coll}}^2$$

$$\approx \frac{q}{10}\left(\frac{1}{\eta} + \xi\right), \tag{6}$$

where

$$q = q(\infty) = \frac{Z_a e Q_0}{\hbar\, v b^2}, \tag{7}$$

and ξ is the adiabaticity parameter

$$\xi = \frac{3\hbar}{\mathscr{I}}\frac{a_0}{v} \tag{8}$$

corresponding to the excitation of the lowest rotational state. We shall assume that both q/η and $q\xi$ are small, in which case χ is constant during the collision and equal to the initial orientation angle:

$$\chi = \theta_0. \tag{9}$$

The angular momentum of the top after the collision is thus equal to

$$\Delta L = p_\chi(\infty)$$

$$= 3\sin 2\theta_0 \frac{Z_a e Q_0}{4}\int_{-\infty}^{\infty}\frac{dt}{r^3}\left[1 + O(q/\eta) + O(q\xi)\right]. \tag{10}$$

The term of order $q\xi$ has been evaluated in Alder and Winther (1966). To evaluate the integral we use energy conservation, noting that the second term in (1) is of order $(q/\eta)[(q/\eta) + q\xi]$ as compared to the third term. We thus obtain

$$dt = \frac{\pm\, dr\sqrt{\frac{1}{2}m_{aA}}}{\sqrt{E - \dfrac{Z_a Z_A e^2}{r} - \dfrac{Z_a e Q_0}{4r^3}(3\cos^2\theta_0 - 1)}} \,, \qquad (11)$$

leading to the result

$$\int_{-\infty}^{\infty}\frac{dt}{r^3} = \frac{2}{3a_0^2 v}\left[1 + O\!\left(\frac{q}{\eta}\right)\right]. \qquad (12)$$

To lowest order in the small parameters we find

$$\Delta L = 2q\hbar\,\sin 2\theta_0. \qquad (13)$$

In order to construct the classical angular-momentum distribution after the collision we have to consider that there are two angles

$$\theta_0^{(1)} = \tfrac{1}{2}\arcsin f \qquad (14)$$

and

$$\theta_0^{(2)} = \tfrac{1}{2}\pi - \tfrac{1}{2}\arcsin f, \qquad (15)$$

which lead to the same angular momentum for

$$f = \frac{\Delta L}{\Delta L_{\max}} < 1. \qquad (16)$$

For $\Delta L > \Delta L_{\max} = 2\hbar\, q$, there are no real solutions, but we shall later use the complex solution

$$\theta_0 = \tfrac{1}{4}\pi \pm i\,\tfrac{1}{2}\operatorname{arccosh} f. \qquad (17)$$

Assuming that the top has an isotropic distribution in space, i.e. a probability distribution

$$p(\theta_0) = \sin \theta_0 \qquad (0 \le \theta_0 \le \tfrac{1}{2}\pi), \qquad (18)$$

the angular-momentum distribution after the collision is given by

$$P(\Delta L) = \sum_i | P_i(\Delta L)| , \qquad (19)$$

with

$$P_i(\Delta L) = p(\theta_0^{(i)}(\Delta L))\frac{d\theta_0}{d(\Delta L)} \Bigg|_{\theta_0 = \theta_0^{(i)}}. \qquad (20)$$

We thus find

$$P(\Delta L) = \frac{\sin(\tfrac{1}{2}\arcsin f)}{4q\hbar \sqrt{1-f^2}} + \frac{\cos(\tfrac{1}{2}\arcsin f)}{4q\hbar \sqrt{1-f^2}}$$

$$= \frac{1}{2\Delta L_{max}} \frac{1}{\sqrt{1-f}} \qquad \text{for} \quad f < 1 \qquad (21)$$

and

$$P(\Delta L) = 0 \qquad \text{for } f > 1. \qquad (22)$$

One may generalize this result to include quantal interference effects by the expression

$$P(\Delta L) = | \sqrt{| P_1 |} \, e^{i\phi_1} + \sqrt{| P_2 |} \, e^{i\phi_2} |^2, \qquad (23)$$

where $P_i = P_i(\Delta L)$ are the classical probabilities used above, and ϕ_i is the action integral

$$\phi = \pm\frac{\pi}{4} - \frac{1}{\hbar} \int_{-\infty}^{\infty} (x\dot{p}_x + r\dot{p}_r) \, dt \qquad (24)$$

for the two trajectories corresponding to the solutions in question. The sign is chosen according to the sign of (20).

The action integral can readily be evaluated to lowest order in q/η. The first term leads to

$$\frac{1}{\hbar} \int_{-\infty}^{\infty} x \dot{p}_x \, dt = 2q\theta_0 \sin 2\theta_0, \tag{25}$$

where (5), (9), and (12) have been used. The last term in (24) is given by

$$\frac{1}{\hbar} \int_{-\infty}^{\infty} r \dot{p}_r \, dt$$

$$= \eta \int_{-\infty}^{\infty} v \left(\frac{1}{r} + \frac{3a_0^2}{r^3} \frac{q}{\eta} (3 \cos^2 \theta_0 - 1) \right) dt \ . \tag{26}$$

where we neglected the first term in (3) since it is of order $(q/\eta)^2$ as compared to the second term. In evaluating the second term in (26) we can neglect the correction in the trajectory, i.e. the q/η term in (12). On the other hand, in evaluating the first term in (26) the quadrupole interaction term in (11) is important. To lowest order in q/η we obtain

$$\eta \int_{-\infty}^{\infty} \frac{v \, dt}{r} = 2\eta \int_{x_0}^{R} \frac{dx}{x \sqrt{1 - \frac{1}{x} - \frac{q}{4\eta x^3} (3 \cos^2 \theta - 1)}}$$

$$= A - \tfrac{4}{3} q (3 \cos^2 \theta_0 - 1), \tag{27}$$

where A is independent of θ_0.

The phase ϕ is thus given by

$$\phi = \pm \tfrac{1}{4}\pi - 2q\theta_0 \sin 2\theta_0 - 2q \cos^2 \theta_0 + \text{const.} \tag{28}$$

For the two trajectories corresponding to the solutions (14) and (15) we find

$$\phi_1 = \tfrac{1}{4}\pi - qf \arcsin f - 2q \cos^2(\tfrac{1}{2} \arcsin f)$$

$$\phi_2 = -\tfrac{1}{4}\pi - qf\pi + qf \arcsin f - 2q \sin^2(\tfrac{1}{2} \arcsin f). \tag{29}$$

Inserting the result (29) and the probabilities appearing in (21) into (23), we obtain the result (7.15) and (7.16).

In the classically forbidden region one should use the complex solutions (17) in the expression (23). One of these solutions gives a probability which increases exponentially in the forbidden region and is to be discarded. The other solution (below the real axis) leads to the result

$$P(\Delta L) = \left| \frac{\sin(\frac{1}{4}\pi - i\,\frac{1}{2}\,\mathrm{arccosh}\,f)}{4q\hbar\,\sqrt{1-f^2}} \right|$$

$$\times \exp\{4q\,Im[\,-i\,\tfrac{1}{2}f\,\mathrm{arccosh}\,f + \cos^2(\tfrac{1}{4}\pi - i\,\tfrac{1}{2}\,\mathrm{arccosh}\,f)]\}, \qquad (30)$$

which reduces to the expression (7.18).

CHAPTER III

ELASTIC SCATTERING

When, in a heavy-ion reaction, the two nuclei come within the range of the nuclear forces, the trajectory will be changed by the attraction which will act between the nuclear surfaces. This surface interaction is a fundamental quantity in all heavy-ion reactions. The corresponding average potential—the ion-ion potential—is a quantity which can be estimated rather accurately by theoretical means from the knowledge of the nuclear densities and effective interactions.

The very short wavelength associated with the relative motion of the interacting ions makes heavy-ion physics, to a large extent, a problem of classical scattering of extended objects with surface degrees of freedom, a picture which has to be supplemented with the superposition principle.

Elastic scattering provides a rich field for exploration of the ion-ion potential and of the semiclassical description. Because of the Coulomb excitation and of the nuclear reactions which take place at the distance where the ion-ion potential is felt, one has also to take into account the depopulation of the elastic channel. This is usually done for "light" heavy-ion reactions through an imaginary potential. Although useful, the vagueness of this quantity was already pointed out in Chap. II (Sec. II.9). Still, we shall be pragmatic and use this quantity in the present chapter in analyzing the experimental data. In the following chapter we shall get a better understanding of this procedure by studying the depopulation due to inelastic processes. In this context we derive an imaginary potential which for the special case of excitation of surface vibrational states only reproduces the elastic scattering amplitude.

For the case of collisions of very heavy ions these questions are less relevant, as one would hardly be able to detect the nuclear effects in elastic scattering because of the depopulation of the elastic channel by Coulomb excitation.

1   ION-ION POTENTIAL

When the two nuclei come to distances within the range of nuclear forces, the nuclear surfaces will attract each other. This attractive surface-surface interaction can be estimated by calculating the matrix element

$$U_{aA}^N(r) = \langle \psi_0^a \psi_0^A | \ V_{aA}^N \ | \psi_0^a \psi_0^A \rangle \,, \tag{1}$$

where V_{aA}^N is the sum of the two-body interactions V_{12} between all nucleons in a and all nucleons in A, while ψ_0^a and ψ_0^A are the wavefunctions describing the ground states of nuclei a and A. We interpret (1) as the matrix element of an effective interaction between model wavefunctions, since we are going to consider only low-energy collisions [cf. e.g. Dover and Vary (1974)].

(a)   Folding potentials

We may then write (1) as

$$U_{aA}^N(r) = \int \rho^a(\mathbf{r}')\rho^A(\mathbf{r}'')V_{12}(|\ \mathbf{r} - \mathbf{r}'' + \mathbf{r}'|)\ d^3r'\ d^3r'' \,. \tag{2}$$

In this double folding integral one may perform the integration over $\mathbf{r}''$ and thereby express the ion-ion potential in terms of the nucleon-nucleus potential U_{nA} acting on a nucleon scattered off the nucleus A, i.e.

$$U_{aA}^N(r) = \int \rho^a(\mathbf{r}')U_{nA}^N(|\ \mathbf{r} + \mathbf{r}'|)\ d^3r' \,, \tag{3}$$

where

$$U_{nA}^N(r) = \int \rho^A(r')V_{12}(|\ \mathbf{r} - \mathbf{r}'|)\ d^3r' \,. \tag{4}$$

For estimates of the nucleon-nucleus potential by means of (4) one may consult Hodgson (1971).

Considerable uncertainty is attached to the evaluation of the potentials $U_{aA}^N(r)$ and $U_{nA}^N(r)$. Part of it is due to the effective two-body interaction, for which one may use the one used in Hartree-Fock calculations of the static nuclear properties, since the relative velocity of the two ions is small compared to the Fermi velocity. As we shall see below, the elastic scattering of two ions is determined mainly by the potential at large relative distances where U_{aA}^N is of the order of 0.5 MeV or less. The potential is thus sensitive to the matter density at distances where it is not very well known.

By utilizing Eq. (3) one avoids the uncertainties associated with the choice of the effective interaction, but is still dependent on the knowledge of the matter density in the far tail. The two functions entering in (3) are often parametrized in terms of Fermi distributions [cf. e.g. Bohr and Mottelson (1969)]. One may thus use for the matter distribution

$$\rho^a = \frac{\rho_0}{1 + \exp\{(r - R_a^d)/a_d\}}, \tag{5}$$

where

$$\rho_0 = 0.17 \text{ fm}^{-3}, \tag{6}$$

$$a_d = 0.54 \text{ fm}, \tag{7}$$

and

$$R_a^d = (1.12A_a^{1/3} - 0.86A_a^{-1/3}) \text{ fm}. \tag{8}$$

For the isoscalar part of the optical potential one may use

$$V_{na}(r) = \frac{-V_0}{1 + \exp\{(r - R_A^p)/a_p\}} \tag{9}$$

with

$$V_0 = (52 - 0.3E) \text{ MeV}, \tag{10}$$

where E is the incident energy of the nucleon, and

$$a_p = 0.65 \text{ fm} ,$$

$$R_A^p = 1.25 A_A^{1/3} \text{ fm} .$$

(11)

The evaluation of the integral (3) is essentially simplified by using instead of a Fermi distribution a folded Yukawa parametrization [cf. Krappe and Nix (1974) and Randrup (1975)] for both the density and the potential. For the density this parametrization is given by

$$\rho(r) = \rho_0 F(\kappa_d, R^d)$$

with

(12)

$$F(\kappa, R) = \frac{\kappa^2}{4\pi} \int d^3 r' \, \vartheta(R - r') \frac{e^{-|\mathbf{r} - \mathbf{r}'| \kappa}}{|\mathbf{r} - \mathbf{r}'|} ,$$

where the function F is obtained by folding a Yukawa function together with a step function ϑ. As is shown in Appendix A, the function (12) is rather similar to the Fermi distribution $(1 + \exp\{(r-R)/a\})^{-1}$ in the tail region. The advantage of this parametrization is that one may evaluate explicitly multiple folded integrals, e.g. of (12) with another Yukawa function or of two distributions of type (12). We can thus evaluate the integral (3) analytically. For the region $r > R_a^d + R_A^p = R$ one finds the result (cf. Appendix A)

$$U_{aA}^N(r) \approx - \frac{\pi \rho_0 V_0}{\kappa_d^{-2} - \kappa_p^{-2}} \frac{R_a^d R_A^p}{r}$$

$$\times (\kappa_d^{-4} e^{-(r-R)\kappa_d} - \kappa_p^{-4} e^{-(r-R)\kappa_p}) .$$

(13)

Since the two quantities κ_d and κ_p are similar, one may write approximately

$$U_{aA}^N(r) = - 2\pi \rho_0 V_0 \bar{\kappa}^{-2} \frac{R_a^d R_A^p}{r} e^{-(r-R)\bar{\kappa}} \left(1 + \frac{(r - R)\bar{\kappa}}{4} \right) ,$$

(14)

where

$$\bar{\kappa} = \tfrac{1}{2}(\kappa_d + \kappa_p) \ . \tag{15}$$

In order to evaluate (13) we need the parameters ρ_0, κ_d and R^d, which are obtained by fitting measured particle densities with functions of the shape (12). The results of a separate fit for the neutron and proton densities are given in Appendix A. For the total particle density the result is

$$\kappa_d = (1.40 + 0.60A^{-1/3}) \ \text{fm}^{-1} \ ,$$
$$R^d = (1.20A^{1/3} - 0.35) \ \text{fm} \ , \tag{16}$$
$$\rho_0 = 0.14(1 + 0.86A^{-1/3}) \ \text{fm}^{-3} \ .$$

No systematic fit of single-particle potentials has been made to folded Yukawa shapes, and we therefore use the simple translation of the parameters (10) and (11) by means of (A.10) and (A.11). The result is

$$V_0 = (52 - 0.3E) \ \text{MeV} \ ,$$
$$\kappa_p = (1.03 + 0.64A^{-1/3}) \ \text{fm}^{-1} \ , \tag{17}$$
$$R^p = (1.24A^{1/3} + 0.23 + 0.16A^{-1/3}) \ \text{fm} \ .$$

Inserting these results in (15), we find

$$\bar{\kappa} = [1.21 + 0.62(A_a^{-1/3} + A_A^{-1/3})] \ \text{fm}^{-1}. \tag{18a}$$

The expression (14) for the ion-ion potential is not quite symmetric under the interchange of a with A. Since however the radii R^p and R^d enter only as $R^p + R^d$ or $R^p R^d$, we may approximately use a common R which is the average

$$R_i = \tfrac{1}{2}(R_i^p + R_i^d)$$
$$\approx 1.21A_i^{1/3} \ . \tag{18b}$$

The r-dependence contained in the last factor of (14) is weak and to some extent canceled by the r-dependence of the denominator,

$$\frac{1}{r} \approx \frac{1}{R_a + R_A}\left(1 - \frac{r - R_a - R_A}{R_a + R_A}\right) .$$

Neglecting this dependence, we find

$$U_{aA} = -30 \frac{R_a R_A}{R_a + R_A} e^{-\bar{\kappa}(r - R_a - R_A)} , \qquad (19)$$

with $\bar{\kappa}$ given by (18a) and R_i by (18b). This expression is in very good agreement with the empirical potential given in Eqs. (36)–(38) below.

The conclusion of the above discussion is that one can reproduce the ion-ion potential, for the large distances important for elastic scattering, in terms of the folding given in Eq. (3) of the single-particle potential of one ion with the density of the other.

Similar conclusions have been reached in attempts to calculate the ion-ion potential from an effective two-body force through the "double folding" (2) [cf. Love (1977) and references cited therein; also Satchler (1979) and Akyüz and Winther (1980)].

(b)   Proximity potentials

An alternative way to derive the ion-ion potential which is convenient for systems of arbitrary shape utilizes the fact that the diffuseness a is much smaller than the radii of the two nuclei. To the extent that higher-order terms in this ratio can be neglected, one can calculate the ion-ion potential in terms of the interaction energy per unit area $e(s')$ between two semiinfinite nuclei with flat parallel surfaces. For two curved nuclear surfaces we expand the distance s' between the surfaces (the position of which we may define as the half-density points) around the point of closest approach.

We thus have

$$s' = s + \tfrac{1}{2}\kappa^{\parallel}x^2 + \tfrac{1}{2}\kappa^{\perp}y^2 , \qquad (20)$$

where s is the distance of closest approach and x and y are the coordinates on a plane perpendicular to $\mathbf{s}$. The coeficients $\kappa^{\parallel}$ and $\kappa^{\perp}$ are related to the radii of curvature of the two surfaces. The total interaction energy is then

$$
\begin{aligned}
U_{aA}^N(s) &= \int\int dx\, dy\, e(s') \\
&= \frac{2}{\sqrt{\kappa^{\parallel}\kappa^{\perp}}} \int\int e(s + x'^2 + y'^2)\, dx'\, dy' \\
&= \frac{2\pi}{\sqrt{\kappa^{\parallel}\kappa^{\perp}}} \int_s^{\infty} e(s')\, ds' \; .
\end{aligned}
\tag{21}
$$

If we assume the density of semiinfinite nuclear matter to be given by [cf. also (A.7)]

$$
\rho(z) = \frac{\rho_0 \kappa_d^2}{4\pi} \int \vartheta(Z_d - z') \frac{e^{-|\mathbf{r}-\mathbf{r}'|\kappa_d}}{|\mathbf{r}-\mathbf{r}'|} d^3 r'
$$

$$
= \rho_0 \times
\begin{cases}
\frac{1}{2} e^{-(z-Z_d)\kappa_d} , & z > Z_d , \\[2mm]
1 - \frac{1}{2} e^{(z-Z_d)\kappa_d} , & z < Z_d ,
\end{cases}
\tag{22}
$$

and the single-particle potential to be of the same form, we can calculate the interaction energy per unit area utilizing the results of Appendix A. One finds

$$
e(s') = \frac{-\rho_0 V_p}{2(\kappa_p^{-2} - \kappa_d^{-2})} (\kappa_p^{-3} e^{-s'\kappa_p} - \kappa_d^{-3} e^{-s'\kappa_d}) ,
\tag{23}
$$

where $s' = Z_p - Z_d$. The proximity potential is thus, according to (21),

$$
U_{aA}^N(r) = \frac{-\pi\rho_0 V_p}{\kappa_p^{-2} - \kappa_d^{-2}} \frac{1}{\sqrt{\kappa^{\parallel}\kappa^{\perp}}} (\kappa_p^{-4} e^{-s\kappa_p} - \kappa_d^{-4} e^{-s\kappa_d}) .
\tag{24}
$$

For two spherical nuclei we find

$$\kappa^{\|} = \kappa^{\perp} = R_a^{-1} + R_A^{-1} \tag{25}$$

and

$$s = r - R_a - R_A . \tag{26}$$

The expression (24) thus coincides with (13a) except that the factor r^{-1} is replaced by $(R_a + R_A)^{-1}$. The expression (24) can also be used for the interaction between nonspherical nuclei (cf. Chap. IV below).

The argument given in connection with Eq. (21) can also be used to estimate the force between two curved surfaces. One finds

$$F(s) = -\frac{\partial U_{aA}^N(r)}{\partial r} = \frac{2\pi}{\sqrt{\kappa^{\|}\kappa^{\perp}}} \int_s^\infty \left(-\frac{\partial e}{\partial s'} \right) ds'$$

$$= \frac{2\pi}{\sqrt{\kappa^{\|}\kappa^{\perp}}} e(s) . \tag{27}$$

For the interaction energy per unit area, $e(s)$, we expect in general that it shows a maximum when the two semiinfinite densities (22) are at the distance $s = 0$, where the total density is constant across the boundary. Since at this distance two unit areas of surface have disappeared, per unit area we have

$$e(0) = 2\gamma , \tag{28}$$

where γ is the surface tension. For two spheres we thus find

$$\left(\frac{\partial U_{aA}^N}{\partial r} \right)_{max} = 4\pi\gamma \frac{R_a R_A}{R_a + R_A} \tag{29}$$

at the distance $r = R_a + R_A$, where R_a and R_A are the half-density radii of the two nuclei. This result is rather general and should include the effects of exchange and isospin. The empirical expression for the surface tension is

$$\gamma_i = 0.95 \left[1 - 1.8 \left(\frac{N_i - Z_i}{A_i} \right)^2 \right] \text{ MeV fm}^{-2}.$$

For the quantity γ appearing in (29) one should use the average γ of the two nuclei. In fact, folding calculations including isovector interactions indicate that the best averaging would be

$$\gamma = 0.95 \left[1 - 1.8 \left(\frac{N_a - Z_a}{A_a} \right) \left(\frac{N_A - Z_A}{A_A} \right) \right] \text{ MeV fm}^{-2}. \quad (30)$$

On the basis of the proximity approximation, Błocki et al. (1977) have given the following parametrization of the ion-ion potential:

$$U^N = 4\pi\gamma \bar{\mathscr{R}} b \Phi \left(\frac{r - \mathscr{R}}{b} \right), \quad (31)$$

with

$$\Phi(\zeta) = \begin{cases} -\frac{1}{2}(\zeta - 2.54)^2 - 0.0852(\zeta - 2.54)^3 & \text{for } \zeta \leq 1.2511, \\ -3.437 \exp(-\zeta/0.75) & \text{for } \zeta > 1.2511. \end{cases} \quad (32)$$

The reduced radius $\bar{\mathscr{R}}$ is defined in terms of the effective sharp radii of the two nuclei $\mathscr{R}_i$ by

$$\bar{\mathscr{R}}_{aA} = \frac{\mathscr{R}_a \mathscr{R}_A}{\mathscr{R}_a + \mathscr{R}_A}, \quad (33)$$

with

$$\mathscr{R}_i = (1.28 A_i^{1/3} - 0.76 + 0.8 A_i^{-1/3}) \text{ fm}. \quad (34)$$

For the diffuseness parameter b a value of

$$b = 1 \text{ fm} \quad (35)$$

may be used, while $\mathscr{R} = \mathscr{R}_a + \mathscr{R}_A$. The function (32) is illustrated in Fig. 1.

In the following we shall often use an exponential potential adjusted to fit available elastic-scattering data [Christensen and Winther (1976)] and which retains the form of the potential (19) or (32) (for $\zeta \geq 1.25$). This empirical potential is given by

$$U^N(r) = -S_0 \bar{R}_{aA} \exp \left(-\frac{r-R}{a} \right) , \qquad (36)$$

with

$$S_0 = 50 \text{ MeV fm}^{-1} ,$$
$$a = 0.63 \text{ fm} , \qquad (37)$$
$$R_i = (1.233 A_i^{1/3} - 0.98 A_i^{-1/3}) \text{ fm}.$$

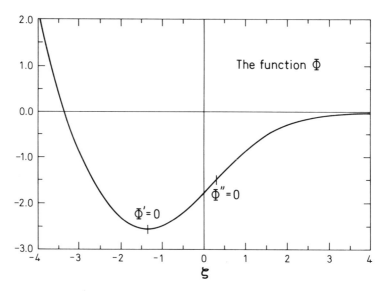

Fig. 1 The function Φ describing the shape of the proximity potential given in Eq. (1.32). The maximum slope is $\Phi' = 1$, which is attained at $r = \mathscr{R} + 0.58$ fm. This radius is considerably larger than the one expected from the discussion in connection with Eqs. (1.28) and (1.29).

112

A more recent improved analysis of the data [Akyüz and Winther (1981)] leads to

$$S_0 = 65.4 \text{ MeV fm}^{-1} ,$$

$$1/a = 1.16[1 + 0.48(A_a^{-1/3} + A_A^{-1/3})] \text{ fm}^{-1} , \qquad (38)$$

$$R_i = (1.20A_i^{1/3} - 0.35) \text{ fm} .$$

The quantity $\bar{R}_{aA}$ is given by (33) with R_i instead of $\mathscr{R}_i$, while $R = R_a + R_A$. The two radii (34) and (37) are plotted as functions of mass number in Fig. 2. The potential (36) should be used only for large values of r. The minimal distance r_{cr} for which (36) is valid can be estimated by evaluating

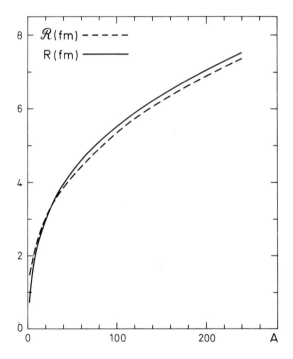

Fig. 2 Mass dependence of the radius of the ion-ion potential. The nuclear radii $\mathscr{R} = (1.28A^{1/3} - 0.76 + 0.8A^{-1/3})$ fm and $R = (1.233A^{1/3} - 0.98A^{-1/3})$ fm appearing in the proximity and in the standard potential respectively [cf. Eqs. (1.34) and (1.37)] are plotted as functions of the mass number A.

the nuclear attractive force $-\partial U^N/\partial r$ and setting it equal to the maximum force (29), leading to the condition

$$r > r_{cr} = R_a + R_A + 1.2 \text{ fm} . \tag{39}$$

In order to improve on this limitation so as to be able to use the potential down to a distance of $r \approx R_a + R_A$ one may use a Saxon-Woods parametrization of the ion-ion potential:

$$U^N_{aA}(r) = \frac{-V_0}{1 + e^{(r-R_0)/a}} . \tag{40}$$

It is possible to adjust the parameters V_0, R_0, and a so that the tail agrees with the tail of the standard potential (36) and so that the maximum force is equal to (29). This leads to

$$V_0 = 16\pi\gamma\bar{R}_{aA}a . \tag{41}$$

Based on the parameter set (37), one finds

$$a = 0.63 \text{ fm} , \tag{42}$$
$$R_0 = R_a + R_A + 0.29 \text{ fm} ,$$

with

$$R_i = 1.233A_i^{1/3} - 0.98A_i^{-1/3} . \tag{43}$$

A least-squares fit of (40) to the experimental data with the subsidiary condition (29) leads to the alternative expression

$$1/a = 1.17[1 + 0.53(A_a^{-1/3} + A_A^{-1/3})]\text{fm}^{-1} , \tag{44}$$
$$R_0 = R_a + R_A ,$$

with

$$R_i = (1.20A_i^{1/3} - 0.09) \text{ fm} . \tag{45}$$

This is close to what would be obtained adjusting the parameters in (40) to agree with the parametrization (38) of the function (36) in the tail region. The parametrization (44) offers a compromise between the empirical potentials (37) and (42), mainly determined from lighter ion scattering and the proximity potential describing the interaction between very heavy systems.

The expressions (31)–(45) apply to the case of spherical nuclei. For deformed nuclei one may use the same expressions with a slightly modified radius as discussed in Chap. IV [cf. Eq. (IV.4.21)].

The function (36) is the isoscalar part of the ion-ion potential. The total interaction between the two ions is expected to contain also an isovector-isovector coupling term. As will be discussed in Sec. IV.2, this coupling is given by [cf. IV.2.36)]

$$U_{iso}^N = -U^N(r) \, 2 \, \frac{(N_a - Z_a)(N_A - Z_A)}{A_a A_A} \, . \tag{46}$$

This estimate is a factor of 4 larger than one would estimate from the isovector potential derived from proton scattering and used in nuclear-structure calculations. The value given has been estimated by Akyüz and Winther (1980) and is due to the increased neutron excess in the surface region (by about a factor of 2) over the value $(N-Z)/A$ appropriate for the bulk of the nucleus. The potential (46) is at most 5% of U^N. Note that the total nuclear interaction

$$U_{tot}^N(r) = U^N(r) + U_{iso}^N(r) \tag{47}$$

has about the same dependence on isospin as (31), using the symmetrization (30) for γ.

2   GENERAL FEATURES OF ELASTIC SCATTERING

The total interaction determining the elastic scattering of two heavy ions is

$$U(r) = U^C(r) + U^N(r) \, , \tag{1}$$

where U^N is the ion-ion potential, while

$$U^C(r) = \frac{Z_a Z_A e^2}{r} \tag{2}$$

is the Coulomb field. The corrections to the Coulomb potential due to the finite size of the charge distributions should, in principle, be applied for distances less than $r^c = R_a^c + R_A^c$, where $R_a^c = 1.2A_a^{1/3}$ fm and similarly for R_A. The Coulomb interaction between two charge densities of the type (1.12) can be evaluated by the technique of Appendix A (for $\kappa \to 0$). For two homogeneously charged spheres of radii R_a^c and R_A^c the interaction has been calculated by De Vries and Clover (1975) and Poling et al. (1976). For small values of the overlap between the two charges the potential (2) is reduced by the factor

$$1 - \frac{3}{16}\left(\frac{s^2}{R_a^c R_A^c}\right)^2 + \cdots , \tag{3}$$

where $s = r - R_a^c - R_A^c < 0$. For all practical purposes this correction can be neglected in grazing collisions.

The most conspicuous feature of the potential $U(r)$ is that it displays a maximum, viz. the Coulomb barrier. Utilizing the empirical potential (1.36), its position is determined by

$$\frac{dU(r)}{dr} = -\frac{Z_a Z_A e^2}{r_B^2} + \frac{50}{a} R_{aA} \exp\left(-\frac{r_B - R}{a}\right) = 0 . \tag{4}$$

The solution of this equation, to an accuracy better than 0.1 fm, is given by

$$r_B = [1.07(A_a^{1/3} + A_A^{1/3}) + 2.72] \text{ fm} . \tag{5}$$

The height of the barrier can, through (1) and (4), be written as

$$E_B = \frac{Z_a Z_A e^2}{r_B}\left(1 - \frac{a}{r_B}\right) . \tag{6}$$

Using the expression (5) for r_B in Eq. (6), one obtains values which differ by at most 1% from the exact solution of (4).

The Coulomb barrier is expected to disappear for sufficiently heavy systems because the increased Coulomb repulsion and the expected maximum value (1.29) of the nuclear attraction. The critical situation can be estimated by comparing the Coulomb repulsion with the nuclear attraction at the point of maximum attraction. One finds in this way that the Coulomb barrier should disappear approximately when $A_a \approx A_A \approx 200$.

Because of the very short wavelength associated with the relative motion of heavy ions (cf. Fig. 3) it seems appropriate to calculate the elastic cross-section by classical mechanics. In the classical description one evaluates, on the basis of the Newtonian equations of motion in the

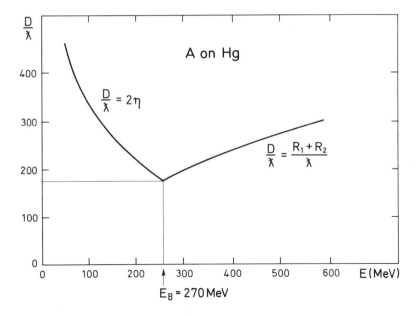

Fig. 3 The ratio between the characteristic length D and the wavelength λ at large distances for Ar projectiles on Hg as a function of the bombarding energy. For energies below the Coulomb barrier E_B, D is taken to be the distance of closest approach in the head-on collision, while for $E > E_B$, D is taken to be the sum of the nuclear radii.

field (1), the connection between the impact parameter ρ and the scattering angle Θ (cf. Appendix B). The functional relation $\Theta = \Theta(\rho)$ is called the deflection function.

Once this function is known, the differential cross-section is seen from Fig. 4 to be given by

$$\frac{d\sigma}{d\Omega} = \left| \frac{\rho}{\sin \Theta} \frac{d\rho}{d\Theta} \right| . \tag{7}$$

A typical example of a deflection function is shown in Fig. I.9. For large impact parameters the deflection function is the one of pure Coulomb scattering, leading to the Rutherford differential cross section

$$\frac{d\sigma_{\text{Ruth}}}{d\Omega} = \left(\frac{a_0}{2} \right)^2 \sin^{-4} \tfrac{1}{2}\Theta . \tag{8}$$

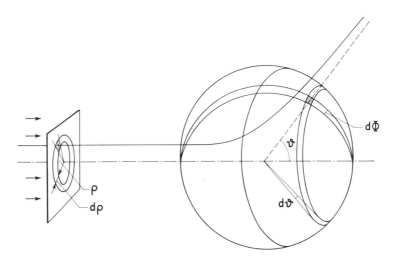

Fig. 4 Classical differential cross-section as evaluated from the connection between the impact parameter ρ and the deflection angle ϑ. The projectiles impinging on the shaded area on the plane perpendicular to the beam direction emerge within the solid angle indicated on the unit sphere centered at the target nucleus in the center-of-mass system. The ratio between the area $\rho \, d\rho \, d\phi$ and the solid angle $\sin \vartheta \, d\vartheta \, d\phi$ is the differential cross section.

For smaller impact parameters, where the distance of closest approach is within the range of the ion-ion potential, the deflection function is bent towards forward angles, and eventually goes asymptotically towards $\Theta = -\infty$. This phenomenon, which is known as the orbiting situation, occurs when the radial motion is stopped at the top of the barrier in the effective potential for the radial motion.

The maximum deflection angle allowed in the classical description is called the rainbow angle Θ_R and is associated with a singularity in the differential cross-section, i.e. $d\rho/d\Theta = \infty$.

For scattering angles smaller than the rainbow angle, two trajectories with the impact parameters ρ_C and ρ_N contribute (cf. Fig. I.9) and the cross-section is given by

$$\frac{d\sigma}{d\Omega} = \sum_i \left| \frac{\rho_i}{\sin\Theta} \frac{d\rho_i}{d\Theta} \right| . \tag{9}$$

In fact in this sum one should also include the small contributions arising from negative scattering angles equal to $\pm\Theta$ modulo 2π.

For impact parameters smaller than the one corresponding to orbiting, the relative motion of the two nuclei will go over the Coulomb barrier. In such cases the nuclear reactions are expected to be so prolific that the two nuclei will not emerge again in the elastic channel. In order to take this feature into account the normal theory of potential scattering should be modified. In a classical description this is done simply by leaving out the corresponding impact parameters, while in a quantal theory one may introduce an incoming-wave boundary condition inside the Coulomb barrier when solving the radial equations.

Even for trajectories which do not penetrate beyond the barrier in the effective potential, nuclear reactions may take place, eliminating flux from the elastic channel.

Outside the range of nuclear interactions, absorption takes place through Coulomb excitation (cf. Fig. I.7 and Sec. II.9). For smaller distances the absorption depends on the overlap of the two nuclear densities, and one may introduce a position dependent mean free path $\lambda(r)$ defined by

$$\frac{dP}{ds} = -\frac{P}{\lambda(r)} .$$

The quantity P is the probability that no nuclear reaction has taken place, while ds is an infinitesimal path length.

Solving this equation with $P(-\infty) = 1$, we obtain

$$P(t) = \exp\left\{-\int_{-\infty}^{t} \frac{v(t')}{\lambda(r(t'))} dt'\right\}, \qquad (10)$$

where $v(t) = ds/dt$ is the velocity of relative motion at time t.

Because of this absorption the cross-section (7) for elastic scattering is modified. One finds

$$\frac{d\sigma}{d\Omega} = P(\infty)\left|\frac{\rho}{\sin\Theta}\frac{d\rho}{d\Theta}\right|. \qquad (11)$$

As an example of how the classical elastic cross-section may be modified by a mean free path, we show in Fig. 5 deflection functions associated with three hypothetical nuclear potentials and the associated angular distributions for different values of $\lambda(r)$. In the figure we have actually used the quantity

$$W(r) = -\frac{\hbar\, v(r)}{2\lambda(r)} \qquad (12)$$

to describe the absorption, and we have parametrized this quantity as $(W_0 > 0)$

$$W(r) = \frac{-W_0}{1 + \exp\left(\dfrac{r - R_W}{a_W}\right)}. \qquad (13)$$

For elastic scattering with not too heavy ions, quantal effects are expected to play an important role, even for the rather short wavelengths encountered. Thus, for scattering angles smaller than the rainbow angle, where at least two trajectories of different impact parameters contribute, interference phenomena will occur. Also, for scattering angles larger than the rainbow angle, the cross section will not vanish suddenly; because of

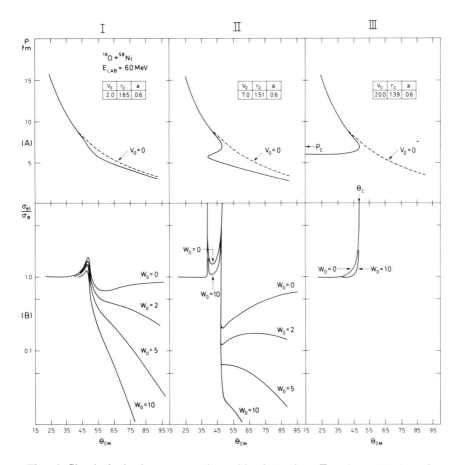

Fig. 5 Classical elastic cross section with absorption. For the scattering of ^{16}O on ^{58}Ni we have used three hypothetical nuclear potentials of the Saxon-Woods shape (1.40) with $R_0 = r_0(A_a^{1/3} + A_A^{1/3})$, where r_0, V_0 and a are given in the figure in units of fm, MeV and fm respectively. The absorption $W(r)$ has the same geometry [cf. (2.13)], and the values of W_0 in MeV label the different cross sections. In case I, where no rainbow scattering occurs, the shape of the cross section is very sensitive to the strength of the absorption, while in cases II and III the cross section is mainly governed by the rainbow angle. This figure is from Broglia et al (1972).

the finite number of partial waves contributing, it will show diffraction and will vanish gradually.

A quantal treatment is straightforward to carry through if one describes the absorptive effects in terms of a mean free path proportional to the velocity, such that the quantity $W(r)$ defined in (12) becomes a function of r only. The function $W(r)$, which is often parametrized in the form (13), has the dimensions of a potential and enters into the quantal treatment as an imaginary part of the potential

$$U_{\text{opt}} = U(r) + iW(r) , \qquad (14)$$

together with the real potential. It thus gives rise to a number of effects, e.g. reflection, which go beyond the concept of a mean free path.

Most experimental data on elastic scattering of heavy ions have been analyzed in this so-called optical model, and one has in almost all cases been able to obtain excellent fits for some values of the six parameters entering into the Woods-Saxon parametrization of the real and imaginary parts of the potential. In fact one can usually fit an experiment with several different sets of parameters.

In this chapter we shall instead use the semiclassical description, which is essentially the same as the classical description but supplemented with quantal interference and diffraction effects. In this way we can treat the absorptive effects in a rather flexible way, where the total absorption in the interior and nonlocal effects in the exterior region can be dealt with independently. Also the semiclassical treatment of elastic scattering gives a simple mechanical picture which can easily be generalized to include inelastic processes and transfer reactions.

3   DEFLECTION FUNCTION

A typical deflection function, calculated numerically on the basis of the expressions given in Appendix B, is shown in Fig. I.9. The characteristic features of this function are the Coulomb dominated region for large impact parameters, the maximum at the so-called rainbow angle ϑ_R and the large negative deflection angles attained at somewhat smaller impact parameters.

(a)   Rainbow scattering

The rainbow angle, which is very important for the discussion of heavy-ion collisions, is determined from the condition

$$\frac{d\Theta(\rho)}{d\rho} = \frac{d\Theta(L)}{dL} = 0 \ . \tag{1}$$

The deflection function for the combined Coulomb repulsion and nuclear attraction of the form (1.36) cannot be calculated in closed form. Since the rainbow angle is usually not very different from the Coulomb scattering angle corresponding to the same impact parameter, we may however use classical perturbation theory as outlined in Appendix C.

In the notation of Appendix C, the deflection function is given by

$$\Theta = 2 \arcsin\frac{1}{\varepsilon^{(0)}} - \phi_0(\infty) \ , \tag{2}$$

where $\varepsilon^{(0)}$ is related to the impact parameter through (C.9), i.e.

$$\varepsilon^{(0)} = \sqrt{1 + \left(\frac{\rho}{a_0}\right)^2} = \sqrt{1 + \left(\frac{L}{\eta_{cl}}\right)^2} \ . \tag{3}$$

The quantity $\phi_0(\infty)$ is, in second-order perturbation theory, given by (C.19). Since the main contribution to $\phi_0(\infty)$ comes from small distances, i.e. $w \approx 0$, we expand the right-hand side of (C.19) to second order in w. Utilizing the potential (1.36) and the expressions

$$\int_{-\infty}^{\infty} e^{-qw^2} dw = \sqrt{\frac{\pi}{q}} \ ,$$

and

$$\int_{-\infty}^{\infty} w^2 e^{-qw^2} dw = \frac{1}{2q}\sqrt{\frac{\pi}{q}} \ ,$$

one finds

$$\phi_0(\infty) = -\sqrt{2\pi}\sqrt{\frac{\varepsilon^{(0)}-1}{\varepsilon^{(0)}}}\,\frac{\varepsilon^{(0)}+1}{\varepsilon^{(0)}}\sqrt{\frac{b(\varepsilon^{(0)})}{a}}\,\frac{U^N(b(\varepsilon^{(0)}))}{2E}$$

$$\times\left(1-\sqrt{2}\,\frac{1+\varepsilon^{(0)}}{\varepsilon^{(0)}}\,\frac{b(\varepsilon^{(0)})}{a}\,\frac{U^N(b(\varepsilon^{(0)}))}{2E}\right). \qquad (4)$$

The rainbow condition (1) for the deflection angle (2) can now be written as

$$\frac{\partial\Theta}{\partial\varepsilon^{(0)}} = -\frac{2}{\varepsilon^{(0)}\sqrt{(\varepsilon^{(0)})^2-1}} - \frac{\partial\phi_0(\infty)}{\partial\varepsilon^{(0)}} = 0 ,$$

or

$$\left(\frac{b}{a}\right)^{3/2}\frac{U^N(b)}{2E}\left(1-\sqrt{2}\,\frac{\varepsilon_R^{(0)}+1}{\varepsilon_R^{(0)}}\,\frac{b}{a}\,\frac{U^N(b)}{E}\right)$$

$$= -\sqrt{\frac{2}{\pi}}\sqrt{\frac{\varepsilon_R^{(0)}}{\varepsilon_R^{(0)}+1}}\,\frac{1}{\varepsilon_R^{(0)}-1} , \qquad (5)$$

where $b = b(\varepsilon_R^{(0)})$ is the unperturbed distance of closest approach [cf. (C.17) or (11) below]. In this expression we have neglected terms of order a/b as compared with $\varepsilon_R^{(0)} - 1$, and the formula should thus only be used for scattering angles smaller than about $\pi/2$.

Using (C.16), we may, within the same approximation, write (5) as

$$\left(\frac{r_R}{a}\right)^{3/2}\frac{U^N(r_R+1.8\Delta)}{2E} = -\sqrt{\frac{2}{\pi}}\sqrt{\frac{\varepsilon_R^{(0)}}{\varepsilon_R^{(0)}+1}}\,\frac{1}{\varepsilon_R^{(0)}-1} , \qquad (6)$$

where r_R is the distance of closest approach for the rainbow scattering, i.e., according to (C.16),

$$r_R^{(1)} = b(\varepsilon_R^{(0)}) + \Delta ,\tag{7}$$

with

$$\Delta = \frac{1 + \varepsilon_R^{(0)}}{\varepsilon_R^{(0)}} \; \frac{U^N(b(\varepsilon_R^{(0)}))}{2E} \, b(\varepsilon_R^{(0)}) .\tag{8}$$

Equation (5) provides an implicit relation for the impact parameter associated with the rainbow scattering.

Using the potential (1.36), one can write (5) in the form

$$\left(\frac{b}{a}\right)^{5/2} e^{-b/a} = 0.0365 \, \frac{Z_a Z_A}{\bar{R}_{aA}} \, e^{-R/a} q(\varepsilon_R^{(0)}) ,\tag{9}$$

where $\bar{R}_{aA}$ is measured in fm and where

$$q(\varepsilon_R^{(0)}) = \frac{\sqrt{\varepsilon_R^{(0)}(\varepsilon_R^{(0)} + 1)}}{\varepsilon_R^{(0)} - 1} \left(1 + \frac{62}{Z_a Z_A} \, \frac{\bar{R}_{aA}}{\varepsilon_R^{(0)}} \left(\frac{b}{a}\right)^2 e^{-(b-R)/a}\right)^{-1}\tag{10}$$

with

$$b \equiv b(\varepsilon_R^{(0)}) = \frac{Z_a Z_A e^2}{2E} (1 + \varepsilon_R^{(0)}) .\tag{11}$$

For the region of interest in heavy-ion scattering the left-hand side of (9) (for $a = 0.63$ fm) can be approximated by $108.5 \exp(-1.369b)$ to an accuracy better than 10%. We can thus solve for b to find

$$b = \left[1.16R + \frac{1}{1.37}\ln\left(\frac{2972\bar{R}_{aA}}{Z_a Z_A}\right) - \frac{1}{1.37}\ln q(\varepsilon_R^{(0)})\right] \text{ fm} .\tag{12}$$

to an accuracy of $a/10$.

Equations (10)–(12) can be solved for $\varepsilon_R^{(0)}$ by iteration, starting e.g. with $q(\varepsilon_R^{(0)}) = 1$ in (12). Using this procedure we have calculated the rainbow angular momentum and the associated angle for the example given in Fig. I.9. We thus obtain after one iteration

$$\varepsilon_R^{(0)} = 4.91. \tag{13}$$

Inserting this value in Eqs. (2)–(4), one finds

$$\Theta_R = 21.7° \quad \text{and} \quad L_R = 39.7\hbar . \tag{14}$$

The connection between $\varepsilon^{(0)}$ and Θ [cf. Eq. (2)] for the rainbow situation [cf. Eq. (6)] can be written to a good approximation as

$$L_R = \eta_{\text{cl}} \cot\left(\tfrac{1}{2}\Theta_R + \frac{\sin^2 \tfrac{1}{2}\Theta_R}{\cos \tfrac{1}{2}\Theta_R} \frac{2aE}{Z_a Z_A e^2} \right). \tag{15}$$

For high energies, where $\varepsilon_R^{(0)} \gg 1$ and where $q(\varepsilon_R^{(0)})$ approaches 1, the quantity b in (12) becomes energy independent. The rainbow angle is then given by

$$\Theta_R = \frac{2}{\varepsilon^{(0)}} + \sqrt{\frac{\pi}{2}} \sqrt{\frac{b}{a}} \frac{U^N(b)}{E}, \tag{16}$$

or

$$\Theta_R E \approx \frac{Z_a Z_A e^2}{b} + \sqrt{\frac{\pi}{2}} \sqrt{\frac{b}{a}} U^N(b) . \tag{17}$$

In this high-energy limit ($\Theta_R \lesssim 30°$) the rainbow angle is thus inversely proportional to the bombarding energy. The distance of closest approach (7) also becomes energy independent, viz.

$$r_R = b = 1.16R + 0.73 \ln\left(\frac{2972 \bar{R}_{aA}}{Z_a Z_A} \right) \text{ fm} . \tag{18}$$

The magnitude of the nuclear attraction at this distance is seen from (5) to be approximately given by

$$U^N(r_R) \approx -\sqrt{\frac{2}{\pi}}\left(\frac{a}{r_R}\right)^{3/2}\frac{Z_a Z_A e^2}{r_R}$$

$$\approx -\sqrt{\frac{2}{\pi}}\left(\frac{a}{r_R}\right)^{3/2} E_B , \tag{19}$$

where E_B is the Coulomb barrier. The rainbow phenomenon is governed by the nuclear potential at rather large distances, i.e. at a distance where its value is about $\sqrt{2a/\pi r_B} \approx \frac{1}{5}$ of the value

$$U^N(r_B) \approx -\frac{a}{r_B} E_B \tag{20}$$

at the Coulomb barrier [cf. Eq. (2.6)].

At lower energies the rainbow angle shifts towards $180°$ and the distance r_R approaches the radius of the Coulomb barrier.

(b) Grazing angular momentum

For impact parameters smaller than the one corresponding to the rainbow scattering, the nuclear attraction bends the trajectory to smaller angles, eventually leading to negative scattering angles.

If the effective potential in the radial motion shows a maximum at the turning point, i.e. $(\partial U_{\text{eff}}/\partial r)_{r=r_0} = 0$ [cf. Eq. (B.11)], both the time and the azimuthal angle ϕ diverge, meaning that the projectile performs a spiral motion towards the distance of closest approach r_0. This phenomenon is called orbiting and corresponds to the situation where the deflection function tends asymptotically towards $-\infty$. In a quantal treatment this is the point where the barrier penetration is equal to $\frac{1}{2}$, and the angular momentum and distance r_0 associated with this situation are known, in heavy-ion physics, as the grazing angular momentum L_g and the grazing distance r_g.

Defining the effective potential for the radial motion by

$$U_{\text{eff}}(r) = U(r) + \frac{L^2}{2m_{aA}r^2} , \tag{21}$$

127

III   Elastic Scattering

the equations determining the grazing parameters are

$$E - (U_{\text{eff}}(r_g))_{L=L_g} = 0 \ , \tag{22}$$

and

$$\frac{\partial U_{\text{eff}}(r)}{\partial r}\bigg|_{r=r_g,\, L=L_g} = 0 \ . \tag{23}$$

Eliminating L_g and using an exponential form of the nuclear potential (1.36), these equations can be solved for r_g, leading to

$$2E - \frac{Z_a Z_A e^2}{r_g} - (50\bar{R} \text{ MeV})\left(\frac{r_g}{a} - 2\right) e^{-(r_g - R)/a} = 0. \tag{24}$$

It is convenient to express r_g in terms of the radius (2.5) of the Coulomb barrier, i.e.

$$r_g = r_B - \delta \ , \tag{25}$$

where

$$r_B = [1.07(A_a^{1/3} + A_A^{1/3}) + 2.72] \text{ fm} \ . \tag{26}$$

To lowest order in δ one finds

$$\delta = a \ln\left(1 + \frac{2(E - E_B)}{E_B}\right) \ , \tag{27}$$

where E_B is the height (2.6) of the Coulomb barrier. The grazing angular momentum can be obtained from (22). One finds the approximate formula

$$L_g = (r_g)_{\text{fm}}\left(\frac{1}{20}\frac{A_a A_A}{A_a + A_A}(E - E_B)_{\text{MeV}}(1 + c)\right)^{1/2} \hbar \ , \tag{28}$$

where

$$c = 2 \frac{a}{r_B} \left[1 - \frac{E_B}{2(E - E_B)} \ln \left(1 + \frac{2(E - E_B)}{E_B} \right) \right] , \quad (29)$$

which for energies close to the Coulomb barrier becomes

$$c \approx \frac{2a}{r_B} \frac{E - E_B}{E_B} . \quad (30)$$

As the energy increases the rainbow distance increases to reach the asymptotic value (18). The grazing distance on the other hand is a monotonically decreasing function of the bombarding energy. For bombarding energies close to the Coulomb barrier both parameters are equal to the radius of the Coulomb barrier.

(c) Nuclear rainbow

As the bombarding energy is increased and the corresponding grazing distance becomes smaller, one eventually reaches a point where the exponential form of the potential assumed in (1.36) is no longer valid. Thus, the force generated by the potential cannot exceed the value (1.29), and the standard potential cannot be used for distances smaller than (1.39).

Since we know the maximum attractive nuclear force, we can estimate the value of the angular momentum L_{cr} for which the minimum in the effective potential for the radial motion disappears. This happens [Wilczynski (1973)] for

$$L_{cr} = \left[m_{aA} R^3 \left(4 \pi \gamma \bar{R} - \frac{Z_a Z_A e^2}{R^2} \right) \right]^{1/2} \quad (31)$$

where $R = R_a + R_A$. For this L-value the effective potential shows an inflection point at $r \approx R_a + R_A$, and the corresponding energy is

$$E_{cr} = \frac{Z_a Z_A e^2}{2R} + 2\pi\gamma R_a R_A \left(1 - \frac{4a}{R} \right) , \qquad (32)$$

where we have used the expression (1.40) for U^N.

For energies below E_{cr}, the shape of the deflection function is characterized quite accurately by the quantities L_R, Θ_R, and L_g. For smaller impact parameters the shape of the deflection function becomes uninteresting, since the projectile will then penetrate beyond the turning point in the effective potential to distances where nuclear reactions are so prolific that no elastic scattering will take place.

For energies above E_{cr} orbiting disappears and the deflection function will instead show a minimum, i.e., there will appear a second rainbow: the so-called nuclear rainbow (cf. Fig. 6). In this situation the projectile does not penetrate very far into the target nucleus, and if the absorption is small in the surface region, the nuclear rainbow phenomenon might be observ-

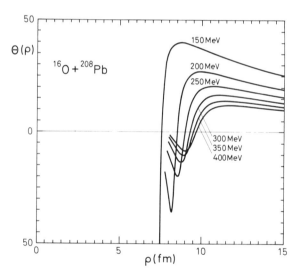

Fig. 6 Deflection functions for ^{16}O + ^{208}Pb for different bombarding energies. For energies below the critical energy $E_{cr} = 150$ MeV [cf. Eq. (3.32)] the deflection functions show orbiting, while for $E > E_{cr}$ the deflection functions show a minimum, which is the so-called nuclear rainbow phenomenon. The calculations were performed utlizing the potential (1.40) – (1.43).

able. There are experimental indications that this is the case, especially for alpha-particle scattering [see Goldberg and Smith (1972, 1974)].

4 PARTIAL WAVES, PHASE SHIFTS AND CROSS-SECTIONS

The angular motion in the central field can be given an accurate treatment classically as well as quantum mechanically, since the angular momentum is a constant of the motion. We shall in these notes often treat the angular momentum quantum mechanically, retaining a classical (or semiclassical) description of the radial motion. In this semiquantal formulation, one completely loses the classical localization in angle, and the radial motion is solved for those impact parameters which correspond to integer values of the orbital-angular-momentum quantum number l.

The radial motion may also show quantal effects due to reflection from, and penetration through, the Coulomb barrier. These quantal effects will be considered in the next sections. In the present section we discuss the conditions under which one may obtain a simple connection between the scattering angle and the angular momentum. We derive in this context the well-known relation between the phase shift and the classical deflection function (Ford and Wheeler, 1959).

We write the scattering amplitude in the form

$$f(\vartheta) = \frac{i}{2k} \sum_{l} (2l+1)(1 - e^{2i\beta_l}) P_l(\cos \vartheta), \tag{1}$$

where l is the orbital-angular-momentum quantum number. The wave-number k is defined by

$$k = \frac{1}{\lambda} = \frac{m_{aA} v}{\hbar}, \tag{2}$$

where v is the relative velocity at infinity. The differential cross section can be written in terms of $f(\vartheta)$ as

$$\frac{d\sigma}{d\Omega} = |f(\vartheta)|^2. \tag{3}$$

The scattering amplitude is governed by the phase shift β_l between the incoming and outgoing waves of angular momentum l.

In the short-wavelength limit the expressions (3) and (2.9) must be equivalent. In order to show this, we utilize the fact that l is large and use the asymptotic form

$$P_l(\cos \vartheta) = \sqrt{\frac{4\pi}{2l+1}} \; \frac{1}{\pi} \; \frac{1}{\sqrt{\sin \vartheta}} \; \cos[(l + \tfrac{1}{2})\vartheta - \tfrac{1}{4}\pi] \tag{4}$$

of the Legendre polynomial. Furthermore we replace the summation in (1) with the sum of integrals over l according to the Poisson formula [cf. Morse and Feshbach (1953)], i.e.

$$f(\vartheta) = \frac{1}{ik\sqrt{2\pi \sin \vartheta}} \sum_{p=-\infty}^{\infty} \int_0^{\infty} dl \, (l + \tfrac{1}{2})^{1/2}$$

$$\times \; e^{2i\beta_l}(e^{-i[(l+\frac{1}{2})\vartheta - 2\pi pl - \frac{1}{4}\pi]}$$

$$+ \; e^{i[(l+\frac{1}{2})\vartheta + 2\pi pl - \frac{1}{4}\pi]}) , \tag{5}$$

where p is an integer and $\vartheta > 0$.

To evaluate the integrals we use the method of steepest descent, where the only contributions to the integral are assumed to come from the regions around the points of stationary phase. These points $\bar{l}$ are determined from

$$2\left(\frac{d\beta_l}{dl}\right)_{l=\bar{l}} \mp \vartheta + 2\pi p = 0 . \tag{6}$$

The quantity

$$2\left(\frac{\partial \beta_l}{\partial l}\right)_{l=\bar{l}} = \Theta(\bar{l}) \tag{7}$$

can thus be identified with the deflection function, in the sense that for a given scattering angle ϑ only those l-values satisfying the relation (6) contribute. This is illustrated in Fig. 7.

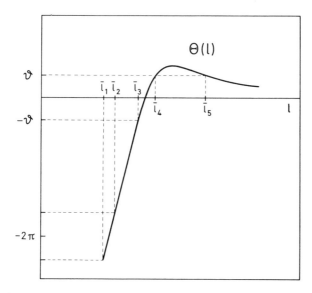

Fig. 7 The stationary-phase points of the scattering amplitude. The values of l contributing to the scattering amplitude are those which satisfy Eq. (4.6) for a given scattering angle $0 < \vartheta < \pi$. The solutions $\bar{l}_2$, $\bar{l}_4$ and $\bar{l}_5$ are the values of l where the first term in (4.5) contributes for $p = 1$, $p = 0$ and $p = 0$ respectively; $\bar{l}_1$ and $\bar{l}_3$ are the values where the second term in (4.5) contributes for $p = 1$ and $p = 0$ respectively.

The contribution from the points of stationary phase are found by expanding the exponent to second order [see e.g. Broglia *et al.* (1974), Eq. (2.21)]. One finds the result

$$f(\vartheta) = \frac{1}{k\sqrt{\sin \vartheta}} \sum_{\bar{l}} (\bar{l} + \tfrac{1}{2})^{1/2} \frac{1}{\sqrt{|\Theta'(\bar{l})|}}$$

$$\times \exp(i\{2\beta_{\bar{l}} - (\bar{l} + \tfrac{1}{2})\Theta(\bar{l}) - \tfrac{1}{2}\pi$$

$$+ \pi p(\bar{l}) + \tfrac{1}{4}\pi[S_1(\bar{l}) + S_2(\bar{l})]\}) , \qquad (8)$$

where

$$S_1(\bar{l}) = \frac{\Theta'(\bar{l})}{|\Theta'(\bar{l})|} , \qquad (9)$$

133

while

$$S_2(\bar{l}) = \begin{cases} +1 & \text{if} \quad 0 < \Theta(\bar{l}) + 2p(\bar{l})\pi = \vartheta < \pi, \\ -1 & \text{if} \quad -\pi < \Theta(\bar{l}) + 2p(\bar{l})\pi = -\vartheta < 0. \end{cases}$$

For the simple situation where only two $\bar{l}$-values contribute, namely one corresponding to scattering in the Coulomb field ($\bar{l} = \bar{l}_5$) and one dominated by the nuclear attraction ($\bar{l} = \bar{l}_4$) (cf. Fig. 7), one finds the cross-section

$$\frac{d\sigma}{d\Omega} = \frac{d\sigma_R}{d\Omega} \left[1 + \frac{d\sigma_N}{d\sigma_R} + 2\sqrt{\frac{d\sigma_N}{d\sigma_R}} \sin(\delta_C - \delta_N) \right], \qquad (10)$$

where

$$\delta_C = 2\beta_{l_5} - (\bar{l}_5 + \tfrac{1}{2})\vartheta, \qquad (11)$$

$$\delta_N = 2\beta_{l_4} - (\bar{l}_4 + \tfrac{1}{2})\vartheta.$$

The cross-section $d\sigma_R$ and $d\sigma_N$ are given by

$$\frac{d\sigma}{d\Omega} = \frac{(\bar{l} + \tfrac{1}{2})}{k^2 \sin\vartheta} \frac{1}{|d\Theta/dl|_{l=\bar{l}}}, \qquad (12)$$

which is identical to the expression (2.7), since

$$(\bar{l} + \tfrac{1}{2})\hbar = m_{aA} v\rho. \qquad (13)$$

For $\bar{l} = \bar{l}_5$ the expression (12) is the Rutherford cross section $d\sigma_R$ [cf. (2.8)]. The interference terms are oscillating functions of the scattering angle. The amplitude of the oscillations is independent of how close the system is to the classical limit. In the classical limit, however, the frequency of the oscillations becomes very large, so that any uncertainty in the bombarding energy needed to form a wave packet well defined in time makes these terms vanish.

As mentioned above, the main quantal effects arise from the angular motion, i.e. the quantization of the angular momentum and the interference between different trajectories. The quantal effects associated with the radial motion which are contained in the phase shifts β_l are less important. In fact one may use for β_l the expression

$$\beta_l = \frac{1}{2} \int_\infty^l \Theta(l) \, dl$$

$$= \frac{m_{aA}v}{2\hbar} \int_\infty^{\rho(l)} \Theta(\rho) \, d\rho \, , \tag{14}$$

as suggested by (7). The quantity $\Theta(\rho)$ is here the classical deflection function, and for the relation between ρ and l we have used (13). The lower limit in the integral (14) is determined from the fact that $\beta_l = 0$ for $l \to \infty$.

Similarly to what was done in the classical description, we assume that all partial waves with angular momentum smaller than the one corresponding to orbiting are absorbed and do not contribute to the partial-wave sum. The corresponding classical trajectories are those that penetrate beyond the barrier of the effective potential.

The classical absorption in the surface, due to a mean free path [Eq. (2.10)], can be included in the scattering amplitude (1) by the substitution

$$e^{i\beta_l} \to \sqrt{P(\infty)} \, e^{i\beta_l} \, . \tag{15}$$

This would correspond to the replacement of the phase shift (14) by the complex phase shift

$$\beta_l = \frac{1}{2} \int_\infty^l \Theta(l') \, dl' + \frac{i}{4} \int_{-\infty}^\infty \frac{v(r) \, dt}{\lambda(r)}$$

$$= \frac{1}{2} \int_\infty^l \Theta(l') \, dl'$$

$$- \frac{i}{\hbar} \sqrt{\frac{m_{aA}}{2}} \int_{r_0}^\infty W(r(t)) \frac{dr}{\sqrt{E - U(r) - (l+\frac{1}{2})^2 \hbar^2 / 2m_{aA}r^2}} . \tag{16}$$

In the last equation we have used the definition (12).

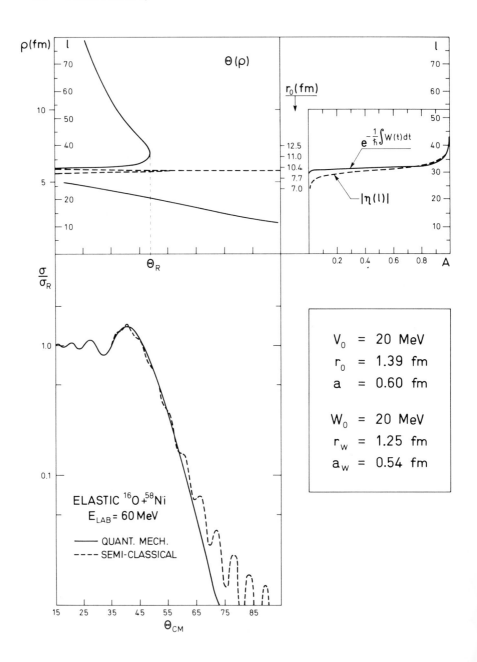

ELASTIC $^{16}O + ^{58}Ni$
$E_{LAB} = 60\,MeV$

—— QUANT. MECH.
---- SEMI-CLASSICAL

$V_0 = 20\ MeV$
$r_0 = 1.39\ fm$
$a = 0.60\ fm$

$W_0 = 20\ MeV$
$r_w = 1.25\ fm$
$a_w = 0.54\ fm$

The deflection function and the cross section for elastic scattering of ^{16}O on ^{58}Ni at 60 MeV using the approximation (16) in the partial-wave sum (1) is illustrated in Fig. 8.

Although these calculations are straightforward, it is useful to be able to make order-of-magnitude estimates from equations like (10). We can thus estimate the frequency of the oscillations caused by the interference between the two trajectories on either side of the rainbow angle (i.e. the one corresponding to the angular momenta $\bar{l}_4$ and $\bar{l}_5$ in Fig. 7). The local variation of the phase in (10) with ϑ is given by

$$\alpha'(\vartheta) = \frac{d}{d\vartheta}(\delta_c - \delta_N) = \bar{l}_4(\vartheta) - \bar{l}_5(\vartheta) , \tag{17}$$

where use has been made of (7).

The local angular wavelength $\Delta\vartheta$ of the oscillations is thus given by

$$\Delta\vartheta = \frac{2\pi}{\alpha'(\vartheta)} = \frac{2\pi}{\bar{l}_4(\vartheta) - \bar{l}_5(\vartheta)} . \tag{18}$$

Utilizing the deflection function in Fig. 8, it is seen that this formula reproduces the wavelength in the oscillation in the elastic scattering for angles less than $45°$ quite well.

Additional oscillations of smaller amplitude may arise especially for higher bombarding energies from the interference with the trajectory of negative scattering angle (i.e. the one corresponding to angular momentum $\bar{l}_3$ in Fig. 7).

Fig. 8 The classical deflection function $\Theta(\rho)$, the absorption coefficient, and the associated angular distribution for the scattering of ^{16}O on ^{58}Ni at 60 MeV. The potentials used were of Saxon-Woods type, with the parameters given. For the deflection function (upper left graph) the impact parameter and the corresponding angular momentum are given on the vertical scale. Also, the corresponding distances of closest approach are indicated. The classical absorption coefficient (upper right) is compared with the corresponding quantal reflection coefficient $\eta_l = \exp\{-\operatorname{Im}\beta_l\}$. The semiquantal angular distribution is indicated in the lower graph by a dashed curve, while the quantal result is given by a solid curve. In both calculations 150 partial waves were utilized. The figure is taken from Broglia et al (1974).

The formulae (8) and (12) cannot be used for angles close to the rainbow angle ϑ_R where $d\Theta/dl \approx 0$. Improved expressions for the scattering amplitudes, which also apply for $\vartheta > \vartheta_R$, will be derived in Sec. 7. Equation (8) also does not apply close to $\vartheta = 0$ and $\vartheta = \pi$. The special phenomena associated with these forward and backward scattering are called the glory effect.

The main discrepancy between the quantal and semiquantal calculations is due to the steep variation of the absorption with impact parameter as shown in Fig. 8. In the two following sections we discuss possible improvements int he expression (16) for the phase shift on the basis of the WKB approximation.

5   SOLUTION OF RADIAL EQUATION

To determine the quantal scattering amplitude (4.1) one needs to evaluate the phase shifts β_l. These quantities are determined from the asymptotic behavior of the regular solution of the radial wave equation

$$\left[\frac{d^2}{dr^2} + k_l^2(r) \right] \chi_l(r) = 0 , \tag{1}$$

where

$$k_l(r) = \left([E - U_{\text{eff}}(r) - iW(r)] \frac{2m_{aA}}{\hbar^2} \right)^{1/2} . \tag{2}$$

We have here introduced an imaginary part in the potential as discussed in Sec. 2.

The radial function χ has the asymptotic behavior

$$\chi_l(r) \sim \sin(kr - \eta \ln 2kr - \tfrac{1}{2}\pi l + \beta_l) , \tag{3}$$

where [cf. (4.2)]

$$k = k_l(\infty) = \frac{m_{aA}v}{\hbar} \tag{4}$$

and [cf. (II.1.5)]

$$\eta = \frac{Z_a Z_A e^2}{\hbar v} .$$

(5)

Only for the case of a pure Coulomb field is there an analytic solution of (1) [see e.g. Mott and Massey (1949)], namely

$$\beta_l^C = \arg \Gamma(l + 1 + i\eta) .$$

(6)

For the general case the solution must be found by numerical integration. It is convenient to write the scattering amplitude (4.1) in the form

$$f(\vartheta) = f^C(\vartheta) + f^N(\vartheta) ,$$

(7)

as a sum of the Coulomb scattering amplitude

$$f^C(\vartheta) = \frac{i}{2k} \sum_l (2l + 1)(1 - e^{2i\beta_l^C})P_l(\cos \vartheta)$$

$$= \frac{a_0}{2} \frac{1}{\sin^2 \tfrac{1}{2}\vartheta} \exp\{-i\eta \ln \sin^2 \tfrac{1}{2}\vartheta + 2i\beta_0^C\} ,$$

(8)

and a nuclear scattering amplitude

$$f^N(\vartheta) = \frac{i}{2k} \sum_l (2l + 1)e^{2i\beta_l^C}(1 - e^{2i\beta_l^N})P_l(\cos \vartheta) ,$$

(9)

where we have introduced the nuclear phase shift β_l^N through

$$\beta_l = \beta_l^C + \beta_l^N .$$

(10)

For the numerical evaluation of the scattering amplitude one needs to consider only angular momenta in a range around the grazing angular momentum (3.28), i.e.

$$l_g + \tfrac{1}{2} = (r_g)_{\text{fm}} \left(\frac{1}{20} \frac{A_a A_A}{A_a + A_A} (E - E_B)_{\text{MeV}}(1 + c) \right)^{1/2} .$$

(11)

139

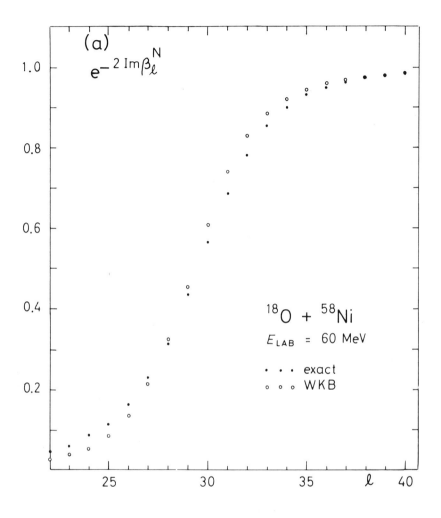

Since the potential has an imaginary part, β_l^N is a complex number. The imaginary part of the phase shift β_l^N describes the flux out of the elastic channel. In fact the total reaction cross section is given by

$$\sigma_R = \frac{\pi}{k^2}\sum_l (2l + 1)(1 - e^{-2\,Im\,\beta_l^N}) . \tag{12}$$

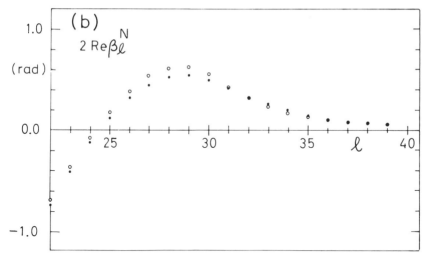

Fig. 9 The nuclear transmission coefficient (a) and phase shift (b) for the elastic scattering of ^{18}O on ^{58}Ni at $E_{lab} = 60$ MeV. The optical potential used was of Saxon-Woods shape [cf. (1.40) and (2.13)] with $V_0 = 90.1$ MeV, $W_0 = 42.9$ MeV, $R_0 = R_w$ 7.92 fm and $a = a_w = 0.5$ fm. The solid dots are the results from integrating the Schrödinger equation numerically. The open circles are calculated in the WKB approximation using the expression (6.15) and the outer turning points in the lower half plane shown in Fig. 15. The figure is taken from Landowne et al (1976).

The quantity $\exp \{- 2 \, Im \, \beta_l^N\}$ is often called the transmission coefficient. An example is shown in Fig. 9 for the scattering of ^{18}O on ^{58}Ni at a bombarding energy of 60 MeV. The figure also shows the quantity $2 \, Re \, \beta_l^N$.

In the quantal description outlined above, the imaginary part of the potential plays two roles. For partial waves close to the grazing angular momentum the tail of W acts as an attenuation, while for low partial waves the effect of W can be replaced by an incoming-wave boundary condition at a radius somewhat inside the Coulomb barrier.

By an incoming-wave boundary condition we mean that the radial wavefunction χ satisfies the equation [Rawitscher (1966), Eisen and Vager (1972)]

$$\left[\frac{d}{dr} \log \chi \right]_{r=R} = - i k(r) . \tag{13}$$

This condition is taken from the WKB approximation (see below), where a clear distinction exists between incoming and outgoing waves at each point in space. Insofar as the point R is chosen well inside the Coulomb barrier, the condition (13) and the condition of regularity of χ at the origin give the same answer. If R is chosen close to the Coulomb barrier, the two prescriptions give somewhat different results. The incoming-wave boundary condition has the advantage that the results are explicitly independent of the potential for $r < R$.

The tail of the absorptive potential can be studied expecially well in elastic scattering close to the Coulomb barrier, where the nuclear potential $U^N + iW$ can be treated as a perturbation. In first-order perturbation theory the nuclear scattering amplitude is given by

$$f^N(\vartheta) = -\frac{m_{aA}}{2\pi\hbar^2} \int d^3r\, \chi_{\text{Coul}}^{(-)*}(\mathbf{k},\mathbf{r})[U^N(r) + iW(r)]\chi_{\text{Coul}}^{(+)}(\mathbf{k},\mathbf{r}), \qquad (14)$$

where

$$\chi_{\text{Coul}}^{(+)}(\mathbf{k},\mathbf{r}) = \chi_{\text{Coul}}^{(-)*}(-\mathbf{k},\mathbf{r}) = \frac{4\pi}{kr}\sum_{lm} i^l e^{i\beta_l^C}\chi_l^C(r)Y_{lm}^*(\hat{k})Y_{lm}(\hat{r}) \qquad (15)$$

are the Coulomb distorted plane waves, $\chi_l^C(r)$ being the solution of (1) for a pure Coulomb field. For $180°$ scattering one finds the approximate result [Traber *et al.* (1977)]

$$\frac{d\sigma_{el}}{d\sigma_{\text{Ruth}}} = 1 + \sqrt{\frac{\pi b}{a}}\,\frac{U^N(b)}{E} + \sqrt{\frac{\pi b}{a_w}}\,2ka_w\frac{W(b)}{E}. \qquad (16)$$

We have used an exponential form of the real and imaginary parts of the nuclear potential, with diffuseness a and a_w, respectively. We have furthermore used that the product $2ka_w$ is large compared to unity. The quantity b is the distance of closest approach in the Coulomb scattering ($b = Z_a Z_A e^2/E$).

The expression (16) can also be derived classically from Eq. (2.11). Thus, the first term in (16) results when the classical cross-section (2.7) is evaluated utilizing (3.2)–(3.5), while the second term arises when the absorption (2.10) is evaluated by the expression

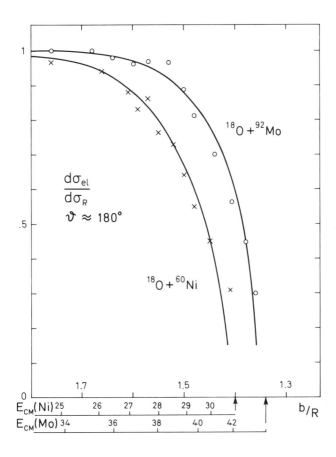

Fig. 10 Ratio of elastic to Rutherford cross section for backward scattering. The quantity $d\sigma_{el}/d\sigma_{Ruth}$ given in Eq. (5.16) is plotted for the scattering of ^{18}O on ^{60}Ni and on ^{92}Mo as a function of the distance of closest approach in the Coulomb field $b = Z_a Z_A e^2/E_{CM}$ in units of $R = R_a + R_A$. The corresponding energies E_{CM} are given on the auxilary scales for the two cases. For $U^N(b)$ we used the empirical potential (1.36)–(1.37), while for $W(b)$ we used $0.5 U^N(b)$ in the case of O+Mo and $0.8 U^N(b)$ for the case of O+Ni. The energy of the Coulomb barrier is indicated by an arrow. The experimental data are taken from Rehm *et al.* (1975).

$$P(\infty) = e^{-(2/\hbar)\int_{-\infty}^{\infty} W(t)\, dt}$$

(17)

to lowest order.

Because of the large value of $2ka_w$, the relation (16) provides a way to determine the imaginary potential from the elastic excitation function at $180°$. An example of this is given in Fig. 10 for the scattering of ^{18}O on ^{92}Mo and on ^{60}Ni. The rather insignificant contribution from the first term in (16) was calculated using the empirical potential (1.36) with the parameters (1.37). The curves shown were calculated with $a_w = a$, but the data would be consistent also with values of a_w ranging from 0.5 fm to 0.7 fm. For the choice $a_w = a$ one obtains $W = 0.5 U^N$ for ^{18}O + ^{92}Mo and $W = 0.8 U^N$ for ^{18}O + ^{60}Ni.

6 WKB APPROXIMATION

For the short wavelengths encountered in heavy-ion collisions the radial equation (5.1) can be solved quite accurately utilizing the WKB approximation [see e.g. Knoll and Schaeffer (1975)].

(a) Single turning point

For a real potential the regular solution takes the form

$$\chi_l(r) = \begin{cases} \sqrt{\dfrac{k}{k_l(r)}} \sin\left(\displaystyle\int_{r_0}^{r} k_l(r')\, dr' + \tfrac{1}{4}\pi \right), & r > r_0, \\[2em] \dfrac{1}{2}\sqrt{\dfrac{k}{\kappa_l(r)}} \exp\left(\displaystyle\int_{r_0}^{r} \kappa(r')\, dr' \right), & r < r_0, \end{cases}$$

(1)

where $k_l(r)$ is defined in (5.2), i.e.

$$k_l(r) = k\left[1 - \frac{U_{\text{eff}}(r)}{E} \right]^{1/2},$$

(2)

144

while

$$\kappa_l(r) = k \left[\frac{U_{eff}(r)}{E} - 1 \right]^{1/2}. \tag{3}$$

The solution (1) applies to the case where there is only one turning point r_0 in the radial motion [cf. Eq. (B.8)].

In order to discuss the situations where absorption occurs, it is convenient to study the solution (1) in the complex r-plane. Thus we may consider the function χ_l for $r < r_0$ as the analytic continuation of the solution for $r > r_0$ beyond the singularity at $r = r_0$. This is seen by writing

$$\chi_l(r) = \begin{cases} \frac{1}{2}[\psi_l^{in}(r) + \psi_l^{out}(r)], & r > r_0, \\ \frac{1}{2}\psi_l^{reg}(r), & r < r_0, \end{cases} \tag{4}$$

where

$$\psi_l^{in}(r) = e^{i \frac{1}{4}\pi} \sqrt{\frac{k}{k_l(r)}} \exp\left[-i \int_{r_0}^{r} k_l(r')\, dr' \right]. \tag{5}$$

The function

$$\psi_l^{reg}(r) = \sqrt{\frac{k}{\kappa_l(r)}} \exp\left[\int_{r_0}^{r} \kappa_l(r')\, dr' \right] \tag{6}$$

is then the analytic continuation of $\psi_l^{in}(r)$ in the upper half plane. This is seen by noting that in the neighborhood of the singularity r_0 the phase integral is given by

$$\int_{r_0}^{r} k_l(r')\, dr' \approx \frac{2}{3} \sqrt{\frac{2m_{aA}}{\hbar^2} \left[-\frac{\partial U_{eff}(r)}{\partial r} \right]_{r_0}} (r - r_0)^{3/2}, \tag{7}$$

and

$$k_l(r) = \sqrt{\frac{2m_{aA}}{\hbar^2} \left[-\frac{\partial U_{eff}(r)}{\partial r} \right]_{r_0}} (r - r_0)^{1/2}. \tag{8}$$

Similarly

$$\psi_l^{\text{out}}(r) = e^{-i\frac{1}{4}\pi} \sqrt{\frac{k}{k_l(r)}} \exp\left[i \int_{r_0}^r k_l(r')\, dr' \right] \tag{9}$$

is obtained by an analytic continuation of ψ_l^{reg} in the lower half plane.

The rules for the analytic continuation can be precisely stated by introducing the Stokes lines, which are the loci where the phase integral

$$S_l(r,r_0) = \hbar \int_{r_0}^r k_l(r')\, dr' \tag{10}$$

is purely imaginary. The analytic structure is illustrated in Fig. 11, where the Stokes lines are indicated by the solid curves emerging from the turning point with an angle of $\frac{2}{3}\pi$ between them. In the first Riemann sheet the function ψ^{in} is exponentially increasing along Stokes lines 2 and 3 and exponentially decreasing along 1. In the second sheet the situation is reversed, and the function along Stokes line 1 becomes proportional to the irregular solution

$$\psi_l^{\text{irreg}}(r) = \sqrt{\frac{k}{\kappa_l(r)}} \exp\left[- \int_{r_0}^r \kappa_l(r')\, dr' \right] . \tag{11}$$

The rules for constructing the WKB solution corresponding to a given boundary condition can now be formulated graphically as shown in Fig. 12. The interpretation of this figure is that if we know that the wavefunction is regular on Stokes line 1, the solution in the whole complex plane is obtained by clockwise continuation in sectors III and I up to Stokes line 3 and anticlockwise continuation through sectors II and I up to Stokes line 2. While the solutions in sectors II and III are given directly by these analytic continuations, the solution in sector I is the sum of the two analytic continuations. If the boundary condition is one of outgoing (incoming) waves only, the same figure applies, but one should consider the behavior of the function in region III (II) as the boundary condition. A more accurate derivation is indicated in Appendix D.

The above rules allow us to generalize the solutions (1) to the situation in which the turning point is complex, as e.g. in the case of complex potentials.

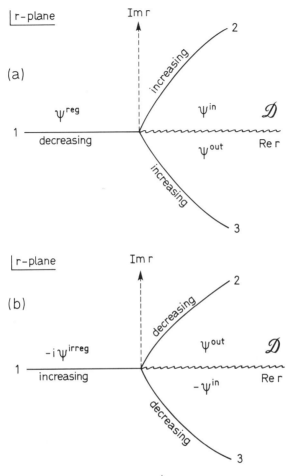

Fig. 11 The analytic continuation of ψ^{in} [Eq. (6.5)]. On the first sheet (a) above the cut indicated by the wavy line, ψ^{in} represents an incoming wave for large values of r. As ψ^{in} is continued anticlockwise around r_0, it becomes exponentially increasing along Stokes line 2 and decreasing along Stokes line 1. Below the cut in region $\mathscr{D}$ the continuation of ψ^{in} is the outgoing wave ψ^{out} given by (6.9). As the anticlockwise continuation proceeds through the cut, the second sheet is entered. The behavior of the function under continuation in the second sheet is shown in (b). The values of the function in the third and fourth Riemann sheets are obtained from those in the first and second sheets, respectively, by an overall change of sign. The further continuation from the fourth sheet anticlockwise leads into the first sheet.

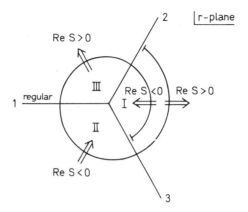

Fig. 12 Analytic continuation of a WKB wavefunction which decreases exponentially along Stokes line 1. The sign of the phase S defined in (6.10) is indicated in each sector by arrows. The analytic continuation is terminated at a Stokes line where one arrives at an exponentially decreasing solution if an exponentially increasing solution was obtained already by analytic continuation in the other sense around the turning point [cf. Fig. 11].

For the situation in which the bombarding energy is below the Coulomb barrier in the effective potential, the position of the turning point and the associated Stokes lines is indicated in Fig. 13. The boundary condition is one of regular or incoming waves (towards the scattering center) in sector III, and the solution is thus, according to Fig. 12, given by

$$\chi_l(r) = \begin{cases} \frac{1}{2}[\psi_l^{\text{in}}(r) + \psi_l^{\text{out}}(r)] & \text{in I}, \\[2mm] \frac{1}{2}\psi_l^{\text{in}}(r) & \text{in III}, \end{cases} \tag{12}$$

where $\psi_l^{\text{in}}(r)$ is given by (5) with

$$k_l(r) = k\left(1 - \frac{U_{\text{eff}} + iW(r)}{E}\right)^{1/2} \qquad (\tfrac{1}{2}\pi > \arg k_l \geq 0). \tag{13}$$

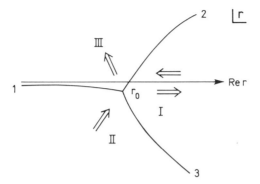

Fig. 13 Stokes lines associated with the outermost turning point for a complex potential and at an energy below the Coulomb barrier in the effective potential. The WKB wavefunction corresponding to the regular solution is indicated with the convention of Fig. 12.

The function ψ_l^{out} is given by (9) and is the analytic continuation of ψ^{in} into sector I below the turning point. The function ψ^{in} in sector III differs from ψ^{reg} only by the presence of the imaginary potential.

From this wavefunction we can deduce the phase shift defined in (5.3), i.e.

$$\beta_l = \int_{r_0}^{r} k_l(r')\, dr' + \tfrac{1}{4}\pi - kr + \eta \ln(2kr) + \tfrac{1}{2}\pi l, \qquad (14)$$

where r is a point at large distance on the real axis. The first integral is a path integral from the singularity at r_0 to r. It may be recast into $\frac{1}{2}$ times the integral from r clockwise around r_0 and back to r, i.e.

$$\beta_l = \tfrac{1}{2}\oint_C k_l(r')\, dr' - kr + \eta \ln 2kr + \tfrac{1}{2}\pi(l + \tfrac{1}{2}) . \qquad (15)$$

This phase shift is a generalization of the phase shift which we obtained from the equivalence of the classical and quantal cross sections in the classical limit in Sec. 4 [cf. Eqs. (4.14) and (4.16)]. This can be shown by utilizing the fact that $|W|$ is small compared to $|U_{\text{eff}} - E|$ along an appropriate path around r_0. Therefore,

$$\beta_l \approx \sqrt{\tfrac{1}{2}m_{aA}} \, \frac{1}{\hbar} \oint_C \sqrt{E - U_{\text{eff}}(r')} \, dr'$$

$$- k(r) \cdot r + \eta \ln 2kr + \tfrac{1}{2}\pi(l + \tfrac{1}{2})$$

$$- \frac{i}{2\hbar} \sqrt{\frac{m_{aA}}{2}} \oint_C \frac{W(r') \, dr'}{\sqrt{E - U_{\text{eff}}(r')}}. \tag{16}$$

The imaginary part of the phase shift is identical to the imaginary part of (4.16). The identity of the real parts is most easily proved by taking the derivative of the real part of (16) with respect to l.

Since we may usually assume that the ion-ion potential is also relatively small, we may expand the integral in (16) in this quantity also to find

$$\beta_l = \beta_l^C - \frac{1}{2\hbar} \int_{-\infty}^{\infty} U^N(r(t)) \, dt - \frac{i}{2\hbar} \int_{-\infty}^{\infty} W(r(t)) \, dt, \tag{17}$$

where β_l^C is the Coulomb phase shift.

In the classical expression (16), W only contributes to the imaginary part of the phase shift. Its only effect is thus to remove flux. In the WKB expression (14), the absorptive potential gives rise also to a real phase shift, leading to reflection and refraction [see e.g. Broglia et al. (1974), p. 12)]. These effects are often small in heavy-ion reactions, since W is usually small in the surface.

(b)   Generalization of simple prescription

The usefulness of (14)–(15) over the classical expression (16) is mainly connected with the fact that it can be used for lower impact parameters than the one corresponding to orbiting, where the bombarding energy is above the Coulomb barrier in the effective potential. In this situation the Coulomb barrier, even for $W = 0$, gives rise to two (complex conjugate)

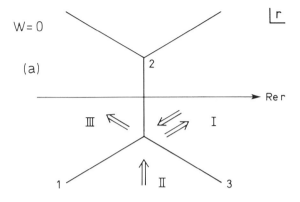

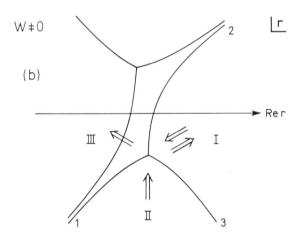

Fig. 14 Stokes lines emerging from the turning points associated with the Coulomb barrier for a bombarding energy above the barrier. For a real potential (a), the two turning points are complex conjugates, and two Stokes lines, one emerging from each of the two turning points, coincide on the interval between them. The WKB solution corresponding to the incoming-wave boundary condition in region III is indicated with arrows following the convention of Fig. 12. In (b) is illustrated the same situation but with $W \neq 0$, in which case the Stokes lines never cross.

turning points. In fact, as the bombarding energy approaches the height of the Coulomb barrier from below, the two outermost turning points from either side of the barrier approach each other and coalesce for the orbiting situation. For higher energies they move out in the complex plane (cf. also Appendix D).

The wavefunction (12) is the WKB solution which corresponds to the boundary condition of incoming waves inside the barrier, provided one chooses as r_0, in (5)–(10), the turning point in the lower half plane. This is illustrated in Fig. 14(a), where we have indicated the solution utilizing the conventions of Fig. 12. The solution associated with the turning point in the upper half plane would correspond to the boundary condition of outgoing waves in the inside region. For the case where the energy is above the barrier and $W \neq 0$, the situation is quite similar except that the turning points are shifted and the Stokes lines emerging from the two turning points do not cross. This is illustrated in Fig. 14(b).

In Fig. 15 we illustrate the turning points for the scattering of ^{18}O on ^{58}Ni at a bombarding energy of 60 MeV. Utilizing (15) with the integration path shown in Fig. 15, one obtains the phase shifts indicated by open circles in Fig. 9. They give a very good approximation to the exact phase shifts. The associated elastic-scattering angular distribution is given in Fig. 16.

A simple classical geometrical interpretation of the WKB results has been given by Miller (1969). The quantal amplitude (as e.g. a wavefunction) connecting a state a with a state b can be written as the product

$$\sqrt{P(a \rightarrow b)}e^{(i/\hbar)S(a \rightarrow b)}$$

of the square root of the classical probability for going from a to b, and an exponential function. The quantity $S(a \rightarrow b)$ is the action integral associated with the states a and b. In many cases there will be several classical paths q connecting the state a with the state b, and the amplitude is then

$$\langle b \mid a \rangle = \sum_q \sqrt{P_q(a \rightarrow b)}\, i^{-n_q}e^{(i/\hbar)S_q(a \rightarrow b)}, \tag{18}$$

where n_q is the number of reflections, i.e. sign changes of the velocity.

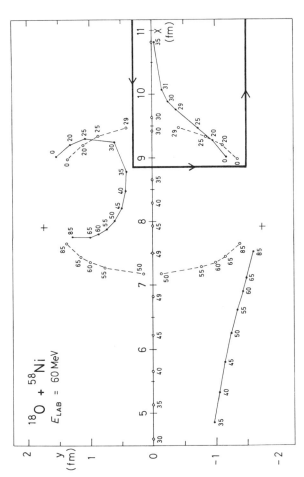

Fig. 15 Complex turning points for $^{18}\text{O} + {}^{58}\text{Ni}$ at a bombarding energy of 60 MeV, calculated with the optical potential given in the caption of Fig. 9. The open circles show the turning points when the imaginary potential is set equal to zero; the solid dots are for the complex potential. The numbers give the angular momentum of the corresponding partial waves in units of $\hbar$. The grazing (orbiting) angular momentum is $29 \leq l_g \leq 30$. For $30 \leq l \leq 49$ there are (for $W=0$) three real turning points. For $l \leq 29$ the two outermost turning points move out in the complex plane as complex conjugates, while for $l \geq 50$, where the bombarding energy is below the minimum in the effective potential, the two innermost turning points become complex. For $W \neq 0$ the transition between real and complex turning points becomes a smooth one. The solid line with arrows shows part of the integration path used in the WKB calculation. It returns to the real axis outside the range of interaction. The plus signs indicate the position of poles in the potential. The figure is taken from Landowne et al (1976.)

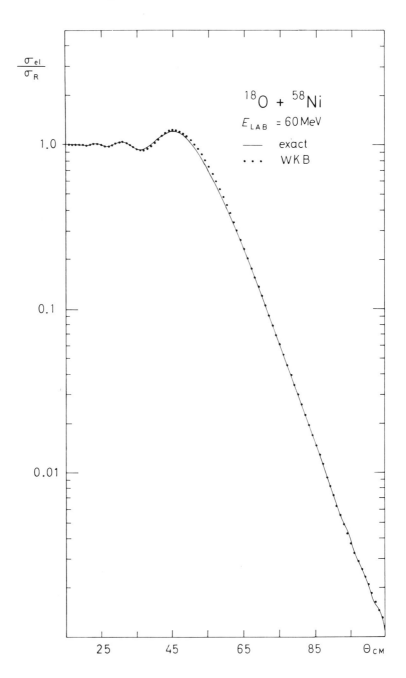

For the one-dimensional motion considered above, a describes a state $|\psi\rangle$ of particles moving towards the nucleus with velocity v and with a density of one particle per unit length. The amplitude to find a particle at the position r receives contributions partly from the path connecting a point far away (r_∞) directly with the point r, and partly from the path which is reflected from the nucleus and passes the point r in the outward motion. For both contributions to the wavefunction the probability $P(r_\infty \rightarrow r)$ is given by

$$P(r_\infty \rightarrow r) = \frac{v}{v_l(r)}, \qquad (19)$$

where v is the velocity at infinity and $v_l(r)$ the radial velocity at the point r. The action integrals are

$$S_1(r_\infty \rightarrow r) = -\hbar \int_{r_\infty}^{r} k_l(r')\, dr'$$

$$= -S_l(r,r_0) + S_l(r_\infty, r_0), \qquad (20)$$

and

$$S_2(r_\infty \rightarrow r) = -\hbar \int_{r_\infty}^{r_0} k_l(r')\, dr' + \hbar \int_{r_0}^{r} k_l(r')\, dr'$$

$$= S_l(r_\infty, r_0) + S_l(r,r_0). \qquad (21)$$

We thus obtain the wavefunction

$$\langle r | \psi \rangle = \sqrt{\frac{v}{v_l(r)}} e^{(i/\hbar)S_l(r_\infty,r_0)} \left(e^{-(i/\hbar)S_l(r,r_0)} - i e^{(i/\hbar)S_l(r,r_0)} \right). \qquad (22)$$

Fig. 16 Ratio of the elastic-scattering differential cross section to the Rutherford cross section for the scattering of ^{18}O on ^{58}Ni at $E = 60$ MeV, calculated with the optical parameters listed in the caption to Fig. 9. The solid curve shows the quantum-mechanical result, while the dots indicate the results obtained with the WKB approximation. The corresponding quantal and WKB phase shifts are given in Fig. 9. The figure is taken from Landowne et al (1976).

Except for a constant factor this solution is equal to the WKB wavefunction (4) in the external region.

The above formulation can be generalized to include complex trajectories, i.e. solutions of the classical equations of motion where the time is allowed to become complex. The action intergral $S(a \rightarrow b)$ is then only dependent on the way the path circumvents the singularities in the complex plane. The quantities S_1 and S_2 thus differ by the clockwise integration aound the singularity at r_0 from r and back again to r. The factor i^{nq} in (18) is obtained automatically when the probability P_q is analytically continued along the same path which was used for the evaluation of S.

All the results obtained above for complex potentials and for reflections from complex turning points can be interpreted in terms of these generalized classical prescriptions.

In general in heavy-ion collisions it is sufficient to consider the reflections from the outermost turning point in the lower half plane, since the inner turning point lies so far inside the nucleus that there is complete absorption. For high bombarding energies it may happen, however, that the three turning points coalesce, the corresponding values of the energy and the angular momentum being given by (3.31)–(3.32). For bombarding energies in the neighborhood of E_{cr} and angular momenta slightly smaller than the grazing angular momentum l_g, the innermost turning point may become important. (Brink and Takigawa (1977)).

The WKB wavefunction for this situation can be constructed on the basis of Fig. 17. One finds, dropping the index l on S,

$$\langle r \mid \psi \rangle = \sqrt{\frac{v}{v_l(r)}} \{ e^{(i/\hbar)S(r_\infty,r)} + e^{(i/\hbar)[S(r_\infty,r_0)+S(r,r_0)]-i\frac{1}{2}\pi}$$

$$+ e^{(i/\hbar)[S(r_\infty,r_1)+S(r,r_1)]-i\frac{1}{2}\pi}$$

$$+ e^{(i/\hbar)[S(r_\infty,r_1)+S(r_2,r_1)+S(r_2,r_1)+S(r,r_1)]-i\frac{3}{2}\pi}\} \tag{23}$$

$$= \sqrt{\frac{v}{v_l(r)}} e^{(i/\hbar)S(r_\infty,r_0)} \{ e^{-(i/\hbar)S(r,r_0)}$$

$$+ e^{(i/\hbar)S(r,r_0)-i\frac{1}{2}\pi}[1 + e^{(i/\hbar)2S(r_0,r_1)}(1 - e^{(i/\hbar)2S(r_1,r_2)})] \}.$$

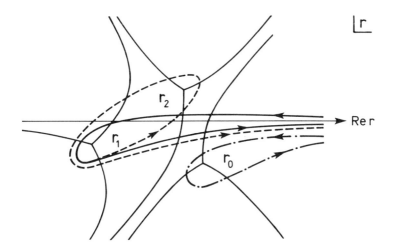

Fig. 17 Stokes lines associated with three turning points. The path associated with the reflection from the outermost turning point is indicated by the dash-dot curve, while the solid curve indicates the path associated with reflection from the innermost turning point. The multiple reflection between the two innermost turning points is shown by dashed curve. The associated wavefunction is given by (6.23).

The first two terms are the same as those appearing in (22) and arise from the dash-dot path of integration in Fig. 17. The third term arises from the reflection in the innermost turning point r_1 (solid curve in Fig. 17). The last term is due to the double reflection first in the turning point r_1 and then in the turning point r_2. The last term is actually the first term in an infinite series of multiple reflections between the turning points r_1 and r_2.

The phase which is calculated from the wavefunction (23) can be written in the form

$$\beta_l = \beta_l^{(0)} + \beta_l' , \tag{24}$$

where $\beta_l^{(0)}$ is given by (15) while β_l' is defined by

$$e^{2i\beta_l'} = 1 + e^{(i/\hbar)2S_l(r_0,r_1)}(1 + e^{(i/\hbar)2S_l(r_2,r_1)})^{-1} . \tag{25}$$

157

Here the multiple reflections have been included by adding up the geometrical series.

It is noticed that this expression contains effects associated with the resonances that occur at energies close to virtual states defined by the well between the two innermost turning points. Thus the phase shift β'_l shows a rapid variation when

$$2S_l(r_2,r_1) = (2n + 1)\pi , \qquad (26)$$

which is the quantization condition for stationary states in this well.

The usefulness of Eq. (25) in heavy-ion scattering is questionable and the derivation above is mainly to be taken as an exercise in the use of the WKB approximation.

7   RAINBOW SCATTERING

In the last section as well as in Appendices D and E we have studied in some detail different semiclassical approximations to the solution of the radial Schrödinger equation. The motivation for this was to obtain expressions for the phase shift β_l as it enters in the scattering amplitude (4.1). In the following chapters we shall use such improved wavefunctions in connection with nuclear reactions.

What we have learned about the phase shift is summarized in Fig. 18, which shows the reflection coefficient for a situation where the effective potential has a barrier at $l = l_g$. In the classical description and in the absence of an absorptive potential the reflection coefficient is a step function corresponding to total absorption for $l < l_g$ and to total reflection for $l > l_g$. The presence of an absorptive potential influences the transmission for $l > l_g$, diminishing η_l by the factor (2.10):

$$P(\infty) = e^{-(2/\hbar)\int_{-\infty}^{\infty} W(t)\,dt} . \qquad (1)$$

In the simple WKB approximation discussed in Sec. 6 there occurs a reflection from the top of the barrier which gives rise to a finite value of η_l for $l < l_g$.

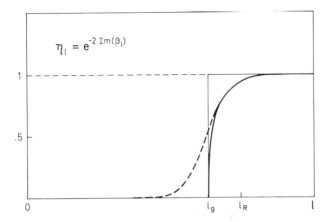

Fig. 18 Reflection coefficient for a bombarding energy above the Coulomb barrier in different approximations. The angular momenta corresponding to orbiting and to the rainbow situation are denoted by l_g and l_R, respectively. The step function at $l = l_g$ is the reflection coefficient in the pure classical limit with total absorption inside the barrier. The solid curve results when a surface absorption, acting as a mean free path, is added. The dashed curve corresponds to the WKB approximation with reflection from the outermost turning point and total absorption inside the nucleus.

When the imaginary potential W is incorporated into the WKB approximation, a smooth reflection coefficient results which has approximately the value $\frac{1}{2}$ at $l = l_g$. For l-values larger than l_R, the curve does not differ significantly from the solid curve. In this region the real part of the phase shift also does not differ appreciably from the value (4.14) obtained on the basis of the classical deflection function. For $l < l_R$, the presence of the imaginary potential usually changes the real part of β_l significantly, and the "deflection function" obtained as the derivative of $2\beta_l$ does not display orbiting. A classical interpretation for such impact parameters can however only be made in terms of complex trajectories.

The results which we have obtained in the last three sections concerning the quantal effects governing the radial motion can be used also to obtain an improved expression for the partial-wave sum (4.8). In fact, the scattering amplitude $f(\vartheta)$ is quite analogous to radial wavefunctions in a situation with one turning point. As a function of ϑ we thus have a classically allowed region $\vartheta < \vartheta_R$, where mainly two trajectories interfere like the incoming and outgoing radial waves in (6.22). At the boundary of the classical region, the simple stationary-phase approximation (4.14) diverges in the same way as the WKB radial wavefunction (6.1).

As the WKB wavefunction has a simple continuation into the classically forbidden region, the stationary-phase approximation (4.8) can be extended into the shadow region for $\vartheta > \vartheta_R$. Some of these solutions, when inserted in (4.8), will lead to scattering amplitudes which increase exponentially beyond the rainbow angle and which have to be discarded.

In order to illustrate the extension of (4.8) into the shadow region, we expand β_l around $l = l_R$ in the following way:

$$2\beta_l = 2\beta_{l_R} + (l - l_R)\Theta_R + \tfrac{1}{6}(l - l_R)^3\Theta_R'' , \qquad (2)$$

where we used that $\Theta'(l_R) = \Theta_R' = 0$. The stationary phase condition (4.6) then takes the form

$$\Theta_R + \tfrac{1}{2}(l - l_R)^2\Theta_R'' = \vartheta , \qquad (3)$$

where we have neglected the contributions from negative scattering angles $(p \neq 0)$. Since $\Theta_R'' < 0$, the solutions of (3) can be written as

$$\bar{l} = l_R \pm iq , \qquad (4)$$

where

$$q = \left[\frac{2(\vartheta - \Theta_R)}{|\Theta_R''|} \right]^{1/2} . \qquad (5)$$

This simple generalization of (4.8) to the case of complex values of $\bar{l}$ leads to the result

$$f(\vartheta) = \frac{1}{k\sqrt{\sin\vartheta}} (l_R \pm iq + \tfrac{1}{2})^{1/2} \frac{1}{\sqrt{q\,|\Theta_R''|}}$$

$$\times \exp(\pm \tfrac{1}{3}q^3 \mid \Theta_R'' \mid)$$

$$\times \exp\{i[2\beta_{l_R} - (l_R + \tfrac{1}{2})(\Theta_R + \tfrac{1}{2}q^2 \mid \Theta_R'' \mid) \pm \tfrac{1}{4}\pi]\} . \qquad (6)$$

In the shadow region we may thus write the cross-section as

$$\frac{d\sigma}{d\Omega} = \frac{1}{k^2 \sin \vartheta} \frac{l_R + \tfrac{1}{2}}{\mid \Theta_R'' \mid^{2/3}} \frac{e^{-2x/3}}{x^{1/3}} , \qquad (7)$$

with

$$x = q^3 \mid \Theta_R'' \mid = \frac{1}{\mid \Theta_R'' \mid^{1/2}} [2(\vartheta - \Theta_R)]^{3/2} . \qquad (8)$$

This equation shows that in the shadow region the slope is mainly determined by the second derivative of the deflection function. This quantity can be estimated from the expressions (3.2)–(3.5). Utilizing also (3.7), one finds approximately

$$\Theta_R'' = -\frac{2a_0}{a} \frac{(l_R + \tfrac{1}{2})}{[\eta^2 + (l_R + \tfrac{1}{2})^2]^{3/2}}(1 - 2\sqrt{2}\,\frac{\Delta}{a}) , \qquad (9)$$

where Δ is defined in (3.8). The expression (7) can only be used for qualitative estimates, since it does not apply very close to the rainbow on account of the factor $x^{-1/3}$, and does not apply far away from the rainbow, where the expression (2) is inaccurate.

To obtain a more accurate expression close to the rainbow angle one may insert (2) directly in the partial-wave expansion (4.5). One can express the integral over l in terms of the Airy function, and one finds the result

$$f(\vartheta) = \frac{1}{ik}\left(\frac{2\pi(l_R + \tfrac{1}{2})}{\sin \vartheta}\right)^{1/2}$$

$$\times \exp\{i[2\beta_{l_R} - (l_R + \tfrac{1}{2})\vartheta - \tfrac{1}{4}\pi]\}$$

$$\times \frac{\mathrm{Ai}(\mid \tfrac{1}{2}\Theta_R'' \mid^{-1/3}(\vartheta - \Theta_R))}{\mid \tfrac{1}{2}\Theta_R'' \mid^{1/3}} . \qquad (10)$$

161

The corresponding cross-section becomes identical to (6) when one utilizes the asymptotic expression of the Airy function for ϑ far away from the rainbow angle.

One can use (10) to estimate the cross section at the rainbow angle. One finds

$$\frac{d\sigma}{d\sigma_{\text{Ruth}}}\bigg|_{\vartheta=\Theta_R} = \frac{4\pi(l_R + \frac{1}{2})\,\sin^{3\frac{1}{2}}\Theta_R}{\eta^2\cos^{\frac{1}{2}}\Theta_R}\,\frac{[\text{Ai}(0)]^2}{(\frac{1}{2}|\,\Theta_R''\,|\,)^{2/3}}$$

$$\approx 1.6\left[\frac{\eta a^2}{b^2}\cot^2\frac{\pi - \Theta_R}{4}\right]^{1/3}, \tag{11}$$

where we have used (9) and where $b = a_0[1 + (\sin\frac{1}{2}\Theta_R)^{-1}] \approx r_R$.

For the example given in (3.13)–(3.14) one finds

$$\frac{d\sigma}{d\sigma_{\text{Ruth}}}\bigg|_{\vartheta=\Theta_R} = 0.55, \tag{12}$$

which is in approximate agreement with Fig. I.8. The rainbow angle occurs at a point somewhat beyond the maximum in the cross-section ratio $d\sigma/d\sigma_R$. The maximum can be estimated from (10) to be at $\vartheta = \Theta_R - 6°$, which is also consistent with Fig. I.8.

The expression (10) for the scattering amplitude is quite analogous to the expression (E.15) for the radial wavefunction. It only applies in the neighborhood of the rainbow angle where the phase shift can be approximated by (2), just as (E.15) only applies in a region close to the turning point where the potential is linear.

This analogy can be understood viewing the scattering amplitude (4.5) as the Fourier transform of the "wavefunction" $e^{2i\beta_l}$ in the corresponding momentum space. Thus, the stationary-phase approximation leads to the simple WKB approximation (4.8) and (6) for $\vartheta < \Theta_R$ and $\vartheta > \Theta_R$ respectively. The linear approximation of the potential in the neighborhood of the turning point leads to the wavefunction (E.15), which is, on the other hand, the Fourier transform of a wavefunction in momentum space of the type $\exp\{i(ap+bp^3)\}$, corresponding to the cubic expansion of the phase shift.

In the same way as one is able to construct a uniform WKB-like approximation to the wavefunction [cf. Eqs. (E.5)–(E.14)], one can perform a mapping of the integrand in (4.5) to obtain an expression for the scattering amplitude which close to the rainbow angle agrees with (10) and far from the rainbow angle agrees with (4.8). Such approximations are discussed in Berry and Mount (1972), da Silveira (1973), and Knoll and Schaeffer (1977).

The WKB approximation in several dimensions can be formulated on the basis of a self-consistent use of the stationary-phase approximation for the evaluation of all Fourier integrals connecting the different quantal amplitudes [Miller (1969)]. For the simple one-dimensional case this leads to the result (6.22). For the scattering amplitude one may also apply the formula (6.18), interpreting P_q as a classical cross-section associated with the trajectory q, while S_q is the classical action integral along the same trajectory. This leads directly to the result (4.8) in the classically allowed region $\vartheta < \vartheta_R$. The amplitudes in the classically forbidden region are obtained considering solutions of the classical equations of motion where the time, and therefore also the coordinates, are allowed to be complex. The action integral along such trajectories will receive an imaginary part which will make the amplitude $\exp\{iS_q/\hbar\}$ vanishingly small if the path is very nonclassical. The nonclassical paths can lead to barrier penetrations but also to reflections from an otherwise attractive potential. The attractive field $U \sim \exp\{(r-R)/a\}$ is thus repulsive if $r \to r + i\pi a$. For a discussion of this method see Schaeffer (1978).

8 COMPARISON WITH EXPERIMENT

The differential cross-section for elastic scattering has been measured for many different combinations of target and projectile where at least one of the partners is relatively light. In collisions where both nuclei are heavy it is more difficult to detect the elastic events because of the energy resolution needed. In this case the elastic differential cross section is essentially the Rutherford cross-section multiplied by the probability of the nuclei staying in their ground state. Thus the probability that the two nuclei will not undergo Coulomb excitation in a collision where they come within the range of nuclear interaction is often very small (see e.g. Fig. I.7). It will be

163

a challenge to detect the nuclear effects in the elastic scattering of such heavy-ion collisions, which are expected to show up far below the Rutherford cross-section.

Since Coulomb-excitation is not expected to change the trajectory of relative motion from the hyperbolic trajectory, the sum of elastic plus Coulomb-excitation cross-sections should give an angular distribution which can be analyzed in a smilar way to the elastic-scattering angular distribution for lighter ions. In this case, however, one does not expect to see interference phenomena, since the cross-section is made up of a large number of incoherent contributions. Such inclusive elastic processes have until now been realized by using bad energy resolution, in which case one includes, besides Coulomb excitation, a number of transfer reactions and nuclear inelastic scattering. The interpretation of such data is made in terms of Coulomb scattering and a strong absorption radius. It is doubtful whether the effect of the ion-ion potential can be seen directly in these measurements.

The situation is rather different in the scattering of relatively light ions (cf. Fig. II.4). In many cases Coulomb excitation and nuclear surface reactions are so weak that the elastic scattering is dominated by the ion-ion potential. One has been able to analyze the experiments in terms of an ion-ion potential which follows naturally from nuclear properties. The absorption can be taken into account by an incoming-wave boundary condition and a weak imaginary potential acting in the nuclear surface.

Examples of such analysis are given in Fig. 19, where we have used the potential (1.36)–(1.37) throughout with the $W(r)$ indicated in the caption. The experiments can actually be fitted by a wide variety of optical potentials. It is found in practice that all real potentials which fit a given experiment, cross each other at a point somewhat outside the Coulomb barrier produced by this potential (cf. Fig. 20). This can be understood by realizing that the angular distribution is mainly governed by the position of the rainbow angle. All ion-ion potentials of exponential form which produce a given rainbow angle can be shown to cross each other at a point slightly inside the distance of closest approach, r_R, for the trajectory corresponding to the rainbow situation (see Appendix F). Since the rainbow distance r_R is a function of the bombarding energy, one may empirically determine the ion-ion potential from elastic scattering over a range of r-values outside the Coulomb barrier. The analysis of a large number of experimental data on this basis has led to the ion-ion potential (1.36) often used in his chapter, which is very similar to the proximity potential (1.31).

The ion-ion potential is important not only for its effect on the trajectory of relative motion, but also because, being a surface-surface interaction, it is the driving force in setting up surface oscillations and rotations, as will be discussed in the following chapter. The success that the calculated ion-ion potential has had in explaining elastic scattering between light ions gives confidence that this potential can be used also in heavy-ion collisions, where it has not been measured directly in elastic scattering.

9   TOTAL REACTION CROSS-SECTIONS

The measured elastic angular distribution is often used to estimate the total reaction cross-section. For light-ion reactions this has been done by using (5.12), where the transmission coefficients $\exp\{-2\,\mathrm{Im}\,\beta_l\}$ are obtained from an optical-model fit of the elastic scattering. Empirically one finds that the total reaction cross-section determined in this way agrees well with the experiment, in spite of the ambiguities of the optical potential. In cases where the absorption is very weak, one could have expected appreciable discrepancies, as elastically scattered particles would emerge at very forward angles, where they would be difficult to detect due to the rather large Rutherford cross-section. However, for typical bombarding conditions and for the impact parameters leading to small angles, the deflection function is very steep. The associated cross section is thus small [see also Schwarzchild *et al.* (1976)].

A widely used prescription is that of relating the total reaction cross-section with the angle $\vartheta_{1/4}$ where the elastic cross section is equal to $\frac{1}{4}$ times σ_{Ruth}. The rule states that the grazing angular momentum where the transmission coefficient is $\frac{1}{2}$ is given by

$$l'_g + \tfrac{1}{2} = \eta \cot \tfrac{1}{2}\vartheta_{1/4} . \tag{1}$$

This rule can be proven for the case in which the elastic scattering can be described as Rutherford scattering for all l-values larger than l'_g and total

III   Elastic Scattering

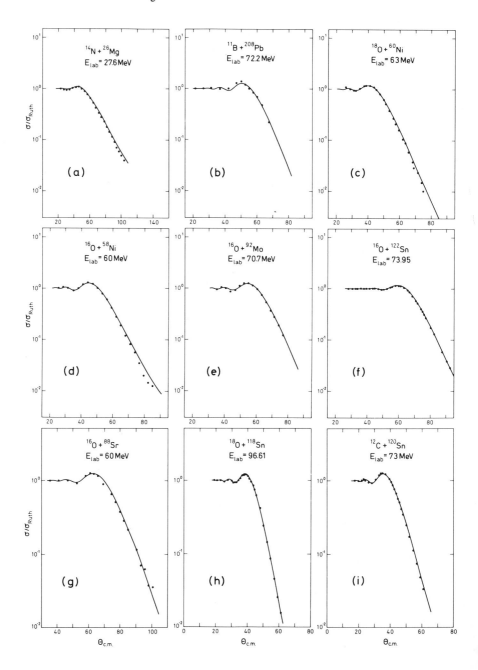

Fig. 19 Ratio of experimental angular distributions for elastic scattering and Rutherford cross section. The curves indicate the theoretical result based on the potential (1.36)-(1.37) and the simple WKB approximation (6.15). The imaginary part of the potential was assumed also to be of exponential form $W(r) = W_0 \exp[-(r-R)/a_w]$, where R is given by (1.37), while W_0 and a_w were adjusted for each case as indicated by the table below:

Case	a	b	c	d	e	f	g	h	i
W_0(MeV)	7.38	31.93	32.80	61.17	33.72	62.71	65.76	31.50	44.89
a_w(fm)	0.806	0.553	0.651	0.467	0.559	0.533	0.489	0.658	0.559

The experimental data indicated by dots were taken from the following references:

(a) Hentschel, E., private communication. (b) Ford, J., *et al.* (1975) *Phys. Rev. C* **8**, 1912.
(c, d, e, f) Rhem, K. E., *et al.* (1975). (g) Christensen, P. R., *et al.* (1973). (h, i) Thorn, C., private communication.

The difference between the theoretical curve and the experimental points is mainly due to the fact that $W(r)$ was determined by a quantal optical-model search routine, which essentially fits the experiments perfectly. The difference thus mainly indicates the difference between the WKB and the quantal calculations. The fitting procedure also accounts for the rather erratic variation of the parameters for $W(r)$. The results are actually not very sensitive to $W(r)$. This figure is due to Andrea Vitturi.

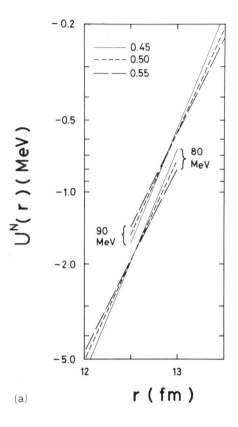

(a)

absorption for $l < l'_g$. The total reaction cross-section is in this case given by

$$\sigma_R = \pi \mathchar'26\mkern-10mu\lambda^2 (l'_g + \tfrac{1}{2})^2 . \tag{2}$$

The elastic-scattering amplitude can be evaluated by the method of stationary phase, and one finds [Frahn (1966)]

$$f(\vartheta) = f^C(\vartheta) \tfrac{1}{2}\{1 + [C(x) + S(x)] + i[C(x) - S(x)]\} , \tag{3}$$

where

$$x = \sqrt{\frac{2\eta}{\pi}} \frac{\sin \tfrac{1}{2}(\vartheta_g - \vartheta)}{\sin \tfrac{1}{2}\vartheta_g} . \tag{4}$$

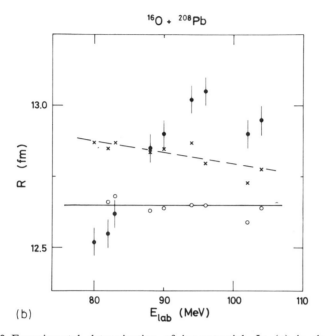

^{16}O + ^{208}Pb

(b)

Fig. 20 Experimental determination of ion-potential. In (a) is shown the tail of the real part of the optical potentials which give equivalent fits to elastic-scattering data of ^{16}O on ^{208}Pb at 80-MeV and 90-MeV bombarding energy respectively. Saxon-Woods parametrizations were used with the diffuseness parameters a as indicated on the figure in fm. It is seen that the potentials cross each other at $R = 12.5$ fm and $R = 12.9$ fm respectively. The values of R obtained by a similar analysis for other bombarding energies are indicated in (b) by dots with an estimated error bar. The open circles indicate the value of the strong-absorption radius as extracted from the quarter-point analysis (9.1) utilizing the relation $R = a_0(1 + \sqrt{[(l + \frac{1}{2})/\eta]^2 + 1})$, valid for Coulomb trajectories. The crosses indicate the strong-absorption radii corresponding to the l-value where the transmission coefficient is $\frac{1}{2}$ in an optical-model fit to the data and utilizing the same prescription to relate l and R. The data as well as the figures are taken from Videbæk *et al.* (1977).

The angle ϑ_g is related to the angular momentum l'_g by

$$l'_g + \frac{1}{2} = \eta \cot \frac{1}{2}\vartheta_g \qquad (5)$$

characteristic for Coulomb scattering. The functions $C(x)$ and $S(x)$ are the

Fresnel integrals. It is seen that for $\vartheta = \vartheta_g$, i.e. $x = 0$, the differential cross section $|f(\vartheta_g)|^2$ is equal to $\frac{1}{4}$ times $|f^c(\vartheta_g)|^2$ and the angle ϑ_g can be identified with $\vartheta_{1/4}$.

The quarter-point prescription cannot be used with confidence for the analysis of actual experiements. In cases of very-heavy-ion reactions which are dominated by Coulomb scattering, the depopulation (to Coulomb excitation) is a smooth function of the angular momentum (cf. Fig. I.7). In this case one would rather estimate the total reaction cross section to be given by the expression (2), with l'_g equal to the l-value of the Rutherford trajectory corresponding to a depopulation of the ground state by $\frac{1}{2}$. On the other hand, for the inclusive elastic scattering with very heavy ions one can estimate the total reaction cross-section, aside from the one counted as "elastic", by using either the quarter-point rule or the one-half rule, since the angular distribution decays very fast from the Rutherford cross-section.

For the collision between lighter ions where the scattering is dominated by the rainbow phenomenon, the total reaction cross-section is determined by (2), where l'_g is approximately equal to the orbiting angular momentum estimated in Sec. 3. As seen from the deflection function, the angle ϑ_g determined by (5) lies in the so-called dark region, beyond the rainbow. The difference between the rainbow angular momentum and the orbiting angular momentum is related to the rate of curvature of the deflection function near the rainbow angle in such a way that a large rate of curvature gives a small difference and vice versa. On the other hand, the exponential slope of the elastic cross section in the classically forbidden region decreases when the rate of curvature increases, implying an increasing difference between ϑ_R and $\vartheta_{1/4}$.

As an example of the uncertainties associated with the use of the quarter-point prescription, we show in Fig. 20(b) an analysis of the reaction $^{16}O + {}^{208}Pb$ at different bombarding energies by Videbæk et al. (1977). They find that the quarter-point analysis leads to total reaction cross-sections which differ by more than 100 mb (i.e. between 10% and 100%) from the reaction cross-sections determined experimentally and derived from the optical-model analysis.

APPENDIX A FOLDING WITH YUKAWA FUNCTIONS

We follow the procedure of Randrup (1975) and define the Yukawa function

$$Y(\kappa) = \frac{e^{-\kappa r}}{r} \tag{1}$$

and the step function

$$\Theta(R) = \vartheta(R - r) = \begin{cases} 1, & r < R , \\ 0, & r > R . \end{cases} \tag{2}$$

The folding of two such functions centered at different points is denoted by the symbol *, e.g.

$$Y(\kappa_1) * Y(\kappa_2) = \int d^3 r' \frac{e^{-\kappa_1 |\mathbf{r}_1 - \mathbf{r}'|}}{|\mathbf{r}_1 - \mathbf{r}'|} \frac{e^{-\kappa_2 |\mathbf{r}_2 - \mathbf{r}'|}}{|\mathbf{r}_2 - \mathbf{r}'|}$$

$$= \frac{4\pi}{\kappa_2^2 - \kappa_1^2} \frac{e^{-\kappa_1 |\mathbf{r}_1 - \mathbf{r}_2|} - e^{-\kappa_2 |\mathbf{r}_1 - \mathbf{r}_2|}}{|\mathbf{r}_1 - \mathbf{r}_2|} . \tag{3}$$

The result is a function of a distance $|\mathbf{r}_1 - \mathbf{r}_2|$ only, as are the functions (1) and (2), and we may therefore write

$$Y(\kappa_1) * Y(\kappa_2) = \frac{4\pi}{\kappa_2^2 - \kappa_1^2} [Y(\kappa_1) - Y(\kappa_2)] . \tag{4}$$

The folding of Y with Θ can also be evaluated exlicitly. One finds

$$Y(\kappa) * \Theta(R) = \int d^3 r' \, \vartheta(R - r') \frac{e^{-\kappa |\mathbf{r} - \mathbf{r}'|}}{|\mathbf{r} - \mathbf{r}'|}$$

$$= \frac{4\pi}{\kappa^2} F(\kappa, R) \tag{5}$$

with

$$
F(\kappa,R) = \begin{cases} 1 - (1 + \kappa R)e^{-\kappa R}\dfrac{\sinh \kappa r}{\kappa r} & \text{for } r < R , \\[2em] (R \cosh \kappa R - \dfrac{1}{\kappa}\sinh \kappa R)\,\dfrac{e^{-\kappa r}}{r} & \text{for } r > R . \end{cases} \tag{6}
$$

For $\kappa R \gg 1$ we find

$$
F(\kappa,R) = \begin{cases} 1 & \text{for } r \ll R , \\[0.8em] 1 - \tfrac{1}{2}e^{-\kappa(R-r)} & \text{for } r \lesssim R , \\[0.8em] \dfrac{R}{2r}\,e^{-\kappa(r-R)} & \text{for } r > R. \end{cases} \tag{7}
$$

The Fermi distribution

$$
f(r) = \frac{f_0}{1 + e^{(r-c)/a}} , \tag{8}
$$

may approximately be written

$$
\begin{aligned} f(r) &\approx f_0 F(\kappa,R) \\ &= \frac{f_0 \kappa^2}{4\pi}Y(\kappa)*\Theta(R) , \end{aligned} \tag{9}
$$

where a fairly good parametrization in the tail region ($r \gtrsim R$) is obtained for [Akyüz and Winther (1981)]

$$
\kappa = \frac{1}{1.5}\left(\frac{1}{a} + \frac{1.2}{c}\right) \tag{10}
$$

and

$$
R = 0.99c + 0.5a - 0.1 + 0.2/c . \tag{11}
$$

Since in a multiple folding one may interchange the order of integrations, we can easily evaluate the folding of two Fermi distributions in the region where the distance between them is larger than the sum of the radii. We thus find

$$F(\kappa_1,R_1)*F(\kappa_2,R_2)$$

$$= \frac{\kappa_1^2\kappa_2^2}{16\pi^2} \int d^3r' \int d^3r_1' \quad \vartheta(R_1 - |\mathbf{r}_1 - \mathbf{r}_1'|) \quad \frac{e^{-\kappa_1|\mathbf{r}_1'-\mathbf{r}'|}}{|\mathbf{r}_1' - \mathbf{r}'|}$$

$$\times \int d^3r_2' \, \vartheta(R_2 - |\mathbf{r}_2 - \mathbf{r}_2'|) \quad \frac{e^{-\kappa_2|\mathbf{r}_2'-\mathbf{r}'|}}{|\mathbf{r}_2' - \mathbf{r}'|}$$

$$= \frac{\kappa_1^2\kappa_2^2}{16\pi^2} \{\Theta(R_1)*[Y(\kappa_1)*Y(\kappa_2)]*\Theta(R_2)\}$$

$$= \frac{\kappa_1^2\kappa_2^2}{4\pi(\kappa_2^2 - \kappa_1^2)} \{\Theta(R_1)*[Y(\kappa_1) - Y(\kappa_2)]*\Theta(R_2)\}$$

$$= \frac{1}{\kappa_2^2 - \kappa_1^2} \{\Theta(R_1)*[\kappa_2^2 F(\kappa_1,R_2) - \kappa_1^2 F(\kappa_2,R_2)]\}. \tag{12}$$

Since for $|\mathbf{r}_1 - \mathbf{r}_2| > R_1+R_2$ only the exterior part of F contributes to the folding integrals, we may use the expression

$$F(\kappa,R) = \left(R \cosh \kappa R - \frac{1}{\kappa} \sinh \kappa R \right) Y(\kappa)$$

$$\approx \tfrac{1}{2} R e^{\kappa R} Y(\kappa) \tag{13}$$

to obtain

$$F(\kappa_1,R_1)*F(\kappa_2,R_2)$$

$$= \frac{1}{\kappa_2^2 - \kappa_1^2} \left(\kappa_2^2 \frac{R_2}{2} e^{\kappa_1 R_2}\Theta(R_1)*Y(\kappa_1) \right.$$

$$\left. - \kappa_1^2 \frac{R_2}{2} e^{\kappa_2 R_2}\Theta(R_1)*Y(\kappa_2) \right)$$

$$= \frac{2\pi R_2}{\kappa_2^2 - \kappa_1^2}\left(\frac{\kappa_2^2}{\kappa_1^2}e^{\kappa_1 R_2}F(\kappa_1,R_1) - \frac{\kappa_1^2}{\kappa_2^2}e^{\kappa_2 R_2}F(\kappa_2,R_1) \right)$$

$$\approx \frac{\pi R_1 R_2}{(\kappa_2^2 - \kappa_1^2)r}\left(\frac{\kappa_2^2}{\kappa_1^2}e^{-\kappa_1(r-R_1-R_2)} - \frac{\kappa_1^2}{\kappa_2^2}e^{-\kappa_2(r-R_1-R_2)} \right) . \qquad (14)$$

A detailed fit of measured neutron (v) and proton (π) densities for a number of nuclei utilizing the folded Yukawa density

$$\rho_i = \rho_{0i}F(\kappa_i,R_i) \qquad (i = v,\pi) \qquad (15)$$

has been made by Akyüz and Winther (1981). The results can approximately be parametrized as functions of A by the expressions

$$\kappa_v = (1.16 + 1.0A^{-1/3}) \text{ fm}^{-1} ,$$

$$\kappa_\pi = (1.65 + 0.2A^{-1/3}) \text{ fm}^{-1} ,$$

$$R_v = (1.20A^{1/3} - 0.43) \text{ fm} , \qquad (16)$$

$$R_\pi = (1.20A^{1/3} - 0.26) \text{ fm} ,$$

$$\rho_{0v} = \frac{N}{A} 0.14(1 + 1.08A^{-1/3})\text{fm}^{-3},$$

$$\rho_{0\pi} = \frac{Z}{A} 0.14(1 + 0.65A^{-1/3}) \text{ fm}^{-3}.$$

APPENDIX B   CLASSICAL DESCRIPTION OF TRAJECTORIES

Separating away the total center-of-mass motion, the relative motion of the two ions is classically determined by the Newtonian equation

$$m_{aA}\ddot{\mathbf{r}} = -\nabla U(r) , \qquad (1)$$

where m_{aA} is the reduced mass

$$m_{aA} = \frac{m_a m_A}{m_a + m_A} . \tag{2}$$

Utilizing the conservation of energy and angular momentum, one finds instead of (1) the equations

$$\tfrac{1}{2} m_{aA} \dot{r}^2 + \frac{L^2}{2 m_{aA} r^2} + U(r) = E \tag{3}$$

and

$$\dot{\phi} = \frac{L}{m_{aA} r^2} . \tag{4}$$

Here E is the total energy of relative motion in the center-of-mass system and L is the total angular momentum. The variable ϕ is the azimuthal angle of relative motion measured in a coordinate system, A, with the z-axis along the total angular momentum (cf. Fig. 21). Instead of the angular momentum L one may specify the initial conditions by the impact parameter ρ, which is given in terms of L by the relation

$$L = \rho \sqrt{2 m_{aA} E} . \tag{5}$$

The connection between the time t and the distance r between the two ions is obtained from (3), i.e.

$$t = \pm \sqrt{\frac{m_{aA}}{2}} \int_{r_0}^{r} \frac{dr'}{\sqrt{E - U_{\text{eff}}(r')}} , \tag{6}$$

where

$$U_{\text{eff}}(r) = \frac{L^2}{2 m_{aA} r^2} + U(r) , \tag{7}$$

is the effective potential for the radial motion. The turning point r_0 in the radial motion is the outermost zero of the equation

$$E - U_{\text{eff}}(r_0) = 0 . \tag{8}$$

(a)

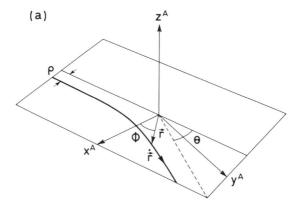

(b)

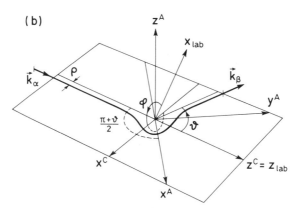

Fig. 21 Coordinate systems used for the description of the classical trajectory. In (a) we show the coordinate system A where the z-axis is perpendicular to the plane of the orbit in the direction of the orbital angular momentum. The x-axis is along the symmetry axis of the trajectory towards the apex (point of closest approach). The azimuthal angle ϕ for a point $\mathbf{r}$ on the trajectory is indicated as well as the impact parameter ρ and the scattering angle ϑ. In (b) we show also the laboratory system where the polar coordinates of the final momentum are ϑ and φ for a case where the deflection function Θ is negative. The coordinate system C is most often used in reaction theory.

Through (6) we have defined the time t to be equal to 0 at the distance of closest approach. For the numerical evaluation of (6) it is useful to introduce a parametric representation

$$r = r_0 + \tfrac{1}{2}\alpha w^2 \,, \tag{9}$$

with

$$\alpha = - \left(\frac{\partial U_{\text{eff}}}{\partial r}\right)_{r_0} \frac{r_0^2}{2E} \,. \tag{10}$$

In terms of the dimensionless parameter w one finds

$$t(w) = \frac{r_0}{\sqrt{\dfrac{2E}{m_{aA}}}} \int_0^w \frac{dw'}{\sqrt{\dfrac{E - U_{\text{eff}}(r(w'))}{-(\partial U_{\text{eff}}/\partial r)_{r_0}(r - r_0)}}} \,. \tag{11}$$

The integrand in this form shows no singularity except if $(\partial U_{\text{eff}}/\partial r)_{r_0} = 0$, which is the so-called orbiting situation.

The coefficient

$$\tau = \frac{dt}{dw}\bigg|_{w=0} = \frac{r_0}{\sqrt{2E/m_{aA}}} = \frac{r_0}{v} \,, \tag{12}$$

where v is the relative velocity at large distances, is a measure of the collision time. The azimuthal angle ϕ is given according to (4) by

$$\phi = \pm \frac{L}{\sqrt{2m_{aA}}} \int_{r_0}^r \frac{dr'}{r'^2 \sqrt{E - U_{\text{eff}}(r')}} \,. \tag{13}$$

In terms of the parameters w one finds

$$\phi(w) = \frac{L}{m_{aA} v r_0} \int_0^w \frac{dw'}{\left(\dfrac{r(w')}{r_0}\right)^2 \sqrt{\dfrac{E - U_{\text{eff}}(r(w'))}{(-\partial U_{\text{eff}}/\partial r)_{r_0}[r(w') - r_0]}}} \,, \tag{14}$$

where

$$\frac{L}{m_{aA}vr_0} = \frac{\rho}{r_0} \ .$$ (15)

In (13) and (14) we have measured the azimuthal angle from an x-axis pointing in the direction of the radius vector at $t = 0$, which is the symmetry axis for the trajectory. This is the coordinate system A indicated in Fig. 21.

For the cases in which one may neglect the nuclear interaction and where we have a pure Coulomb scattering, the integrals in (6), (11), (13) and (14) can be evaluated in closed form. It is, however, in this case more convenient to use instead of (9) the parametrization

$$r = r_0 + \alpha(\cosh w - 1)$$
$$= a_0(\varepsilon \cosh w + 1) \ .$$ (16)

The turning point is equal to

$$r_0 = a_0(1 + \varepsilon) \ ,$$ (17)

where

$$a_0 = \frac{Z_a Z_A e^2}{m_{aA}v^2}$$ (18)

is half the distance of closest approach in a head-on collision. The quantity ε is

$$\varepsilon = \sqrt{1 + \left(\frac{L}{\eta_{cl}}\right)^2}$$ (19)

where

$$\eta_{cl} = \frac{Z_a Z_A e^2}{v}$$ (20)

is the action of the Coulomb field. Note that $\eta_{cl}/\hbar = \eta$ is the Sommerfeld parameter for the Coulomb field. The quantity ε is the eccentricity in the hyerbolic orbit.

With the parametric representation (16) one finds from (11) and (13)

$$t = \frac{a_0}{v}(\varepsilon \sinh w + w) \tag{21}$$

and

$$\phi = \arcsin\left(\frac{\sqrt{\varepsilon^2 - 1}\ \sinh w}{\varepsilon \cosh w + 1}\right). \tag{22}$$

For the general case the deflection angle Θ [cf. Fig. 21(a)] is given by

$$\Theta = \pi - 2\phi(\infty) \tag{23}$$

Utilizing (13), one finds

$$\Theta = \pi - \frac{2L}{\sqrt{2m_{aA}}} \int_{r_0}^{\infty} \frac{dr'}{r'^2\sqrt{E - U_{eff}(r')}}$$

$$= \pi - 2 \int_0^{x_0} \frac{dx}{\sqrt{\dfrac{2m_{aA}E}{L^2} - \dfrac{2m_{aA}U(1/x)}{L^2} - x^2}}, \tag{24}$$

where in the second expression we have introduced

$$x = \frac{1}{r} \tag{25}$$

and $x_0 = 1/r_0$.

For pure Coulomb scattering we find from (22)

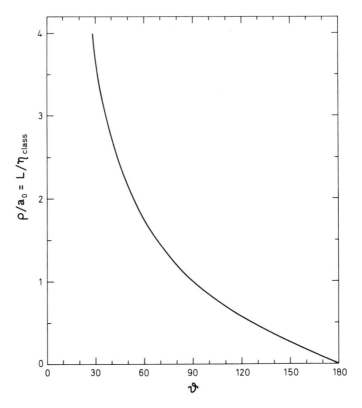

Fig. 22 Coulomb deflection function. The connection between the impact parameter measured in units of the quantity a_0 or the angular momentum measured in units of η_{cl}, and the scattering angle ϑ is given for scattering in the pure Coulomb field.

$$\Theta = \pi - 2 \arcsin \frac{\sqrt{\varepsilon^2 - 1}}{\varepsilon}$$

$$= 2 \arcsin \frac{1}{\varepsilon} = 2 \arctan \frac{a_0}{\rho} \tag{26}$$

This function is shown in Fig. 22.

APPENDIX C   CLASSICAL PERTURBATION OF
COULOMB TRAJECTORY

In this appendix we study the equations which govern the relative motion of two ions under the influence of Coulomb repulsion and relatively weak nuclear attraction.

In the spirit of perturbation theory we parametrize the motion in the coordinate system A in Fig. 21 as follows:

$$r = \frac{Q}{v^2}(\varepsilon \cosh w + 1) ,$$

$$\phi = \arcsin\left(\frac{\sqrt{\varepsilon^2 - 1} \, \sinh w}{\varepsilon \cosh w + 1}\right) + \phi_0 , \tag{1}$$

$$\vartheta = \tfrac{1}{2}\pi ,$$

$$t = \frac{Q}{v^3}(\varepsilon \sinh w + w) + t_0 .$$

For velocity independent interactions the associated momenta are

$$p_r = \frac{m_{aA} v \varepsilon \sinh w}{\varepsilon \cosh w + 1} ,$$

$$p_\phi = m_{aA}\frac{Q}{v}\sqrt{\varepsilon^2 - 1} , \tag{2}$$

$$p_\vartheta = 0 .$$

We have here introduced the abbreviation

$$Q = \frac{Z_a Z_A e^2}{m_{aA}} . \tag{3}$$

For pure Coulomb scattering the quantities ε, v, ϕ_0, and t_0 are constants signifying the eccentricity of the hyperbola, the velocity at large distances, the azimuthal angle at the turning point and the time at this point respectively.

In the more general case, when besides the Coulomb field there is a nuclear attractive force $-\partial U^N/\partial r$, these parameters become functions of w. The equations to be satisfied by the parameters are obtained by inserting (1) and (2) in the Hamiltonian equations of motion

$$\frac{dr}{dw} = \frac{p_r}{m_{aA}} \frac{dt}{dw},$$

$$\frac{d\phi}{dw} = \frac{p_\phi}{m_{aA} r^2} \frac{dt}{dw},$$

$$\frac{dp_r}{dw} = \left(\frac{1}{m_{aA} r^3} p_\phi^2 + \frac{m_{aA} Q}{r^2} - \frac{\partial U^N}{\partial r}\right) \frac{dt}{dw}, \tag{4}$$

$$\frac{dp\phi}{dw} = 0.$$

The result can be written as

$$\frac{dv}{dw} = \frac{v\varepsilon}{\varepsilon^2 - 1} \frac{d\varepsilon}{dw}, \tag{5}$$

$$\frac{d\varepsilon}{dw} = -\frac{\varepsilon\sqrt{\varepsilon^2 - 1}\ \sinh w}{\varepsilon + \cosh w} \frac{d\phi_0}{dw}, \tag{6}$$

$$\frac{d\phi_0}{dw} = \frac{\dfrac{Q}{m_{aA} v^4} \dfrac{\sqrt{\varepsilon^2 - 1}}{\varepsilon}(\varepsilon + \cosh w)\dfrac{\partial U^N}{\partial r}}{1 - \dfrac{Q}{m_{aA} v^4}\left[\dfrac{\varepsilon^2 + 1}{\varepsilon}\cosh w + 2\right]\dfrac{\partial U^N}{\partial r}}, \tag{7}$$

and a similar equation for dt_0/dw. Note that the equation (5) ensures that the angular momentum p_ϕ is conserved.

Assuming that $\partial U^N/\partial r$ is a small quantity, the equations (5)–(7) can be solved by iteration. To zeroth order (i.e. setting $\partial U^N/\partial r = 0$), we find

$$v = v^{(0)} = \sqrt{\frac{2E}{m_{aA}}}, \tag{8}$$

$$\varepsilon = \varepsilon^{(0)} = \sqrt{1 + \left(\frac{L}{\eta_{cl}}\right)^2} = \sqrt{1 + \left(\frac{\rho}{a_0}\right)^2}, \tag{9}$$

and

$$\phi_0 = \phi_0^{(0)} = 0, \tag{10}$$

in terms of the initial conditions of given impact parameter and angular momentum with the choice of coordinate system A.

To first order the equations (5) and (6) can be solved, with the following result:

$$v = v^{(0)} + v^{(1)} = v^{(0)} - \frac{U^N(r^{(0)}(w))}{m_{aA}v^{(0)}}$$

$$= v^{(0)}\left(1 - \frac{U^N(r^{(0)}(w))}{2E}\right), \tag{11}$$

$$\varepsilon = \varepsilon^{(0)} + \varepsilon^{(1)} = \varepsilon^{(0)} - \frac{(\varepsilon^{(0)})^2 - 1}{\varepsilon^{(0)}} \frac{U^N(r^{(0)}(w))}{2E}, \tag{12}$$

and

$$\phi_0 = \phi_0^{(1)} = \frac{a_0}{2E} \frac{[(\varepsilon^{(0)})^2 - 1]^{1/2}}{\varepsilon^{(0)}}$$

$$\int_{-\infty}^{w} (\varepsilon^{(0)} + \cosh w') \frac{\partial U^N(r^{(0)}(w'))}{\partial r^{(0)}} dw', \tag{13}$$

where

$$r^{(0)}(w) = a_0(\varepsilon^{(0)} \cosh w + 1). \tag{14}$$

With these expressions we can calculate the trajectory to first order. We find e.g.

$$r(w) = r^{(0)}(w) + a_0 \frac{U^N(r^{(0)}(w))}{E} \left(\frac{(\varepsilon^{(0)})^2 + 1}{2\varepsilon^{(0)}} \cosh w + 1 \right) , \qquad (15)$$

which shows that the distance of closest approach for a given value of L is

$$r_0 = b(\varepsilon^{(0)}) \left(1 + \frac{1 + \varepsilon^{(0)}}{\varepsilon^{(0)}} \frac{U^N(b(\varepsilon^{(0)}))}{2E} \right) , \qquad (16)$$

where

$$b(\varepsilon^{(0)}) = a_0(1 + \varepsilon^{(0)}) \qquad (17)$$

and $\varepsilon^{(0)}$ is connected with L by the general relation (9). It is noted that the turning point is to any order of perturbation theory reached at $w = 0$. To first order the time t_0 is given by

$$\frac{dt_0^{(1)}}{dw} = \left(3 + (\varepsilon^{(0)})^2 - (\varepsilon^{(0)})^2 \cosh^2 w \right.$$

$$\left. + \frac{3(\varepsilon^{(0)})^2 + 1}{\varepsilon^{(0)}} \cosh w - 3\varepsilon^{(0)}w \sinh w \right)$$

$$\times \frac{Q^2}{v^7} \frac{1}{m_{aA}} \frac{\partial U^N(r^{(0)}(w))}{\partial r} . \qquad (18)$$

Finally we give the expression for the angle ϕ_0 to second order in perturbation theory. One finds by inserting (11), (12) and (15) in (7)

$$\frac{d\phi_0}{dw} = \frac{d\phi_0^{(1)}}{dw} \left\{ 1 + \frac{U^N(r^{(0)}(w))}{2E} \right.$$

$$\times \left[\frac{4(\varepsilon^{(0)})^2 - 1}{(\varepsilon^{(0)})^2} - \frac{(\varepsilon^{(0)})^2 - 1}{\varepsilon^{(0)}(\varepsilon^{(0)} + \cosh w)} \right.$$

$$+ a_0 \left(\frac{\partial^2 U^N(r^{(0)}(w))/\partial r^2}{\partial U^N(r^{(0)}(w))/\partial r} + \frac{\partial U^N(r^{(0)}(w))/\partial r}{U^N(r^{(0)}(w))} \right)$$

$$\times \left(\frac{(\varepsilon^{(0)})^2 + 1}{\varepsilon^{(0)}} \cosh w + 2 \right) \Big] \Big\} . \tag{19}$$

APPENDIX D   STOKES CONSTANTS

The WKB approximation described in this chapter suffers from a number of deficiencies which partly have to do with the inability to reproduce the barrier penetration close to the top of the barrier (orbiting situation), and partly have to do with the singular behavior of the wavefunction close to the turning points. Both of these questions have been dealt with in detail in the literature [Heading (1962), Berry and Mount (1972)].

In this appendix we discuss the deficiency associated with the description of the barrier penetration. For angular momenta close to the orbiting situation the simple description gives, in the absence of absorption, a reflection coefficient of unity. On the other hand, in a proper quantum-mechanical description the reflection is one-half when the energy is equal to the height of the barrier.

In order to improve the WKB approximation we study the analytic continuation of the WKB solutions (6.5) and (6.9) in more detail, keeping track not only of the dominant but also of the subdominant components.

Let us first consider the situation with one turning point, as illustrated in Fig. 13. If the solution to the Schrödinger equation in sector I can be approximated by

$$\psi_\mathrm{I} = b_+ \psi^\mathrm{out} + b_- \psi^\mathrm{in} , \tag{1}$$

we can in general conclude that the analytic continuation of the solution in sector III can be approximated by

$$\psi_\mathrm{III} = b_- \psi^\mathrm{in} + (b_+ + \beta b_-) \psi^\mathrm{out} , \tag{2}$$

where β is the Stokes constant associated with stokes line 2. The function ψ^in is the analytic continuation into sector III of the function (6.5) above the turning point, and represents a wave moving away from the turning point. The function ψ^{out} is similarly the analytic continuation into sector III of the function (6.9). The fact that the coefficient on the wavefunction ψ^{out} is not continuous across Stokes line 2 is because the function ψ^{out} is exponentially decreasing (subdominant) on this Stokes line.

We can determine the Stokes constant β by anticlockwise analytic continuation of the solution (2) into sectors II and I, requiring the solution thus obtained to be identical to ψ_I. In sector II we find

$$\psi_{II} = (b_+ + \beta b_-)\psi^{out}$$
$$+ [b_- + \alpha(b_+ + \beta b_-)]\psi^{in} , \tag{3}$$

and thus, in sector I,

$$\psi_I = [b_-(1 + \alpha\beta) + \alpha b_+]\psi^{out}$$
$$+ [(b_+ + \beta b_-) + \gamma(b_-(1 + \alpha\beta) + \alpha b_+)](-\psi^{in}) . \tag{4}$$

In the last equation we have used the fact that the anticlockwise analytic continuation of ψ^{in} from region I back to region I is equal to ψ^{out}, while the corresponding analytic continuation of ψ^{out} is equal to $-\psi^{in}$. We have introduced the Stokes constant α associated with the discontinuity in the subdominant function ψ^{in} on Stokes line 1 and the corresponding Stokes constant γ associated with Stokes line 3. Requiring that coefficients on ψ^{in} and ψ^{out} in (4) be identical with those in (1) for any choice of b_+ and b_-, we obtain

$$\alpha = 1 \tag{5}$$

and

$$\beta = \gamma = -1 . \tag{6}$$

We thus obtain

$$\psi_{II} = \psi_{III} = b_+\psi^{in} + (b_+ - b_-)\psi^{out} . \tag{7}$$

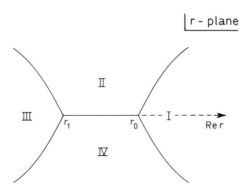

Fig. 23 Stokes lines associated with barrier penetration where the turning points r_0 and r_1 are real.

The special solution used in Sec. 6 [cf. Eq. (6.4)], which is illustrated in Fig. 12, is characterized by $b_+ = b_- = \frac{1}{2}$.

Utilizing this technique, one may find the solution for the situation illustrated in Fig. 23, which shows the turning points and the associated Stokes lines when the total energy is below the Coulomb barrier. We assume that the solution in sector I is given by

$$\psi_{\mathrm{I}} = \psi^{\mathrm{in}} + b_R \psi^{\mathrm{out}} , \qquad (8)$$

and in Sector III by

$$\psi_{\mathrm{III}} = b_T \psi^{\mathrm{in}} \qquad (9)$$

where ψ^{in} and ψ^{out} in (8) are given by (6.5) and (6.9), while ψ^{in} in (9) is the analytic continuation of ψ^{in} from I above the turning points into III. One finds [e.g. Fröman and Fröman (1965)]

$$b_R = \frac{e^{-i\delta}}{(1 + e^{-2S(r_0, r_1)/\hbar})^{1/2}} , \qquad (10)$$

where δ is real and S is the real phase integral defined in (6.10).

From (10) one obtains the reflection coefficient

$$R = \mid b_R \mid^2 = \frac{1}{1 + e^{-2S(r_0,r_1)/\hbar}} \ . \qquad (11)$$

At the top of the barrier, where $S(r_0,r_1) = 0$, one finds $R = \frac{1}{2}$.

APPENDIX E   UNIFORM APPROXIMATION

We shall consider how to improve on the singular behavior of the WKB wavefunction close to the classical turning point. The singularity of the wavefunction is of no direct concern for elastic scattering when we are only interested in the phase shifts. However, the singularity in the classical angular distribution occurring at the boundary between the classically allowed and classically forbidden regions (at the rainbow angle) is of similar nature and sets an important limitation on the usefulness of the classical description.

In order to obtain improved WKB-like solutions of the wave equation

$$\left(\frac{d^2}{dr^2} + k^2(r)\right)\psi(r) = 0 , \qquad (1)$$

we note that the standard WKB solution can be considered as the wavefunction of a "free" particle with slowly changing kinetic energy. Introducing the mapping

$$s(r) = \int^r k(r')\,dr' \qquad (2)$$

and writing ψ as

$$\psi(r) = \left(\frac{ds}{dr}\right)^{-1/2} \phi(s(r)) , \qquad (3)$$

we find that $\phi(\sigma)$ satisfies the equation

$$\left(\frac{d^2}{ds^2} + 1\right)\phi(s) = 0 , \qquad (4)$$

provided that the quantity $k^{-3/2} d^2(k^{-1/2})/dr^2$ is small compared to unity.

We may generalize this method by introducing a general mapping function [cf. e.g. Berry and Mount (1972)]

$$\sigma = \sigma(r) , \qquad (5)$$

which together with the *Ansatz* (3) with $s = \sigma$ will lead to the following wave equation:

$$\left(\frac{d^2}{d\sigma^2} + \Gamma(\sigma) \right) \phi(\sigma) = 0 . \qquad (6)$$

The function Γ is given by

$$\Gamma(\sigma) = \frac{k^2(r(\sigma))}{(d\sigma/dr)^2} \qquad (7)$$

provided that

$$\left(\frac{d\sigma}{dr} \right)^{1/2} \frac{d^2}{dr^2} \left(\frac{d\sigma}{dr} \right)^{-1/2} \ll k^2(r) . \qquad (8)$$

After choosing the function $\Gamma(\sigma)$, one may use (7) as a differential equation to determine the mapping function σ. Thus if we use the *Ansatz* $\Gamma(\sigma) = 1$, corresponding to a constant potential in the σ-variable, we obtain the standard WKB solution (2)–(4).

The simplest potential with one turning point in the σ-variable corresponds to the choice

$$\Gamma(\sigma) = -\sigma , \qquad (9)$$

which leads to the differential equation

$$\sigma^{1/2} \frac{d\sigma}{dr} = [-k^2(r)]^{1/2} \qquad (10)$$

for the mapping function. A solution of this equation is

$$\sigma = - \left(\frac{3}{2} \int_{r_0}^{r} k(r') \, dr' \right)^{2/3} . \tag{11}$$

The equation for $\phi(\sigma)$,

$$\left(\frac{d^2}{d\sigma^2} - \sigma \right) \phi(\sigma) = 0 , \tag{12}$$

is the Airy equation, with the two independent solutions $\mathrm{Ai}(\sigma)$ and $\mathrm{Bi}(\sigma)$ [see Abramowitz and Stegun (1966)]. The approximate solution (3) of the Schrödinger equation is therefore

$$\psi(r) = \left[\frac{-\sigma(r)}{k^2(r)} \right]^{1/4} [\alpha \, \mathrm{Ai}(\sigma(r)) + \beta \, \mathrm{Bi}(\sigma(r))] , \tag{13}$$

where α and β are determined by the boundary conditions.

The regular solution χ_l to the radial Schrödinger equation (5.1) can thus be approximated by

$$\chi_l(r) = \left[- \sigma(r)\pi^2 \frac{k^2}{k^2(r)} \right]^{1/4} \mathrm{Ai}(\sigma(r)) . \tag{14}$$

This expression constitutes the so-called uniform approximation to the Schrödinger equation with one turning point. For values of r far away from the turning point one may use the asymptotic expression for the Airy function, and one finds that (14) coincides with the WKB solution (6.4). In the neighborhood of the turning point r_0, we may write (14) as

$$\chi_l(r) = \left[\frac{\pi^3 k^3}{\left(\frac{dk^2}{dr} \right)_{r_0}} \right]^{1/6} \mathrm{Ai}\left(-\left(\frac{dk^2(r)}{dr} \right)_{r_0}^{1/3} (r - r_0) \right) , \tag{15}$$

which is analogous to (7.10).

APPENDIX F DETERMINATION OF
EMPIRICAL POTENTIAL

As was seen in Sec. 3, the elastic scattering is only sensitive to the ion-ion potential at large distances, where it may be approximated by an exponential function [Igo (1958, 1959)]

$$U^N(r) = Ke^{-r/a} . \qquad (1)$$

Even with this potential one will usually find it possible to obtain fits of equivalent quality for a range of values of the diffuseness parameter a if at the same time the constant K is adjusted.

Since the angular distribution is mainly governed by the position of the rainbow angle which is produced by this potential, we may assume that the adjustment which has to be done is such that the rainbow angle is unchanged. We may express this constraint by asking the question: Where do the potentials cross each other, which have the property that they have the same rainbow angle under variations of a? To answer this we write (1) as

$$U^N(r) = U^N(x)e^{-(r-x)/a} \qquad (2)$$

and ask that for fixed x (the crossing point) the rainbow angle should not change. The infinitesimal changes in the impact parameter or the value of $\varepsilon^{(0)}$ [cf. (3.3)] which are necessary in order to compensate for these changes are unimportant, since Θ_R is stationary in $\varepsilon^{(0)}$ [cf. (3.5)]. The condition for an unchanged value of Θ_R is thus

$$\left. \frac{\partial \Theta_R}{\partial a} \right|_{\varepsilon^{(0)}=\varepsilon_R^{(0)}} = \left. \frac{\partial \phi_0}{\partial a} \right|_{\varepsilon^{(0)}=\varepsilon_R^{(0)}} = 0 , \qquad (3)$$

where $\varepsilon^{(0)}$ is kept fixed, equal to $\varepsilon_R^{(0)}$. Utilizing the expression (3.4) for ϕ_0, we find that (3) leads to the equation

$$-\frac{1}{2a} + \frac{\partial}{\partial a}\left(-\frac{b(\varepsilon_R^{(0)}) + \sqrt{2}\,\Delta - x}{a} \right) = 0 , \qquad (4)$$

where Δ is given by (3.8), i.e.

$$x\left(1 - \sqrt{2}\ \frac{\Delta}{a}\right) = b(\varepsilon_R^{(0)})\left(1 - \sqrt{2}\frac{\Delta}{a}\right) - \frac{a}{2} + \sqrt{2}\Delta\ . \quad (5)$$

To first order in Δ/a the result may be written

$$x = r_R - \frac{a}{2} + \left(\frac{\sqrt{2}}{2} - 1\right)\Delta\ , \quad (6)$$

where we have used the expression (3.7) for the distance of closest approach r_R in the rainbow scattering.

In some cases experiments on the elastic angular distribution have been analyzed with several ion-ion potentials, and it has been possible to verify that the fitting potentials cross each other at a position compatible with the expression (6) [cf. Fig. 20]. In most cases only one fitting potential has been quoted. By calculating r_R and the quantity x for the potential given, the value of the potential in the point $r = x$ is the most relevant number to be extracted from the experiment. Treating available data on elastic scattering of heavy ions in this fashion, Christensen and Winther (1976) have determined the parameters in the potential (1.37) by a least-squares fit. A more recent improved analysis of the data [Akyüz and Winther (1981)] leads to the parameters (1.38).

CHAPTER IV

INELASTIC SCATTERING

For collisions between very heavy ions it is exceptional that the nuclei come to distances where nuclear interactions take place without already being excited by the Coulomb field. At such distances further inelastic processes take place through the nuclear fields, leading to the excitation of new modes of high excitation energy like giant resonances. These excitations, together with the multiple excitation of the low-energy surface modes, have a significant influence on the dynamics of the collision through the associated energy loss and deformations of the nuclear surfaces. The inelastic processes are thus one of the most conspicuous features of heavy-ion reactions from grazing collisions through deep-inelastic reactions.

Specific information about the inelastic response function for the colliding nuclei is contained in the grazing collisions, where one can study the population of individual excited states. The region of the spectrum which can be studied is limited by the adiabatic cutoff, which is controlled by the collision time. While at energies close to the Coulomb barrier one populates mainly the low-lying surface modes, at higher bombarding energies (> 20 MeV/nucleon) one may also excite the giant resonances in grazing collisions. At these energies Coulomb excitation of giant resonances also becomes important. For more violent collisions where the two ions overlap strongly, the inelastic channels become even more important, but either are doorways to fusion or lead to deep-inelastic processes, where the individual states can hardly be disentangled.

An essential element in evaluating cross sections for grazing collisions is the absorptive potential. Because the trajectory of relative motion is determined by the position of the nuclear surfaces and because these in turn show quantal zero-point fluctuations, the excitation of definite states lead to quantal effects that go beyond those discussed in the previous chapter. In order to deal with these effects in a self-consistent way, we first calculate the inelastic processes, utilizing coupled equations with real potentials and form factors, and from there determine the depopulation which should be used to construct the imaginary potential, which in turn governs the (complex) trajectory of relative motion.

1   GENERAL FEATURES OF INELASTIC SCATTERING

Inelastic scattering through the combined action of Coulomb and nuclear interactions is governed by the same coupled equations (cf. Sec. II.1) which apply to Coulomb excitation:

$$i\hbar \, \dot{a}_{nn'}(t) = \sum_{mm'} \langle \psi_n^a \psi_n^A | \, V_{aA} - U_{aA}(r) | \, \psi_m^a \psi_m^A \rangle$$

$$\times \, e^{(i/\hbar)\Delta E \, t} a_{mm'}(t) \, , \tag{1}$$

where

$$\Delta E = E_n^A - E_m^A + E_{n'}^a - E_{m'}^a \, . \tag{2}$$

In the matrix element the interaction V_{aA} between all nucleons in a and all nucleons in A includes, besides the Coulomb interaction (discussed in Sec. II.1), also the nuclear interactions. The quantity $U_{aA}(r)$ is the expectation value of this interaction in the entrance channel where both nuclei are in their ground states, i.e.

$$U_{aA}(r) = \frac{Z_a Z_A e^2}{r} + U_{aA}^N(r) \, , \tag{3}$$

where $U_{aA}^N(r)$ is the ion-ion potential discussed in Sec. III.1. The diagonal matrix element in (1) in the ground state thus vanishes [cf. (III.1.1)].
   The nondiagonal matrix elements can be written as

$$\langle \psi_n^a \psi_n^A | \, V_{aA} - U_{aA} | \, \psi_m^a \psi_m^A \rangle$$

$$= \langle \psi_n^a | \, V_E(a,r) | \, \psi_m^a \rangle + \langle \psi_n^A | \, V_E(A,r) | \, \psi_m^A \rangle$$

$$+ \langle \psi_n^a \psi_n^A | \, V_{aA}^N | \, \psi_m^a \psi_m^A \rangle \, , \tag{4}$$

where the electric interactions are given in (II.1.8). The nuclear inelastic matrix elements will be calculated in Sec. 2 below. In (4) we neglect the electric multipole-multipole interaction which is generally a good approximation for grazing reactions. The multipole-multipole interaction is important for the collision between two deformed nuclei.

The nuclear states which are most strongly excited are the surface modes. For the excitation of an isoscalar one-phonon state of multipolarity λ, μ in the target nucleus, the monopole-multipole nuclear matrix element in (4) can be estimated by (cf. Sec. 3 below)

$$\langle \psi_{n'}^a \psi_n^A | V_{aA}^N | \psi_n^a \psi_m^A \rangle = \frac{(-1)^\lambda}{\sqrt{2\lambda + 1}} f_\lambda^N(r) Y_{\lambda\mu}^*(\hat{r}) \qquad (5)$$

with the radial form factor

$$f_\lambda^N(r) = -\sqrt{\frac{\hbar \omega_\lambda(n)}{2C_\lambda(n)}} R_A^{(0)} \frac{\partial U_{aA}^N}{\partial r} (-1)^\lambda \sqrt{2\lambda + 1} . \qquad (6)$$

The quantity ω_λ is the frequency of the mode, and C_λ is the restoring force (cf. Sec. II.4).

In the tail region (important for grazing reactions) the radial dependence of the matrix element (6) is of the form

$$\frac{\partial U_{aA}^N}{\partial r} \sim e^{-(r-R)/a} . \qquad (7)$$

This is to be contrasted to the radial dependence of the electric matrix element for the excitation of the same state

$$\langle \psi_n^A | V_E(A,\vec{r}) | \psi_m^A \rangle = f_\lambda^c(r) Y_{\lambda\mu}^*(\hat{r}) \frac{(-1)^\lambda}{\sqrt{2\lambda + 1}} , \qquad (8)$$

which is given by the form factor

$$f_\lambda^c(r) = (-1)^\lambda \frac{3}{\sqrt{2\lambda + 1}} \frac{Z_a Z_A e^2}{R_A^{(0)}} \sqrt{\frac{\hbar \omega_\lambda(n)}{2C_\lambda(n)}} \left(\frac{R_A^{(0)}}{r}\right)^{\lambda+1} , \qquad (9)$$

where we have used (II.1.8), (II.4.10) and (II.4.13).

The distances over which the Coulomb excitation takes place is measured by $r \approx R_a + R_A$. The high multipole excitations are thus reduced by the factor $[R_A/(R_a + R_A)]^{\lambda+1}$, while for the nuclear interaction the r-dependence is the same for all λ. Nuclear inelastic scattering is thus much more sensitive to high multipole components of the interaction than Coulomb excitation.

The radial dependence of the form factors also influences the time dependence of the field producing the excitation. Thus the time over which the electric field changes by a factor of 2 can be estimated by (cf. Appendix III.B)

$$\tau_{coll} \approx \sqrt{\frac{2(2^{1/(\lambda+1)} - 1)(R_a + R_A)}{\ddot{r}_0}} \approx \frac{R_a + R_A}{v}, \qquad (10)$$

where $\ddot{r}_0$ is the acceleration at the turning point. For the nuclear field the corresponding time is given by

$$\tau_{char} \approx \sqrt{\frac{2a \ln 2}{\ddot{r}_0}} \approx \sqrt{\frac{a}{\ddot{r}_0}} \approx \frac{\sqrt{2(R_a + R_A)a}}{v} \qquad (11)$$

where we used that $\ddot{r}_0$ is approximately given by $E/(m_{aA}r_0) = v^2/(2r_0)$ [cf. (9.32) below]. This time is typically a factor of 4 smaller than τ_{coll}.

While in Coulomb excitation the slow variation of the field in time gave rise to a severe limitation on the possibility of exciting states of high energy, nuclear excitation can take place to states of much higher energy. In fact the adiabaticity parameter ξ for Coulomb excitation above the Coulomb barrier is

$$\xi_C = \frac{1}{2} \frac{\tau_{coll}}{\hbar} \Delta E = \frac{R_a + R_A}{2v} \frac{\Delta E}{\hbar} \qquad (12)$$

[cf. II.2.4)], while for nuclear excitations it is

$$\xi_N = \frac{1}{2} \frac{\sqrt{2a(R_a + R_A)}}{v} \frac{\Delta E}{\hbar}. \qquad (13)$$

For bombarding energies close to the Coulomb barrier, the adiabatic cutoff ($\xi_N > 1$) appears typically at ΔE of the order of 10 MeV (cf. Sec. 7 below).

The question of which states will be excited in this energy region is determined by the strength of the interaction. This is measured by the parameter χ, which is the action of the external field in units of $\hbar$ (cf. Sec. II.3). For collisions close to grazing the strength parameter can be estimated by

$$\chi_N^{(\lambda)} = \frac{1}{\hbar} f_\lambda^N(r_g)\tau_{\text{coll}}^N , \qquad (14)$$

where r_g is the grazing distance [cf. (III.3.25)], while τ_{coll}^N is the collision time associated with the nuclear interaction. For grazing collisions one may estimate τ_{coll}^N by (11), i.e.

$$\tau_{\text{coll}}^N \approx \tau_{\text{char}} . \qquad (15)$$

For such collisions the strength for exciting a given (vibrational) state by Coulomb interaction is (cf. Sec. II.3)

$$\chi_C^{(\lambda)} = \frac{1}{\hbar}\tau_{\text{coll}} f_\lambda^C(r_g) . \qquad (16)$$

The ratio of $\chi_N^{(\lambda)}$ to $\chi_C^{(\lambda)}$ is accordingly

$$\frac{\chi_N^{(\lambda)}}{\chi_C^{(\lambda)}} \approx -\sqrt{\frac{2a}{r_g}} \frac{R_A^{(0)}\dfrac{\partial U^N}{\partial r}\bigg|_{r_g}}{\dfrac{Z_a Z_A e^2}{r_g}\left(\dfrac{R_A^{(0)}}{r_g}\right)^\lambda}\frac{2\lambda+1}{3}$$

$$\approx -\sqrt{\frac{2a}{r_g}}\left(\frac{r_g}{R_A^{(0)}}\right)^{\lambda-1}\frac{2\lambda+1}{3} , \qquad (17)$$

where we have used the fact that $(\partial U^N/\partial r)_{r_g} \approx E_B/r_g$ [cf. (III.3.20)].

It is seen from this expression that the strength of the nuclear excitation becomes larger than the strength of Coulomb excitation for multipole orders above 3. In evaluating the actual excitation amplitude of the state, one should multiply the strength parameters by factors which take into account the reduction due to the adiabaticity of the excitation. For Coulomb excitation this factor is $R_\lambda(\vartheta,\xi)$ (cf. Sec. II.3), which for bombarding energies close to the barrier is important already for states of an energy of 1 MeV and strongly hinders the excitation of states above 3 MeV. The corresponding reduction factor for nuclear excitations plays a similar role at an excitation energy which is about 4 times higher.

The two excitation mechanisms occur simultaneously and try to cancel each other because the formfactors (6) and (9) have opposite sign. For excitation energies above 1 MeV or for multipolarities larger than 3 the nuclear excitation dominates for grazing collisions, while for $\lambda = 2$ and low excitation energies the nuclear and the Coulomb excitation tend to cancel each other and the process is mostly dominated by Coulomb excitation.

Because of the rainbow phenomenon the nuclear excitation, which takes place on a trajectory with $l = l_N < l_R$ [cf. Fig. I.9)], will interfere with the pure Coulomb excitation, which takes place on the trajectory with $l = l_c > l_R$, leading to the same angle in much the same way as for elastic scattering. The magnitude of the oscillations in the angular distribution gives a measure of the importance of nuclear excitation (see Fig. 1).

The equations (1) can also be used to follow the reaction beyond the barrier. Due to the large energy and angular-momentum loss which will occur, the conditions (II.1.19) and (II.1.20) for the applicability of the semiclassical approximation are violated. However, if the loss of energy and angular momentum happens in many small steps, the trajectory can be continuously modified to take the average effect of these losses into account. For such collisions, where one follows the development of the equations of motion beyond the barrier, the interaction time τ_{coll}^N in (14) is rather of the order of (10). Such reactions are the subject of Chap. 8.

2 MICROSCOPIC FORM FACTORS

The inelastic scattering is induced by the nondiagonal matrix elements of V_{aA} appearing in (1.1). We may write V_{aA} in second quantization as

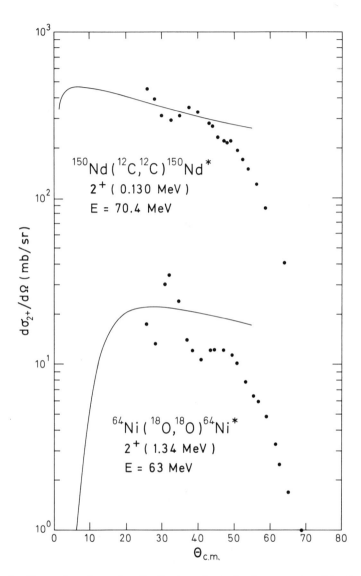

Fig. 1 Examples of angular distributions for inelastic scattering. The O + Ni reaction offers a typical example where Coulomb and nuclear excitation are of similar order of magnitude and where strong interference effects appear. The C + Nd reaction is dominated by Coulomb excitation. The curves indicate the pure Coulomb-excitation cross section. The experimental data are taken from Hillis *et al.* (1977) and Rehm *et al.* (1975).

199

$$\hat{V}_{aA} = \int d^3 r_{1C} \, d^3 r_{2c} \, \hat{\rho}^A(\mathbf{r}_{1C}) \hat{\rho}^a(\mathbf{r}_{2c}) V_{12}(|\,\mathbf{r}_{cC} + \mathbf{r}_{2c} - \mathbf{r}_{1C}|) \, , \qquad (1)$$

where V_{12} is the effective two-body interaction, while the density operators are given by

$$\hat{\rho}^A(\mathbf{r}_{1C}) = \sum_{\substack{a_1 m_1 \\ a_2 m_2}} \int (\phi^{(C)}_{j_2 m_2}(a_2;\mathbf{r}_{1C}\zeta_1))^* \phi^{(C)}_{j_1 m_1}(a_1;\mathbf{r}_{1C}\zeta_1)$$

$$\times \, a^{\dagger}_{j_2 m_2}(a_2) a_{j_1 m_1}(a_1) \, d\zeta_1 \, , \qquad (2)$$

and correspondingly for $\hat{\rho}^a$. The operator $a^{\dagger}_{j_1 m_1}(a_1)$ creates a particle in the state $a_1 m_1$ in nucleus A, where a_1 stands for the radial-, orbital- and total-angular-momentum quantum numbers n_1, l_1 and j_1 respectively, m_1 being the magnetic quantum number, while ζ_1 is the spin coordinate.

The single-particle wavefunction

$$\phi^{(C)}_{j_1 m_1}(a_1;\mathbf{r}_{1C},\zeta_1) = R^{(C)}_{a_1}(r_{1C})[Y_{l_1}(\hat{r}_{1C})\chi(\zeta_1)]_{j_1 m_1} \qquad (3)$$

describes the state a_1, m_1 of particle 1 moving around the core C of nucleus A. The radial wavefunction is $R^{(C)}_{a_1}(r_{1C})$, while $\chi(\zeta_1)$ indicates the spin function. In (1) we have neglected the spin dependence of the effective interaction.

The diagonal matrix elements of (1) are identical to the "double folding" expression (III.1.2) for the ion-ion potential. The nondiagonal matrix elements are of three types: the monopole-multipole matrix elements, where the intrinsic state of either nucleus a or nucleus A is unchanged, and the multipole-multipole matrix elements, where both nuclei are excited simultaneously. In this section we neglect the latter and obtain, for target excitation,

$$\hat{V}^N_{aA} = \int U^N_{1a}(r_{1a}) \hat{\rho}^A(\mathbf{r}_{1A}) \, d^3 r_1 \, , \qquad (4)$$

where U^N_{1a} is the isoscalar nucleon-nucleus potential acting on a nucleon scattered off nucleus a [cf. (III.1.4)]. The isovector interaction is discussed at the end of this section.

Since V_{aA}^N and $U_{1a}^N(r_{1a})$ are scalars, we may write (4) in the form

$$\hat{V}_{aA}^N = \sum_{\substack{a_1 a_2 \\ \lambda\mu}} (-1)^{\lambda+\mu+\pi_1} \frac{\sqrt{2j_2+1}}{2\lambda+1}$$

$$\times f_{\lambda\mu}^{a_2 a_1}(\mathbf{r}_{aA})[a_{j_2}^\dagger(a_2)b_{j_1}^\dagger(a_1)]_{\lambda-\mu}, \qquad (5)$$

where we have coupled the angular momenta j_2 and j_1 to the total transferred angular momentum λ, $-\mu$, i.e.

$$[a_{j_2}^\dagger(a_2)b_{j_1}^\dagger(a_1)]_{\lambda-\mu}$$

$$= \sum_{m_1 m_2} \langle j_2 m_2 j_1 m_1 | \lambda - \mu \rangle a_{j_2 m_2}^\dagger(a_2)b_{j_1 m_1}^\dagger(a_1)$$

$$= (-1)^{j_1-j_2+\pi_1-\pi_2+\mu}[a_{j_1}^\dagger(a_1)b_{j_2}^\dagger(a_2)]_{\lambda\mu}^\dagger. \qquad (6)$$

The operator

$$b_{j_1 m_1}^\dagger(a_1) = (-1)^{j_1+m_1+\pi_1}a_{j_1-m_1}(a_1) \qquad (7)$$

creates a hole in the orbital a_1, m_1, where π_1 is the even or odd parity of the state, i.e. $(-1)^{\pi_1} = (-1)^{l_1}$.

(a)   Evaluation of single-particle form factor

The quantity $f_{\lambda\mu}^{a_2 a_1}(\mathbf{r}_{aA})$ is the single-particle form factor

$$f_{\lambda\mu}^{a_2 a_1}(\mathbf{r}_{aA}) = \frac{(2\lambda+1)^{3/2}}{2j_2+1}(-1)^{\lambda+\mu}$$

$$\times \sum_{m_1 m_2} \langle j_1 m_1 \lambda - \mu | j_2 m_2 \rangle$$

$$\times \int d^3 r_1 d\zeta_1 (\phi_{j_2 m_2}^{(C)}(a_2;\mathbf{r}_{1C}\zeta_1))^*$$

$$\times \phi_{j_1 m_1}^{(C)}(a_1;\mathbf{r}_{1C}\zeta_1)U_{1a}^N(\mathbf{r}_{1a}), \qquad (8)$$

which is a tensor of rank λ, μ. We may therefore conveniently express it in the "intrinsic" frame where the z-axis is chosen along the vector $\mathbf{r}_{aA}$. We thus obtain

$$f_{\lambda\mu}^{a_2 a_1}(\mathbf{r}_{aA}) = \sum_{\nu} D_{\mu\nu}^{\lambda}(\hat{z} \rightarrow \hat{r}_{aA})[f_{\lambda\nu}^{a_2 a_1}(r_{aA})]_{\text{intr}} \tag{9}$$

where $[f_{\lambda\nu}^{a_2 a_1}(r_{aA})]_{\text{intr}}$ is given by the same expression (8) except that the wavefunctions are evaluated in the intrinsic frame. Since U_{1a} is assumed to be symmetric around $\mathbf{r}_{aA}$, the intrinsic form factor is only nonvanishing for $\nu = 0$, and we may therefore write

$$f_{\lambda\mu}^{a_1 a_2}(\mathbf{r}) = f_{\lambda}^{a_2 a_1}(r) Y_{\lambda\mu}(\hat{r}) . \tag{10}$$

The radial single-particle form factor is thus given by

$$\begin{aligned}
f_{\lambda}^{a_2 a_1}(r) = \sqrt{\pi} \, &\sqrt{(2\lambda + 1)(2j_1 + 1)} \, (-1)^{j_2 - 1/2} \\
&\times \langle j_2 \tfrac{1}{2} j_1 - \tfrac{1}{2} | \, \lambda 0 \rangle \delta(\pi_1 + \pi_2 + \lambda, \text{ even}) \\
&\times \int_0^{\infty} r_1^2 \, dr_1 \int_{-1}^{1} d(\cos \vartheta)(R_{a_2}^{(C)}(r_1))^* \\
&\times U_{1a}(\sqrt{r_1^2 + r^2 - 2r_1 r \cos \vartheta}) \\
&\times R_{a_1}^{(C)}(r_1) P_{\lambda}(\cos \vartheta) .
\end{aligned} \tag{11}$$

Since we employ real radial wavefunctions, the radial form factors are real and they satisfy the symmetry relation

$$f_{\lambda}^{a_1 a_2}(r) = (-1)^{j_1 - j_2} \left(\frac{2j_2 + 1}{2j_1 + 1} \right)^{1/2} f_{\lambda}^{a_2 a_1}(r) , \tag{12}$$

where we have used (6).

In order to calculate the matrix elements for nuclear inelastic scattering appearing in (1.1) one thus evaluates the matrix element of (5) between the nuclear states, utilizing the definition (10)–(11). For the idealized situation of a single-particle excitation only one term in (5) would contribute. Examples of such single-particle formfactors are given in Fig. 2. Since even small admixtures of collective states would change the form factors drastically, these form factors can hardly be used directly to analyze inelastic scattering.

(b) Excitation of two-particle configurations

As another example we may consider the excitation of a system of two particles outside a closed shell. We assume that the core of the nucleus remains inert during the process of excitation. The generalization to the case where ground-state correlations are included is discussed below. The ground state is thus described by the wavefunction

$$| I_A M_A \rangle \equiv | 00 \rangle = \sum_{a_1} X(a_1 a_1; I_A = 0) \frac{[a_{j_1}^\dagger(a_1) a_{j_1}^\dagger(a_1)]_{00}}{\sqrt{2}} | 0 \rangle, \tag{13}$$

where X is the amplitude of the different two-particle configurations and the square bracket indicates the vector coupling of the two angular momenta to $I_A = M_A = 0$. The state $|0\rangle$ indicates the closed-shell system, which is also the vacuum of the operators $a_{jm}^\dagger$. The wavefunction of the excited state is given by

$$| I_A' M_A' \rangle \equiv | \lambda\mu \rangle = \sum_{a_1 \geq a_2} X(a_1 a_2; I_A = \lambda) \frac{[a_{j_1}^\dagger(a_1) a_{j_2}^\dagger(a_2)]_{\lambda\mu}}{\sqrt{1 + \delta(a_1, a_2)}} | 0 \rangle. \tag{14}$$

The matrix element of the operator (5) between the two states defined above can then be evaluated to give

$$\langle \lambda\mu | V_{aA}^N | 0 \rangle = \sqrt{2} \sum_{a_2 \geq a_1} \frac{\sqrt{2j_1 + 1}}{2\lambda + 1} \frac{X(a_1 a_2; \lambda)}{\sqrt{1 + \delta(a_1, a_2)}}$$

$$\times \left\{ \frac{X(a_1 a_1; 0)}{\sqrt{2j_1 + 1}} + (-1)^\lambda \frac{X(a_2 a_2; 0)}{\sqrt{2j_2 + 1}} \right\} f_{\lambda\mu}^{a_1 a_2 *}(\mathbf{r}). \tag{15}$$

An example of a form factor calculated according to (15) is given in Fig.

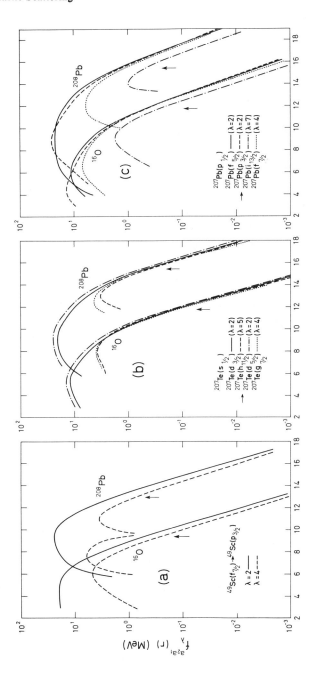

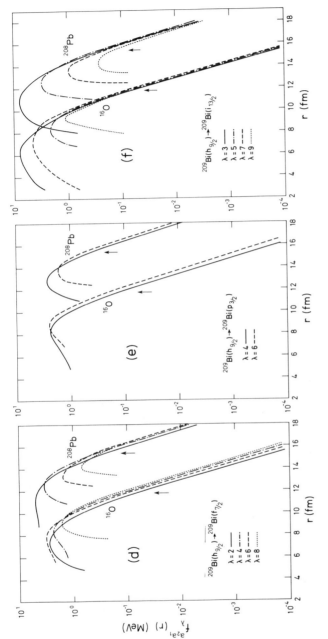

Fig. 2 Radial single-particle form factors for nuclear inelastic scattering. The expression (2.11) for $f_\lambda^{a_2 a_1}$ has been evaluated (in MeV) for inelastic scattering induced by ^{16}O and ^{208}Pb. The arrow indicates the radius r_B of the Coulomb barrier (III.2.5). The single-particle radial wavefunctions were determined utilizing a Saxon-Woods shell-model potential of the form (III.1.9) with a depth adjusted to fit the single-particle binding energy. The same potential with $V_0 = 50$ MeV was used for the quantity $U_{1a}(r)$. The figure is due to Andrea Vitturi.

205

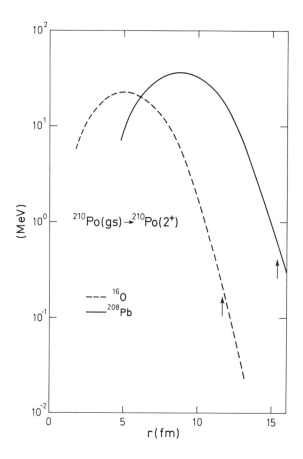

Fig. 3 Form factors for the excitation of two-particle configurations in ^{210}Po. The form factors were evaluated on the basis of Eq. (2.15) for ^{16}O and ^{208}Pb projectiles. The wavefunctions used for the ground state and the 2^+ excited state of ^{210}Po were

$$|0^+\rangle = 0.88\,|\,(h^2_{9/2})_0\rangle + 0.36\,|\,(f^2_{7/2})_0\rangle - 0.29\,|\,(i^2_{13/2})_0\rangle + 0.11\,|\,(p^2_{3/2})_0\rangle,$$

$$|2^+\rangle = 0.95\,|\,(h^2_{9/2})_2\rangle + 0.13\,|\,(f^2_{7/2})_2\rangle - 0.11\,|\,(i^2_{13/2})_2\rangle.$$

The single-particle wavefunctions were determined [cf. Bortignon *et al.* (1977)] utilizing a Saxon-Woods shell-model potential with a depth adjusted to fit the single-particle binding energies (cf. Fig. 2). The same potential (with $V_0 = 60$ MeV) was used for the quantity U_{1a}. The arrow indicates the radius r_B of the Coulomb barrier (III.2.5).

3. The same reservation applies to the use of this formula as to the single-particle form factor (11).

(c)   Phonon excitation

The inelastic form factors were derived in terms of wavefunctions expressed in a fermion basis. It has proved useful to describe the nucleus in terms of elementary modes of excitation where both fermion and boson degrees of freedom are utilized [cf. Bohr and Mottelson (1975)]. The two types of bosons entering into this picture are those corresponding to correlated particle-hole states (wavy line in Fig. 4) and correlated two-particle states (double-arrowed line in Fig. 4). The field generated by the projectile can, in this picture, either change the state of motion of a particle [as for instance in graphs (a) and (f) of Fig. 4] or create a particle-hole boson [as in graphs (b) and (d)].

The form factor (15) is the matrix element associated with the process depicted by graph (a). The quantity $X(a_1a_1;0)$ is the amplitude for the ground-state boson on the configuration (a_1,a_1), while $X(a_1a_2;\lambda)$ is the amplitude with which the configuration (a_1,a_2) enters in the excited state λ [cf. Eqs. (13) and (14)].

In order to evaluate graph (b) we should recast the operator (5) in terms of the boson operators. In the random-phase approximation (RPA) the boson operator is defined as

$$\Gamma^\dagger_{\lambda\mu}(n) = \sum_{a_i a_k} \{X_n(a_k a_i;\lambda)\Gamma^\dagger_{\lambda\mu}(a_k a_i) + (-1)^\mu Y_n(a_k a_i;\lambda)\Gamma_{\lambda-\mu}(a_k a_i)\},$$

$$(16)$$

acting on the correlated ground state $|\tilde{0}\rangle$ defined by the relation

$$\Gamma_{\lambda\mu}(n)|\tilde{0}\rangle = 0 . \qquad (17)$$

In the following we use the convention that the index k is used for particle states and the index i for hole states. The quantities X and Y are the forward-going and backward-going amplitudes and are obtained by diagonalizing the residual Hamiltonian in RPA. The index n indicates the corresponding root-number. The operator $\Gamma^\dagger_{\lambda\mu}(a_1a_1)$ is defined by

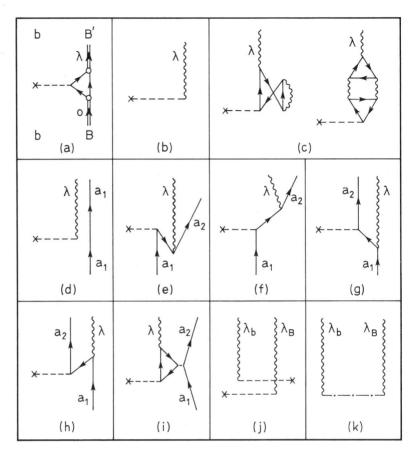

Fig. 4 Graphical representation of inelastic scattering. Graphs (a)–(i) represent the excitation of the target nucleus through the field of the projectile, while (j) and (k) represent excitation of both systems. The excitation of the pairing vibration in (a) takes place through a particle excitation after the phonon has been decomposed into a two-particle state. Graph (b) indicates the excitation of a surface-vibrational phonon in a closed-shell system. Two of the graphs representing the higher-order corrections to this process are depicted in (c). Graphs (d)–(i) indicate the excitation of a surface-vibrational mode in an odd system. The lowest-order graph in (d) is dominant if the phonon is strongly collective. Graph (j) indicates the second-order process in which the phonon is excited in both target and projectile (simultaneous excitation), while (k) indicates mutual excitation.

208

$$\Gamma^\dagger_{\lambda\mu}(a_k a_i) = [a^\dagger_{j_k}(a_k) b^\dagger_{j_i}(a_i)]_{\lambda\mu} . \tag{18}$$

From the commutation relations

$$[\Gamma_{\lambda\mu}(n), \Gamma^\dagger_{\lambda\mu}(m)] = \delta(n,m) , \tag{19}$$

and

$$[\Gamma_{\lambda\mu}(n), \Gamma_{\lambda\mu}(m)] = [\Gamma^\dagger_{\lambda\mu}(n), \Gamma^\dagger_{\lambda\mu}(m)] = 0 , \tag{20}$$

we can express the particle-hole operator (18) in terms of the collective operators $\Gamma^\dagger_{\lambda\mu}(n)$ and $\Gamma_{\lambda\mu}(n)$. One thus obtains

$$\Gamma^\dagger_{\lambda\mu}(a_k a_i) = \sum_n \{X_n(a_k a_i; \lambda)\Gamma^\dagger_{\lambda\mu}(n) - (-1)^\mu Y_n(a_k a_i; \lambda)\Gamma_{\lambda-\mu}(n)\}. \tag{21}$$

Observing that the interaction (5) can be rewritten as

$$\hat{V}_{aA} = \sum_{\substack{a_1 \geq a_2 \\ \lambda\mu}} \frac{(-1)^{\lambda+\mu}}{(2\lambda + 1)\{1 + \delta(a_1, a_2)\}}$$

$$\times \{(-1)^{\pi_1}(2j_2 + 1)^{1/2} f^{a_2 a_1}_{\lambda\mu}(\mathbf{r})[a^\dagger_{j_2}(a_2) b^\dagger_{j_1}(a_1)]_{\lambda-\mu}$$

$$+ (-1)^{\pi_2}(2j_1 + 1)^{1/2} f^{a_1 a_2}_{\lambda\mu}(\mathbf{r})[a^\dagger_{j_1}(a_1) b^\dagger_{j_2}(a_2)]_{\lambda-\mu}\}, \tag{22}$$

and utilizing the relations (6) and (12), we obtain the following expression for the form factor in terms of the boson operator (16):

$$(\hat{V}_{aA})_B = \sum_{\lambda\mu n} \frac{1}{\sqrt{2\lambda + 1}}(-1)^{\lambda+\mu} f^n_{\lambda-\mu}(\mathbf{r})[\Gamma^\dagger_{\lambda\mu}(n) + (-1)^\mu \Gamma_{\lambda-\mu}(n)] \tag{23}$$

where

$$f^n_{\lambda\mu}(\mathbf{r}) = \sum_{a_i a_k}(-1)^{\pi_i} \sqrt{\frac{2j_k + 1}{2\lambda + 1}} f^{a_k a_i}_{\lambda\mu}(\mathbf{r})$$

$$\times [X_n(a_k a_i; \lambda) - Y_n(a_k a_i; \lambda)] . \tag{24}$$

To lowest order the graph in Fig. 4(b) or the corresponding matrix element of (23), $\langle \lambda \mu | V_{aA} | \tilde{0} \rangle$, describes the excitation of vibrational quanta. In a closed-shell system the corrections to this description arise from graphs of type (c) in Fig. 4. For an odd system the corresponding corrections are given by graphs (e)–(i), the lowest-order graph being given by (d).

To calculate systematically the form factors within the framework of the nuclear field theory [see e.g. Bortignon *et al.* (1977)], one thus has to utilize both the boson representation (23) of the form factor and the fermion representation (5), i.e.,

$$(\hat{V}_{aA})_F = \sum_{\substack{a_1 a_2 \\ \lambda \mu}} (-1)^{\lambda + \mu + \pi_1} \frac{\sqrt{2j_2 + 1}}{2\lambda + 1} f_{\lambda - \mu}^{a_2 \ a_1}(\mathbf{r}) [a_{j_2}^\dagger(a_2) b_{j_1}^\dagger(a_1)]_{\lambda \mu}. \quad (25)$$

The generalization of (23) and (25) to the case of pairing deformed (superfluid) and shape deformed nuclei has been worked out by Broglia *et al.* (1978).

Note that all form factors can be made to include components responsible for Coulomb excitation by using instead of (2) the corresponding charge density. We may expand the Coulomb part of the single-particle potential as

$$U_{1a}^C(| \mathbf{r}_{1C} - \mathbf{r} |) = Z_a e \sum_{\lambda \mu} \frac{4\pi e_1}{2\lambda + 1} Y_{\lambda \mu}^*(\hat{r}_{1C}) Y_{\lambda \mu}(\hat{r}) r_{1C}^\lambda r^{-\lambda - 1}, \quad (26)$$

where e_1 is the charge of the nucleon and Z_a is the charge number of the projectile. This expansion leads to the single-particle form factor

$$(f_\lambda^{a_2 a_1}(r))^C = \frac{4\pi Z_a e}{\sqrt{(2\lambda + 1)(2j_2 + 1)}} (-1)^\lambda \langle j_2 \| \mathscr{M}(E\lambda) \| j_1 \rangle r^{-\lambda - 1} \quad (27)$$

where

$$\mathscr{M}(E\lambda) = e_1 r_{1C}^\lambda Y_{\lambda \mu}(\hat{r}_{1C}). \quad (28)$$

Inserting (27) in (24), we find for the Coulomb excitation of a phonon

$$(f^n_{\lambda\mu}(\mathbf{r}))^C = \frac{4\pi Z_a e}{2\lambda + 1}(-1)^\lambda \langle n\lambda \| \mathscr{M}(E\lambda) \| 0 \rangle r^{-\lambda-1}Y_{\lambda\mu}(\hat{r})$$

$$= (-1)^{\lambda-\mu}\sqrt{2\lambda+1}\ \langle \lambda -\mu | V_E(a,\mathbf{r}) | 00 \rangle, \qquad (29)$$

where the reduced matrix element of the electric multipole operator is given by

$$\langle n\lambda \| \mathscr{M}(E\lambda) \| 0 \rangle$$

$$= \sum_{a_k a_i}(-1)^{\pi i}\{X_n(a_k a_i;\lambda) - Y_n(a_k a_i;\lambda)\}$$

$$\times \langle j_k \| e_1 r^\lambda_{1c} Y_\lambda \| j_i \rangle. \qquad (30)$$

These quantities can be determined experimentally (cf. also Sec. II.1) in a model independent way and thus provide a test of the description of the nuclear states.

In Fig. 5 we give examples of inelastic form factors associated with both collective and noncollective vibrational states of ^{208}Pb. Noncollective refers, in the present context, to states which are dominated by few components or states whose wavefunctions display destructive interference.

At large distances the formfactors display the smooth $r^{-\lambda-1}$ dependence characteristic for Coulomb excitation. The nuclear part shows a rapid variation with r. The two contributions have opposite sign and the form factor thus vanishes at a point close to the radius of the Coulomb barrier [cf. Eq. (3.24) below].

Quasielastic heavy-ion reactions are sensitive to the form factors only in the external region for r-values larger than $R_a + R_A + 2$ fm, where all the nuclear form factors shown are in fact almost identical except for the scale.

(d) Excitation through the isovector interaction

The expressions derived above apply to the excitation of states through the isoscalar interaction. The form factors for the excitation through the

IV   Inelastic Scattering

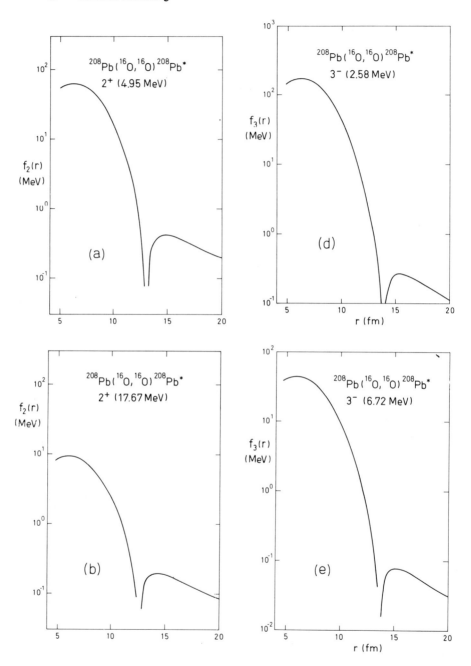

212

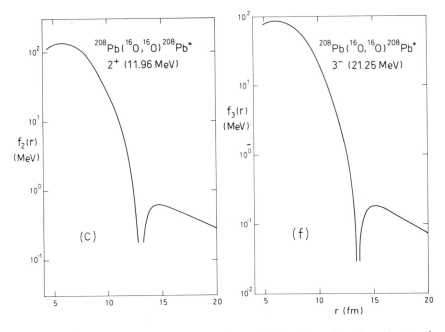

Fig. 5 Microscopic form factors for the excitation of quadrupole and octupole vibrational states. The figures (a) and (d) show form factors associated with the lowest state, (b) and (e) with a noncollective state and (c) and (f) with the $\Delta N = \lambda$ giant resonance respectively. The figs. (a)–(c) pertain to quadrupole states in ^{208}Pb; while (d)–(f) indicate the form factors for the octupole states in ^{208}Pb. The form factors were calculated on the basis of Eq. (2.24). The X and Y coefficients were determined by diagonalizing, in the random-phase approximation, a separable multipole interaction with radial dependence $r\partial U/\partial r$ including both an isoscalar and an isovector term. The unperturbed particle-hole excitations and the radial wave functions were taken from a Hartree-Fock calculation utilizing the Skyrme III interaction. The strength of the separable residual interaction was the same for all multipolarities and was fixed by requiring that the isoscalar dipole state be at zero energy [cf. Bertsch and Tsai (1975)]. The ratio between the isoscalar and isovector strengths was set equal to $\kappa(\lambda, \tau = 1)/\kappa(\lambda, \tau = 0) = -0.45(3 + 2\lambda)$ [c.f. Bes *et. al* (1975)]. The single-particle form factors were determined by Eq. (2.8), the wavefunctions ϕ being replaced by harmonic-oscillator wavefunctions. In order to generate these we used for protons the oscillator parameter $\nu_p = 0.167$ fm^{-2}. For neutrons we required that the mean square radius be equal to that of the protons, i.e. $\nu_n = 0.197$ fm^{-2}. The single-particle potential U_{1a} was that of Eq. (III.1.9). The figure is due to Pier Francesco Bortignon and Andrea Vitturi.

isovector interaction are obtained by adding the isovector-isovector interaction to (4).

This interaction can be estimated from nucleon scattering [Bohr and Mottelson (1975), Vol. I, Eq. (2-177)]. In heavy-ion scattering, however, one should take into account that the neutron excess in the surface of the projectile is larger than the neutron excess in the bulk of nuclear matter [cf. Akyüz and Winther (1981)]. A rough estimate based on the densities (III.A.15) shows that this enhancement is about a factor of 2. We thus find the isovector field to be

$$ \hat{V}_{aA}^{\text{iso}} \approx -\frac{2T_a}{A_a} \int U_{1a}(r_{1a}) \hat{\rho}_{\text{iso}}^{A}(r_{1A}) \, d^3 r_1, \tag{31} $$

where $2T_a = N_a - Z_a$. The isovector density is

$$ \hat{\rho}_{\text{iso}}^{A} = \hat{\rho}_{\text{neutron}}^{A} - \hat{\rho}_{\text{proton}}^{A} , \tag{32} $$

where $\hat{\rho}_{\text{neutron}}^{A}$ and $\hat{\rho}_{\text{proton}}^{A}$ are defined by (2), the summation being carried out over neutrons and over protons, respectively. We may write (31) analogously to (5) as

$$ \hat{V}_{aA}^{\text{iso}} = -\frac{2T_a}{A_a} \sum_{\substack{a_1 a_2 \\ \lambda\mu}} (-1)^{\lambda+\mu+\pi_1} \, 2t_z(a_1) $$

$$ \times \frac{\sqrt{2j_2 + 1}}{2\lambda + 1} f_{\lambda\mu}^{a_2 a_1}(\mathbf{r}_{aA}) [a_{j_2}^\dagger(a_2) b_{j_1}^\dagger(a_1)]_{\lambda-\mu} , \tag{33} $$

where $t_z(a_1) = \pm\frac{1}{2}$ if a_1 (and a_2) label a neutron or a proton state respectively.

As an example, the form factor for the excitation of a phonon state through the isovector field is given by an expression analogous to (24), i.e.

$$ f_{\lambda\mu}^n(\mathbf{r}) = -\frac{2T_a}{A_a} \sum_{a_i a_k} 2t_z(a_i) \sqrt{\frac{2j_k + 1}{2\lambda + 1}} (-1)^{\pi_i} f_{\lambda\mu}^{a_k a_i}(\mathbf{r}) $$

$$ \times [X_n(a_k a_i; \lambda) - Y_n(a_k a_i; \lambda)] . \tag{34} $$

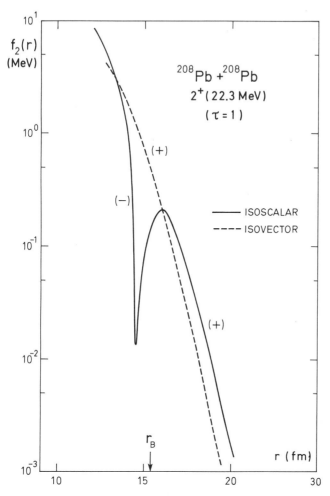

Fig. 6 Nuclear form factor for the excitation of a component of the quadrupole giant isovector resonance in ^{208}Pb by a Pb projectile. The solid curve shows the form factor (2.24) for the excitation of the state through the isoscalar field. The dashed curve is the form factor (34) for the excitation through the external isovector field. The nuclear-structure calculation giving the coefficients X and Y as well as the excitation energy is described in the caption of Fig. 5. The state carries about 20% of the isovector sum rule, i.e. $f_n(2, \tau = 1) = 0.2$ (cf. Sec. 6 below). This figure is due to Andrea Vitturi.

Examples of form factors associated with the isovector field are given in Fig. 6.

Except for the giant dipole resonance, which has a pure isovector character, the collective excitations in heavy nuclei can only approximately be associated with an isospin quantum number. An "isovector" giant resonance can thus be excited both by the isoscalar field and by the isovector field of a projectile with $N > Z$, often with similar strength (cf. Fig. 6).

Assuming again the neutron excess in the surface to be about twice the value it has in the bulk of the nucleus, we find

$$\langle \psi_0^A | \hat{\rho}_{iso}^A | \psi_0^A \rangle = \frac{4T_A}{A_A} \langle \psi_0^A | \hat{\rho}^A | \psi_0^A \rangle \ . \qquad (35)$$

This leads to the following diagonal matrix element of (31)

$$\langle \psi_0^a \psi_0^A | \hat{V}_{aA}^{iso} | \psi_0^a \psi_0^A \rangle = U_{iso}^N(\mathbf{r}_{aA})$$

$$= -\frac{8T_a T_A}{A_a A_A} U^N(\mathbf{r}_{aA}) \ , \qquad (36)$$

as was assumed in Eq. (III.1.44).

3   MACROSCOPIC FORM FACTORS FOR VIBRATIONS

The strongly collective isoscalar states of spherical nuclei can be interpreted as vibrational states corresponding to excitations of surface modes (cf. Sec. II.4). In this macroscopic description the nuclear radius depends on the direction in which it is measured through the expression

$$R_i(\hat{r}) = R_i^{(0)} \left(1 + \sum_{n,\lambda\mu} \alpha_{\lambda\mu}^{(i)}(n) Y_{\lambda-\mu}(\vartheta,\varphi)(-1)^\mu \right) \ , \qquad (1)$$

where i stands for the label (a or A) identifying the nucleus. The quantity $R_i^{(0)}$ is the average radius of the nucleus, while $\alpha_{\lambda\mu}^{(i)}(n)$ indicates the amplitude of the vibrational mode n of multipolarity λ,μ.

The Hamiltonian associated with these degrees of freedom is assumed to be

$$H_i = \sum_{n,\lambda,\mu} \left\{ \frac{1}{2D_\lambda^{(i)}(n)} \mid \pi_{\lambda\mu}^{(i)}(n) \mid^2 + \tfrac{1}{2} C_\lambda^{(i)}(n) \mid \alpha_{\lambda\mu}^{(i)}(n) \mid^2 \right\} , \qquad (2)$$

where $\pi_{\lambda\mu}^{(i)}(n)$ is the momentum conjugate to $\alpha_{\lambda\mu}^{(i)}(n)$. The quantities $D_\lambda^{(i)}(n)$ and $C_\lambda^{(i)}(n)$ are the mass parameter and the restoring-force parameter, respectively. In terms of the boson creation and annihilation operators $(c_{\lambda\mu}^{(i)}(n))^\dagger$ and $c_{\lambda\mu}^{(i)}(n)$ the deformation parameters can be written as

$$\alpha_{\lambda\mu}^{(i)}(n) = \sqrt{\frac{\hbar \, \omega_\lambda^{(i)}(n)}{2C_\lambda^{(i)}(n)}} \left\{ (c_{\lambda\mu}^{(i)}(n))^\dagger + (-1)^\mu \, c_{\lambda-\mu}^{(i)}(n) \right\}, \qquad (3)$$

where $\omega_\lambda^{(i)}(n) = \sqrt{C_\lambda^{(i)}(n)/D_\lambda^{(i)}(n)}$ is the frequency of the mode.

The parameters entering into the definition of the collective variable (3) are determined from the energy of the physical state and from the electromagnetic transition probability which determines the matrix element of the multipole operator $\mathcal{M}(E\lambda,\mu)$. This operator is related to the deformation amplitude by

$$\mathcal{M}_i(E\lambda,\mu) = \sum_n \frac{3Z_i e(R_i^c)^\lambda}{4\pi} \alpha_{\lambda\mu}^{(i)}(n) . \qquad (4)$$

This expression holds also for diffuse charge distributions of the Yukawa folding type [cf. Appendix B, Eq. (B.12)] with R_i^c equal to the equivalent sharp radius. The standard value of R_i^c,

$$R_i^c \approx 1.2 A_i^{1/3} \text{ fm} , \qquad (5)$$

is thus in good agreement with the value R_π in (III.A.15). The quantities A_i and Z_i are the mass and charge numbers of nucleus i.

In the macroscopic description the nuclear interaction V_{aA}^N is assumed to be a function of the shortest distance between the nuclear surfaces. Neglecting terms quadratic in the $\alpha_{\lambda\mu}$'s, this distance is given by

$$s = r_{aA} - R_a(-\hat{r}_{aA}) - R_A(\hat{r}_{aA}) , \qquad (6)$$

where R_a and R_A are given by (1). For strongly deformed nuclei the dependence of V^N on the radii of curvature can also be of significance (cf. Sec. 4 below).

In the following we consider only target excitation and drop the indices $i = A$ and n. The matrix element associated with the excitation of the state $| 1_{\lambda\mu} \rangle$ with one quantum in the mode $\lambda\mu$ is then given by

$$\langle \lambda\mu | V_{aA} | 00 \rangle = \langle 1_{\lambda\mu} | V^N_{aA} | 0 \rangle + \langle 1_{\lambda\mu} | V_E(a,\mathbf{r}) | 0 \rangle, \qquad (7)$$

where $| 0 \rangle$ indicates the ground state. The matrix element corresponding to the Coulomb part can be written [cf. also (II.1.8)]

$$\langle 1_{\lambda\mu} | V_E(a,\mathbf{r}) | 0 \rangle = \frac{4\pi Z_a e}{(2\lambda + 1)^{3/2}} \langle \lambda || \mathcal{M}_B(E\lambda) || 0 \rangle$$

$$\times (-1)^\mu \frac{Y_{\lambda -\mu}(\hat{r})}{r^{\lambda+1}}. \qquad (8)$$

The nuclear part of (7) can be related to the nuclear part of the ion-ion potential $U_{aA}(r)$. This quantity is defined as

$$U^N_{aA}(r) = \langle 0 | V^N_{aA} | 0 \rangle. \qquad (9)$$

From (3) and the corresponding definition for the conjugate operator

$$\pi_{\lambda\mu} = -i\hbar \frac{\partial}{\partial \alpha_{\lambda\mu}} = -i\hbar \sqrt{\frac{C_\lambda}{2\hbar \omega_\lambda}} (c_{\lambda\mu} - (-1)^\mu c^\dagger_{\lambda -\mu}), \qquad (10)$$

we can express the phonon creation and annihilation operators in terms of $\alpha_{\lambda\mu}$ and $\pi_{\lambda\mu}$, e.g.

$$c_{\lambda\mu} = \sqrt{\frac{C_\lambda}{2\hbar \omega_\lambda}} (-1)^\mu \alpha_{\lambda-\mu} + \sqrt{\frac{\hbar \omega_\lambda}{2C_\lambda}} \frac{\partial}{\partial \alpha_{\lambda\mu}}. \qquad (11)$$

The nuclear matrix element is given by

$$\langle 1_{\lambda\mu} | V_{aA}^N(s) | 0 \rangle = \langle 0 | [c_{\lambda\mu}, V_{aA}^N] | 0 \rangle$$

$$= \sqrt{\frac{\hbar \omega_\lambda}{2C_\lambda}} \langle 0 | \frac{\partial V_{aA}^N(s)}{\partial \alpha_{\lambda\mu}} | 0 \rangle. \tag{12}$$

For the isoscalar modes described by (1), we find, utilizing (6),

$$\frac{\partial V_{aA}^N(s)}{\partial \alpha_{\lambda\mu}} = - R_A^{(0)} Y_{\lambda\mu}^*(\hat{r}) \frac{\partial V_{aA}^N(s)}{\partial r}. \tag{13}$$

Since furthermore

$$\langle 1_{\lambda\mu} | \alpha_{\lambda\mu} | 0 \rangle = \sqrt{\frac{\hbar \omega_\lambda}{2C_\lambda}}, \tag{14}$$

we obtain

$$\langle 1_{\lambda\mu} | V_{aA}^N | 0 \rangle = - \sqrt{\frac{\hbar \omega_\lambda}{2C_\lambda}} R_A^{(0)} Y_{\lambda\mu}^*(\hat{r}) \langle 0 | \frac{\partial V_{aA}^N}{\partial r} | 0 \rangle$$

$$= - \langle 1_{\lambda\mu} | \alpha_{\lambda\mu} | 0 \rangle R_A^{(0)} Y_{\lambda\mu}^*(\hat{r}) \frac{\partial U_{aA}^N}{\partial r}. \tag{15}$$

This expression is not valid for $\lambda = 1$, since the corresponding nuclear deformation (1) only corresponds to a shift of the center of mass. The excitation of the isovector modes was discussed in connection with the microscopic description.

The term with $\lambda = 0$ corresponds to the compression mode. This monopole mode cannot be excited in Coulomb excitation, but can readily be excited by the nuclear interaction, as is evident from the above derivation, which also applies for $\lambda = 0$. A discussion of the mass and restoring-force parameters for the different compression modes is given in Sec. 5 below.

The quantity V_{aA}^N is the ion-ion potential in the absence of zero-point fluctuations of the nuclear surfaces, i.e. the surface-surface interaction on a time scale short compared to the collective vibrations. If we assume that V_{aA}^N at large distances is of the exponential form

$$V_{aA}^N(s) = S e^{-s/a} , \qquad (16)$$

we find according to (9)

$$U_{aA}^N(r) = S e^{-(r - R_a^{(0)} - R_A^{(0)})/a} e^{\sigma^2/2a^2} \qquad (17)$$

with

$$\sigma^2 = \sum_{in\lambda} \frac{2\lambda + 1}{4\pi} (R_i^{(0)})^2 \frac{\hbar \, \omega_\lambda^{(i)}(n)}{2C_\lambda^{(i)}(n)} , \qquad (18)$$

i.e.

$$V_{aA}^N(s(\alpha = 0)) = e^{-\sigma^2/2a^2} U_{aA}^N . \qquad (19)$$

We have calculated the macroscopic form factors as discussed above for some of the reactions studied in Fig. 5, and the results are given in Fig. 7 together with the corresponding microscopic form factors.

As is seen from the figure, the macroscopic and microscopic form-factors agree well outside the distance $R_a + R_A + 2$ fm, inside which the absorption is expected to dominate.

At shorter distances there is a marked difference between the two form factors in that the macroscopic form factor reaches a minimum at a radius closer to the surface than the microscopic form factor. This is associated with the repulsive element in the ion-ion potential arising from the increase of the local density at the contact point over the average nuclear density. A corresponding correction might be applied to the microscopic form factor due to particle exchange.

The agreement between the two form factors for collective states in the surface region is a natural consequence of the definition of the collective state through the random-phase approximation as discussed e.g. by Broglia et al. (1978).

In the following we shall calculate the form factor using the expression (III.1.36) for the ion-ion potential. The inelastic matrix element is given by

$$
\langle \lambda\mu | V_{aA} | 00 \rangle
$$

$$
= \sqrt{\frac{\hbar\,\omega_\lambda}{2C_\lambda}} Y^*_{\lambda\mu}(\hat{r}) \left\{ \frac{3Z_a A_A e^2}{(2\lambda+1)R^c_A} \left(\frac{R^c_A}{r} \right)^{\lambda+1} - R_A \frac{\partial U^N_{aA}}{\partial r} \right\}. \quad (20)
$$

For the Coulomb-excitation matrix element we have used (8) and (4). The radius R^c_A, which is approximately given by (5), can be calculated more accurately by utilizing the fact that the charge distribution (III.1.5) is deformed according to (1). One finds [cf. Bohr and Mottelson (1969), p. 160]

$$
\mathcal{M}_A(E\lambda\mu) \approx - \int \frac{\partial\rho}{\partial r}(r')^{\lambda+2}\, dr'\, R^d_A \alpha_{\lambda\mu}
$$

$$
= R^d(\lambda+2) \int \rho(r')(r')^{\lambda+1}\, dr'\, \alpha_{\lambda\mu}
$$

$$
= \frac{3Z_A e(R^c_A)^\lambda}{4\pi}\alpha_{\lambda\mu}, \quad (21)
$$

where, for the Fermi distribution (III.1.5), one finds

$$
(R^c_A)^\lambda = (1.12A_A^{1/3})^\lambda \times \begin{cases} (1 + 0.74A_A^{-2/3})\text{ fm} & \text{for } \lambda = 2, \\ (1 + 3.0A_A^{-2/3})\text{ fm} & \text{for } \lambda = 3,\,(22) \\ (1 + 6.0A_A^{-2/3})\text{ fm} & \text{for } \lambda = 4. \end{cases}
$$

As mentioned in connection with Eq. (5), one finds for Yukawa folded distributions that the multipole moment is identical to the one for the equivalent sharp distribution (cf. Appendix B) and we find $R^c_A \approx 1.2A_A^{1/3}$ fm independent of λ. When one uses the exponential ion-ion potential (III.1.36), the definition of the radius parameter is somewhat uncertain.

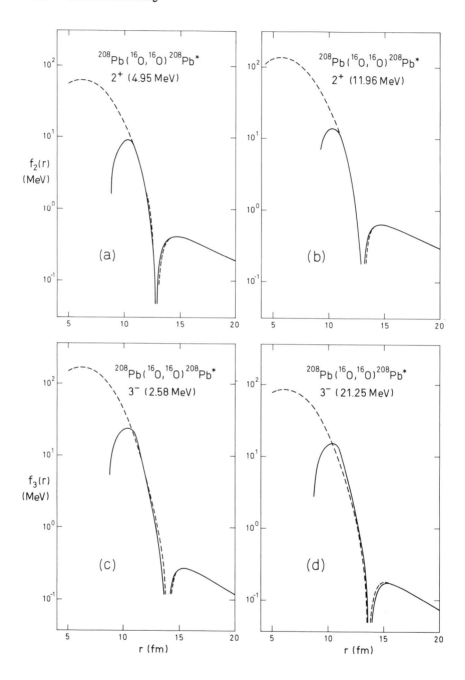

The value appearing in the second term in (20) should be the effective sharp radius [cf. (III.1.34)], which is not very different from the radius (III.1.38) appearing in the empirical potential. The uncertainty in this number is of minor importance, since it appears only linearly in (20).

Since the two terms in (20) appear with opposite signs, the form factor vanishes at a distance r_F determined by

$$\left. \frac{\partial U_{aA}}{\partial r} \right|_{r_F} \approx \frac{3}{2\lambda + 1} \frac{Z_a Z_A e^2}{r_F^2} \left(\frac{R_A^c}{r_F} \right)^{\lambda - 1} . \tag{23}$$

For $\lambda = 1$ this equation is identical to the equation determining the radius r_B of the Coulomb barrier [cf. Eq. (III.2.4)]. For other values of λ one finds

$$r_F = r_B + a \left(1 + (\lambda + 1) \frac{a}{r_B} \right) \ln \left\{ \left(\frac{r_B}{R_A^c} \right)^{\lambda - 1} \frac{2\lambda + 1}{3} \right\} , \tag{24}$$

where one may use [cf. (III.2.5)]

$$r_B = 1.07(A_a^{1/3} + A_A^{1/3}) + 2.72 \text{ fm} . \tag{25}$$

The form factor reaches a minimum when the attractive force between the surfaces reaches the maximum value (III.1.29). One finds approximately

$$(\langle 1_{\lambda\mu} | V_{aA} | 0 \rangle)_{\min}$$
$$= -\sqrt{\frac{\hbar \omega_\lambda}{2C_\lambda}} \left[R_A 4\pi\gamma\bar{R}_{aA} - \frac{3E_B}{2\lambda + 1} \left(\frac{R_A}{r_B} \right)^\lambda \right] Y_{\lambda\mu}^*(\hat{r}) , \tag{26}$$

Fig. 7 Macroscopic form factors for some of the reactions studied in connection with Fig. 5. They were calculated using Eq. (3.15) and the proximity potential [cf. Eq. (III.1.31)], adjusting the matrix elements of $\alpha_{\lambda\mu}$ to agree with the microscopic result for the matrix element of the multipole operator, which also agrees with the experimental values. For comparison we also display with dashed lines the corresponding microscopic form factors taken from Fig. 5. The figure is due to Andrea Vitturi.

where E_B is the Coulomb barrier. Even for very heavy nuclei this is a negative number, which means that the form factor always changes sign.

4   FORM FACTORS FOR DEFORMED NUCLEI

For deformed nuclei the macroscopic description of the form factors can be worked out on the basis of (2.4), introducing the ion-ion potential as a function of the deformation parameters:

$$V_{aA}(\mathbf{r},\alpha_{\lambda\mu}) = \int \rho^A(\mathbf{r'},\alpha_{\lambda\mu})U_{1a}(\,|\,\mathbf{r'} - \mathbf{r}\,|\,)\,d^3r'. \qquad (1)$$

In evaluating this integral we use a coordinate system in which the z-axis is along the symmetry axis of the equilibrium deformation. We thus assume that the density can be written as

$$\rho^A(\mathbf{r},\alpha_{\lambda\mu}) = \rho^A(r - R_A^{(0)}[1 + \alpha'_{20}Y^*_{20} + \alpha'_{22}(Y_{22} + Y^*_{2\,-2})$$
$$+ \sum_{\lambda \geq 3,\nu} \alpha'_{\lambda\nu}Y^*_{\lambda\nu}]) \qquad (2)$$

in terms of the deformation parameters $\alpha'_{\lambda\nu}$ in the intrinsic system [cf. Bohr and Mottelson (1975)]. The quantity $\langle \alpha'_{20} \rangle = \beta_0$ is the equilibrium deformation. The vibrational amplitudes α'_{20} and α'_{22} give rise to β- and γ-vibrations respectively [cf.(II.8.3–5)].

The form factor for excitation of rotational and vibrational states are found by evaluating matrix elements of (1) between rotational wave-functions. We shall consider even nuclei, in which case

$$|\,\psi^A_{I_A M_A}\,\rangle \equiv |\,n_{\lambda\nu}KI_A M_A\,\rangle$$

$$= \sqrt{\frac{2I_A + 1}{16\pi^2[1 + \delta(K,0)]}}\,[D^{I_A}_{M_A K}(\theta_i)|\,n_{\lambda\nu}K\,\rangle$$

$$+ (-1)^{I_A+K+\pi}\,D^{I_A}_{M_A-K}(\theta_i)|\,\overline{n_{\lambda\nu}K}\,\rangle]\,, \qquad (3)$$

where $| n_{\lambda\nu}K \rangle$ indicates a state with $n_{\lambda\nu}$ quanta of multipolarity λ and component ν, i.e. an eigenstate of the vibrations in the variable $\alpha'_{\lambda\nu}$. The quantum number K is equal to the sum of the ν's. The state $|\overline{nK}\rangle$ is the time reversed state to $| nK \rangle$, and π its parity.

The form factor for excitations within the ground-state rotational band is given by

$$\langle 00I_A M_A | U_{aA}(\mathbf{r},\alpha_{\lambda\mu})| 0000 \rangle$$
$$= \frac{\sqrt{2I_A + 1}}{(8\pi^2)} \int d^3\theta_i \, D^{I_A*}_{M_A 0}(\theta_i) U^{\text{int}}_{aA}(r,\theta',\beta_0), \qquad (4)$$

where the ion-ion potential in the intrinsic frame is defined by

$$U^{\text{int}}_{aA}(r,\theta',\beta_0) = \int d^3r' \, \langle 00| \rho(r')| 00 \rangle U_{1a}(| \mathbf{r}' - \mathbf{r} |). \quad (5)$$

This quantity is a function of r and the angle θ' between $\mathbf{r}$ and the intrinsic symmetry axis 3.

The expression (4) can be calculated making use of the relation

$$D^{I_A}_{M_A 0}(\theta_i) = \sum_{M'} D^{I_A}_{M_A M'}(\varphi,\vartheta,\psi)(D^{I_A}_{0M'}(\theta'_i))^* \qquad (6)$$

where the three Eulerian angles (φ,ϑ,ψ) indicate the rotation from the laboratory system to a system with z-axis along $\mathbf{r}$, while the Eulerian angles θ'_i indicate the orientation of this axis with respect to the intrinsic coordinate system (cf. Fig. 8).

Inserting (6) in (4), the integration over two of the Eulerian angles can be performed and we find

$$\langle 00I_A M_A | U_{aA}(\mathbf{r},\alpha_{\lambda\mu})| 000 \rangle = f_{I_A}(r) Y^*_{I_A M_A}(\hat{r}) , \qquad (7)$$

with

$$f_I(r) = \sqrt{\pi} \int \sin \theta' \, d\theta' \, P_I(\cos \theta') U^{\text{int}}_{aA}(r,\theta',\beta_0). \qquad (8)$$

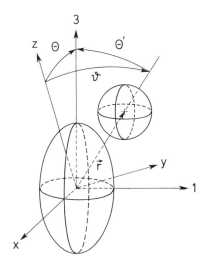

Fig. 8 Coordinate systems for the evaluation of form factors for deformed nuclei. The intrinsic coordinate system $(1,2,3)$ for the axially symmetric deformed density indicated is chosen such that the $(1,3)$ plane contains the vector $\mathbf{r}$. The laboratory system (x,y,z) is also indicated. The Eulerian angles indicating the orientation of the rotor are (ϕ,θ,ψ) while the direction of $\mathbf{r}$ is given by the polar angles ϑ and φ. The orientation of the rotor with respect to $\mathbf{r}$ is specified by the Eulerian angles (ϕ',θ',ψ').

The form factor (8) can be evaluated analytically if we use the proximity approximation for the ion-ion potential in (5), i.e.

$$U_{aA}^{N}(r) = - S\bar{R}e^{-(r-R_a-R_A)/a} , \tag{9}$$

where [cf. III.1.37)] $S \approx 50$ MeV/fm and $a \approx 0.63$ fm. The quantity $\bar{R}$ is given by

$$\bar{R} = (\kappa_a^{\parallel} + \kappa_A^{\parallel})^{-1/2}(\kappa_a^{\perp} + \kappa_A^{\perp})^{-1/2} , \tag{10}$$

where the κ's are related to the radii of curvature at the point of contact between a and A [cf. (III.1.20)]. We shall assume that a is spherical, i.e.

$$\kappa_a^{\parallel} = \kappa_a^{\perp} = 1/R_a \ .$$ (11)

For the radius R_A of nucleus A we assume the angular dependence corresponding to an axially deformed equilibrium shape, i.e., in the intrinsic frame,

$$R_A = R_A^{(0)} \left(1 + \sqrt{\frac{5}{4\pi}} \beta_0 P_2(\cos\theta') \right) \ .$$ (12)

The quantities $\kappa_A^{\parallel}$ and $\kappa_A^{\perp}$ are then the principal rates of curvature at the point of contact. We shall use the expression

$$\bar{R} = \frac{R_a R_A^{(0)}}{R_a + R_A^{(0)}} [1 - B_2 P_2(\cos\theta')]$$ (13)

with

$$B_2 = \frac{2R_a}{R_a + R_A^{(0)}} \sqrt{\frac{5}{4\pi}} \beta_0 \ ,$$ (14)

which is correct to first order in B_2.

Inserting (12) and (13) in (9), we find the result

$$U_{aA}^{\text{int}}(r,\theta') = -S \frac{R_a R_A^{(0)}}{R_a + R_A^{(0)}} [1 - B_2 P_2(\cos\theta')]$$

$$\times \exp\left\{ -\frac{r - R_a - R_A^{(0)}[1 + \sqrt{5/4\pi}\ \beta_0 P_2(\cos\theta')]}{a} \right\}$$ (15)

for the ion-ion potential (5) in the intrinsic frame.

In evaluating the integral (8) it is not appropriate to expand the exponential in (15) in powers of β, since the expansion parameter

$$c = \sqrt{\frac{5}{4\pi}} \frac{R_A^{(0)}}{a} \beta$$

is often of the order of magnitude of unity. The integral can, however, be evaluated in terms of error functions [see Alder and Winther (1975), p. 219]

$$q_I(c) = \frac{1}{2} \int_{-1}^{1} dx\, P_I(x) e^{cP_2(x)}$$

$$= \frac{\Gamma\left(\dfrac{I+1}{2}\right)}{2\Gamma\left(\dfrac{2I+3}{2}\right)} e^c (\tfrac{3}{2}c)^{I/2}\, {}_1F_1\left(\frac{I+2}{2},\ \frac{2I+3}{2},\ -\tfrac{3}{2}c\right), \quad (16)$$

where ${}_1F_1$ is a confluent hypergeometric function. The functions $q_I(c)$ are given in Fig. 9 for $I = 0$, 2 and 4.

In order to evaluate the integral (8) in terms of the functions $q_I(c)$, we note that the term proportional to B_2 can be expressed in terms of $dq_I(c)/dc$. We may write the result to first order in B_2

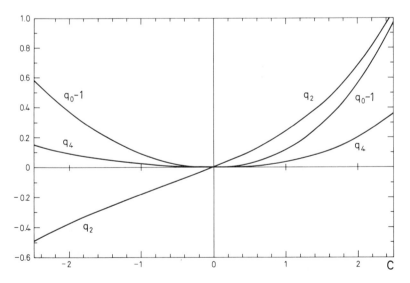

Fig. 9 The functions $q_I(c)$ defined in (4.16). They describe the strength of the ion-ion potential and of the form factors for excitation within the ground-state rotational band in deformed nuclei and are plotted as functions of the parameter c.

228

$$f_I(r) = \sqrt{4\pi}(U_{aA}^N(r))_{\beta=0}\, q_I(c')$$ (17)

with

$$c' = c\left(1 - \frac{2aR_a}{R_A^{(0)}(R_a + R_A^{(0)})}\right).$$ (18)

The fact that the form factor (17) has the same radial shape independent of I is a consequence of the exponential form of the ion-ion potential, which is expected to be a rather good approximation for grazing reactions. It is seen from (18) that the curvature effect proportional to B_2 in (15) tends to cancel the first term, as is to be expected from the fact that when the distance between the surfaces is small, the rate of curvature is large, and vice versa. Corrections to (17) are expected to arise from a possible variation of a with angle and from a change in the value of $R_A^{(0)}$ as compared to the value for nondeformed nuclei (III.1.37), i.e.

$$R_i^{(0)} = 1.233A_i^{1/3} - 0.98A_i^{-1/3}.$$ (19)

One might thus expect a correction due to volume conservation which, for the spheroidal shape used, would lead to a radius parameter

$$R_A'^{(0)} = R_A^{(0)}\left(1 - \frac{1}{4\pi}\beta^2\right)$$ (20)

to lowest order in β^2. Including this correction, the expression (17) for the form factor would read

$$f_I(r) = \sqrt{4\pi}(U_{aA}^N(r))_{\beta=0}\, q_I(c')$$

$$\times \left(1 - \frac{a}{5R_A^{(0)}}c^2\right).$$ (21)

For many deformed nuclei the ion-ion potential $[U_{aA}^N(r) = (4\pi)^{-1/2} f_0(r)]$ according to this formula is almost the same as the potential between the corresponding spherical nuclei. These conclusions are based

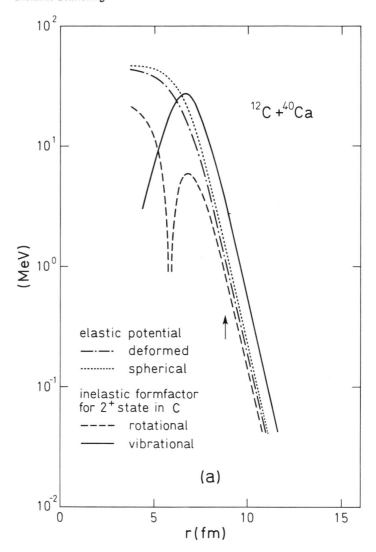

Fig. 10 Form factors and ion-ion potentials for deformed nuclei. In (a) is given the ion-ion potential for the scattering of ^{12}C on ^{40}Ca based on Eq. (4.8) and including the curvature correction with a deformation parameter $\beta_0 = -0.6$ for ^{12}C$[U_{aA}(r) = (4\pi)^{-1/2}f_0(r)]$, calculated on the basis of (III.1.40). For comparison the potential assuming $\beta_0 = 0$ is shown. The inelastic form factor for excitation of the 2^+ state at 4.43 MeV in ^{12}C calculated according to (4.8) is compared with the

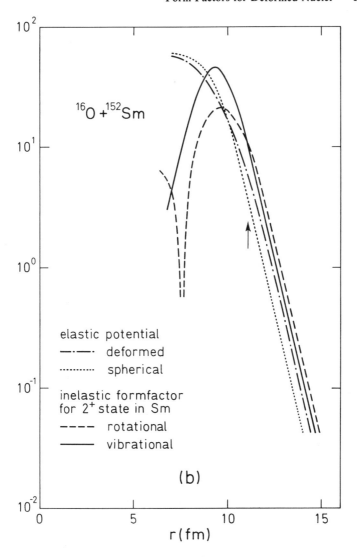

form factor (3.15) with $\langle 1_{2\mu} | \alpha_{2\mu} | 0 \rangle = \beta_0 / \sqrt{5}$. Note that both the potential and the form factor are smaller in the tail region for the deformed case. In (b) is shown the corresponding quantities for the scattering of ^{16}O on ^{152}Sm. The deformation of Sm was taken to be $\beta_0 = 0.306$. Here the potential and form factor are largest for the deformed case. The arrow indicates the position of the Coulomb barrier r_B [cf. Eq. (III.2.5)]. This figure is due to Andrea Vitturi.

on the exponential form of the potential, and are only expected to be correct for

$$r \gtrsim R_a + R_A^{(0)} \left(1 + \sqrt{\frac{5}{4\pi}} |\beta_0| \right) + 1.5 \text{ fm.} \qquad (22)$$

In Fig. 10 are displayed the ion-ion potential and the form factors for the excitation of the lowest 2^+ state for two deformed nuclei in comparison with the corresponding quantities for spherical nuclei. They were calculated numerically from the expression (4), and agree outside the radius r_B with the result (21).

Before we conclude this section we note that the vibrational states which have been identified in deformed nuclei offer an interesting field of study for inelastic scattering, since they lie at a high excitation energy (≈ 1 MeV) where one expects large Coulomb-nuclear interference effects.

The form factor for the excitation of vibrations associated with the deformation parameters $\alpha'_{\lambda\nu}$ in (2) can be evaluated in a similar way to that in Sec. 3. We thus find in the intrinsic frame

$$\langle 1_{\lambda\nu}\nu | V_{aA} | 00 \rangle = \langle 00 | [c'_{\lambda\nu}, V_{aA}] | 00 \rangle$$

$$= \langle 1_{\lambda\nu} | \alpha'_{\lambda\nu} | 0 \rangle \langle 00 | \frac{\partial}{\partial \alpha'_{\lambda\nu}} V_{aA} | 00 \rangle$$

$$= - \langle 1_{\lambda\nu} | \alpha'_{\lambda\nu} | 0 \rangle R_A^{(0)} Y_{\lambda\nu}^*(\theta',\phi') \frac{\partial}{\partial r} U_{aA}^{\text{int}}(r,\theta',\beta_0), \qquad (23)$$

where we have used the fact that $U_{aA}^{\text{int}} = \langle 00 | V_{aA} | 00 \rangle$ depends on the deformation parameters only through the distance between the surfaces. Using this result, we can evaluate the form factor by calculating the matrix element of V_{aA} between the rotational wavefunctions (3). Utilizing (6) we find

$$\langle 1_{\lambda\nu}KI_AM_A | V_{aA} | 0000 \rangle$$

$$= - R_A^{(0)} \frac{\langle 1_{\lambda\nu} | \alpha'_{\lambda\nu} | 0 \rangle}{\sqrt{8\pi}}$$

$$\times \sum_{J}(2J + 1)\sqrt{2\lambda + 1}(-1)^{\nu} \times \begin{pmatrix} I_A & \lambda & J \\ 0 & 0 & 0 \end{pmatrix} \begin{pmatrix} I_A & \lambda & J \\ \nu & -\nu & 0 \end{pmatrix}$$

$$\times [1 + (-1)^{J}]\frac{\partial}{\partial r} [f_{J}(r)] Y^{*}_{I_A M_A}(\hat{r}) \,, \tag{24}$$

where we have used the fact that the parity π is equal to λ.

The factor $1 + (-1)^{J}$ ensures that the form factor vansihes if $I_A + \lambda$ is odd. The states of odd spin in the γ-vibrational band or the states of even spin in octupole vibrational bands can therefore only be excited indirectly via states of other spins in the two bands.

The form factor (23) in the intrinsic frame can be tested directly in inelastic scattering, since the rotational angular momentum transferred to the nucleus can be related to the orientation of the axis at the time of contact. In particular one expects that for γ-vibrational states where the form factor (23) vanishes at $\theta' = 0$, mainly high members of the rotational band are excited.

An alternative way to treat the interaction between deformed nuclei utlizing Yukawa-folding techniques is discussed in Appendix B.

5   FORM FACTORS FOR COMPRESSION MODES

Besides the surface modes, one expects through nuclear interaction also to excite volume modes. In the framework of the schematic hydrodynamic model, where the density is assumed to be constant throughout the volume, the properties of these modes [Bohr and Mottelson (1975)] are determined by a single constant, viz. the nuclear compressibility coefficient b_{comp} defined by

$$b_{comp} = \rho^{2} \frac{\partial^{2}(E/A)}{\partial \rho^{2}} \,, \tag{1}$$

where ρ is the nuclear density and E/A is the energy per nucleon. The available evidence indicates that b_{comp} is approximately equal to 15 MeV.

The eigenmodes of the compression degree of freedom with

233

amplitude $\alpha'_{n\lambda\mu}$, for a spherical nucleus of constant equilibrium density ρ_0, correspond to the density variation

$$\delta\rho = \rho_0 j_\lambda(k_{n\lambda}r) Y^*_{\lambda\mu}(\hat{r})\alpha'_{n\lambda\mu}(t),$$ (2)

and the associated velocity field

$$\mathbf{v} = k_{n\lambda}^{-2} \nabla \left(j_\lambda(k_{n\lambda}r) Y^*_{\lambda\mu}(\hat{r})\right)\dot{\alpha}'_{n\lambda\mu}(t) .$$ (3)

The quantity j_λ is a spherical Bessel function, and $k_{n\lambda}$ is determined by the condition that there is no excess pressure at the surface, i.e.

$$j_\lambda(k_{n\lambda}R_0) = 0.$$ (4)

The quantum number n indicates the number of nodes. For the lowest modes one finds

$$k_{n\lambda}R_0 = \begin{cases} 3.14 & \text{for} \quad \lambda = 0, \quad n = 1, \\ 4.49 & \text{for} \quad \lambda = 1, \quad n = 1, \\ 5.76 & \text{for} \quad \lambda = 2, \quad n = 1, \\ 6.28 & \text{for} \quad \lambda = 0, \quad n = 2. \end{cases}$$ (5)

These vibrations give rise to a deformation of the surface of the type (3.1) with amplitude

$$\alpha_{\lambda\mu} = \alpha'_{n\lambda\mu} \frac{j'_\lambda(k_{n\lambda}R_0)}{k_{n\lambda}R_0},$$ (6)

where j'_λ is the derivative of the Bessel function. The change in density due to this surface deformation is

$$\delta\rho = -\frac{\partial\rho}{\partial r} \frac{j'_\lambda(k_{n\lambda}R_0)}{k_{n\lambda}} Y^*_{\lambda\mu}(\hat{r})\alpha'_{n\lambda\mu}.$$ (7)

Since (2) is zero at the surface, we may combine Eqs. (2) and (7) to obtain the total density variation for a nucleus with a diffuse surface. For small amplitudes one finds

$$\delta\rho = \left(\rho(r)j_\lambda(k_{n\lambda}r)\,\frac{k_{n\lambda}R_0}{j'_\lambda(k_{n\lambda}R_0)} - R_0\frac{\partial\rho}{\partial r} \right) Y^*_{\lambda\mu}(\hat{r})\alpha_{n\lambda\mu}. \quad (8)$$

We have here introduced the $\alpha_{n\lambda\mu}$ defined as

$$\alpha_{n\lambda\mu} = \frac{j'_\lambda(k_{n\lambda}R_0)}{k_{n\lambda}R_0}\,\alpha'_{n\lambda\mu}, \quad (9)$$

so that for $n = 0$ we may identify them with the amplitudes for surface vibrations discussed in Sec. 3. Thus, for $n = 0$ we use $k_{0\lambda} = 0$, whereby the first term in (8) vanishes and $\delta\rho$ coincides with the result of Sec. 3. The definition (9) provides a convenient normalization of the amplitudes for the excitation of compression modes in surface reactions, where the amplitude of surface deformations plays a more important role than the amplitude in the interior.

For $n \neq 0$ the total electric multipole moment associated with the deformation (8) is

$$\mathscr{M}(\lambda\mu) = \int \delta\rho\; r^\lambda Y_{\lambda\mu}\, d^3r$$

$$= \alpha_{n\lambda\mu}\int_0^\infty \rho(r)\left(j_\lambda(k_{n\lambda}r)\frac{k_{n\lambda}R_0}{j'_\lambda(k_{n\lambda}R_0)} + (\lambda+2)\frac{R_0}{r} \right)r^{\lambda+2}dr. \quad (10)$$

For a nucleus of sharp surface this integral vanishes identically, since

$$\int_0^1 x^{\lambda+2}j_\lambda(k_{n\lambda}R_0x)\, dx = \frac{1}{k_{n\lambda}R_0}j_{\lambda+1}(k_{n\lambda}R_0)$$

$$= -\frac{1}{k_{n\lambda}R_0}j'_\lambda(k_{n\lambda}R_0). \quad (11)$$

For a nucleus with diffuse surface the same result should be enforced by adjusting the values of $k_{n\lambda}$ in (5). While for $\lambda = 0$ this condition is equivalent to the conservation of particles, for $\lambda > 0$ it is a definition of the mode. The radial density variation for the lowest monopole vibration in ^{208}Pb is shown in Fig. 11 as calculated from Eq. (8) utilizing this prescription. The change of k_{10} by about a factor of 2 as compared to (5) indicates the limitation of the schematic model.

The mass parameter and restoring force for compression modes ($n \geq 1$) are evaluated by calculating the total kinetic energy associated with the velocity field (3):

$$
\begin{aligned}
T &= \int \tfrac{1}{2} M \rho(r) \mathbf{v}^2 \, d^3 r \\
&= - \sum_{\substack{nn' \\ \lambda\lambda'\mu\mu'}} \tfrac{1}{2} \frac{M}{k_{n\lambda}^4} \rho_0 \int d^3 r \, j_\lambda(k_{n'\lambda'} r) Y_{\lambda'\mu'}(\hat{r}) \\
&\qquad \times \nabla^2 [j_\lambda(k_{n\lambda} r) Y^*_{\lambda\mu}(\hat{r})] \dot{\alpha}'^*_{n'\lambda'\mu'} \dot{\alpha}'_{n\lambda\mu} \\
&= \tfrac{1}{2} \sum_{n\lambda\mu} D_{n\lambda} | \dot{\alpha}_{n\lambda\mu} |^2,
\end{aligned} \tag{12}
$$

where M is the nucleon mass and

$$
D_{n\lambda} = \frac{3}{8\pi} A M R_0^2 . \tag{13}
$$

In deriving (12) we used a sharp nuclear density. Under the same assumption we may derive the restoring force. Since the energy density is given by

$$
\varepsilon = \frac{1}{2} \frac{b_{\text{comp}}}{\rho} (\delta\rho)^2 \tag{14}
$$

we find using (2)

$$
\begin{aligned}
E &= \int \frac{1}{2} \frac{b_{\text{comp}}}{\rho} (\delta\rho)^2 \, d^3 r \\
&= \tfrac{1}{2} \sum_{n\lambda\mu} C_{n\lambda} | \alpha_{n\lambda\mu} |^2
\end{aligned} \tag{15}
$$

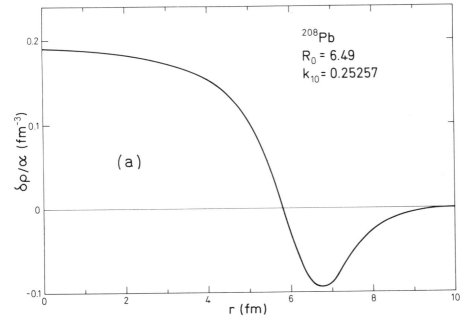

Fig. 11 The radial density variation for the lowest monopole compression mode ($n = 1$, $\lambda = 0$) according to Eq. (5.8) for ^{208}Pb. The value of k_{10} was adjusted to conserve the particle number.

with

$$C_{n\lambda} = \frac{3}{8\pi} A b_{\text{comp}}(k_{n\lambda}R_0)^2. \qquad (16)$$

The excitation energies of the compression modes are thus

$$\hbar \omega_{n\lambda} = \sqrt{\frac{b_{\text{comp}}}{M}} k_{n\lambda} \hbar. \qquad (17)$$

In calculating the form factor for inelastic scattering it should be noted that the first term in (8) is small compared to the last term until well inside the nuclear radius R_0. For grazing collisions we may thus neglect this term. The evaluation of the form factor is thus identical to the

237

evaluation in Sec. 3 of the form factor for surface vibrations, and we find the result

$$\langle 1_{n\lambda\mu} | V_{aA}^N | 0 \rangle = - \sqrt{\frac{\hbar \, \omega_{n\lambda}}{2 C_{n\lambda}}} \, R_A^{(0)} \frac{\partial U_{aA}^N}{\partial r} \, Y_{\lambda\mu}^*(\hat{r}) \tag{18}$$

The study of the compression modes is a new very active field of research. Experimentally, there exists strong evidence for the monopole compression mode [see e.g. Bertrand (1976)].

6   SUM RULES

The energy weighted sum rules mentioned in Sec. II.4 play an important role in inelastic scattering, since they can provide an upper limit to the energy which can be transferred to the nucleus under the action of an external field.

The expectation value of the excitation energy is given by

$$\langle E \rangle = \sum_n (E_n - E_0) P_n , \tag{1}$$

where P_n is the excitation probability of the state $| n \rangle$. The index 0 indicates the ground state. In perturbation theory we evaluate this probability as

$$P_n = \left| -\frac{i}{\hbar} \int_{-\infty}^{\infty} dt \, \langle n | \tilde{V}(t) | 0 \rangle \right|^2$$

$$= \left| -\frac{i}{\hbar} \int_{-\infty}^{\infty} dt \, \langle n | V(t) | 0 \rangle e^{i\omega_n t} \right|^2 , \tag{2}$$

where $\omega_n = (E_n - E_0)/\hbar$ and $V(t)$ is the external field at time t, while $\tilde{V}$ is the same quantity in the interaction representation.

The excitation probabilities show the adiabatic cutoff for high excitation energies which arise from the factor $e^{i\omega_n t}$ in the integral (2). An upper limit to $\langle E \rangle$ is obtained by neglecting this factor in (2). We obtain then

238

$$\langle E \rangle_{max} = \frac{1}{\hbar^2} \int_{-\infty}^{\infty} dt \int_{-\infty}^{\infty} dt'$$

$$\times \sum_n (E_n - E_0) \langle n| V(t)| 0 \rangle \langle n| V(t')| 0 \rangle^*. \tag{3}$$

If $V(t)$ is separable, i.e. if the matrix elements can be written in the form

$$\langle n| V(t)| 0 \rangle = \langle n| M| 0 \rangle F(t), \tag{4}$$

then we obtain

$$\langle E \rangle_{max} = |q|^2 \sum_n (E_n - E_0) \Big| \langle n| M| 0 \rangle \Big|^2, \tag{5}$$

where

$$q = \frac{1}{\hbar} \int_{-\infty}^{\infty} F(t)\, dt. \tag{6}$$

For such separable interactions the energy loss can thus be written as the square of an orbital integral times an energy weighted sum of the form encountered in Sec. II.4.

The result (3) does not actually depend on perturbation theory. In the sudden approximation, we may instead use the general expression for the solution of the coupled equations (1.1) (cf. Appendix II.B)

$$|f\rangle \equiv |\psi(\infty)\rangle = e^{-i\chi}|0\rangle, \tag{7}$$

with

$$\chi = -\frac{1}{\hbar} \int_{-\infty}^{\infty} V(t)\, dt. \tag{8}$$

The expectation value of the energy in the final state $|f\rangle$ can then be written as

$$\langle f \mid H_0 \mid f \rangle = \langle 0 \mid e^{-i\chi} H_0 e^{i\chi} \mid 0 \rangle, \tag{9}$$

where H_0 is the unperturbed Hamiltonian. We can evaluate (9) by the expansion

$$e^{-i\chi} H_0 e^{i\chi} = H_0 - i[\chi, H_0]$$
$$+ \frac{(-i)^2}{2!}[\chi, [\chi, H_0]] + \cdots. \tag{10}$$

Under the assumption that $V(t)$ only depends on the coordinates of the nucleons, the commutator appearing in (10) receives contributions solely from the kinetic-energy part of the Hamiltonian H_0, and the double commutator is a c-number. This means that all higher-order commutators vanish in the sum, and we obtain the expression

$$\langle E \rangle_{max} = \langle f \mid H_0 \mid f \rangle - \langle 0 \mid H_0 \mid 0 \rangle$$
$$= \sum_n (E_n - E_0) \mid \langle 0 \mid \chi \mid n \rangle \mid^2, \tag{11}$$

which is identical to (3).

On the other hand, we may evaluate the double commutator explicitly, obtaining the result

$$\langle E \rangle_{max} = \frac{\hbar^2}{2M} \langle 0 \mid \sum_k (\nabla_k \chi)^2 \mid 0 \rangle. \tag{12}$$

We have labeled the nucleons in the target by the index k. The two different ways of evaluating the expectation value of the double commutator, (11) and (12), form the basis for the standard evaluation of energy weighted sum rules. The result (12) has a simple classical interpretation in that the quantity

$$\hbar \nabla_k \chi = - \int_{-\infty}^{\infty} \nabla_k V \, dt \tag{13}$$

indicates the momentum transferred to the kth nucleon by the external field in a sudden collision. The total energy transferred is the sum of the associated kinetic energies.

It is noteworthy that the result (11) holds also for finite collision times if all degrees of freedom can be considered as harmonic vibrations and the interaction is linear in the amplitudes. In this case the S-matrix is still given by (8) if we replace $V(t)$ by (cf. Appendix II.B)

$$\tilde{V}(t) = e^{iH_0t/\hbar}\, V(t) e^{-iH_0t/\hbar} \; . \tag{14}$$

In this case the expression (11) gives the exact value of the expectation value of the energy loss. This result coincides with (1) with P_n given by (2). The formula is therefore correct to all orders of perturbation theory, although the quantities P_n may become larger than unity.

The energy loss can be expressed, through (12), in terms of an expectation value in the entrance channel.

Explicit calculations of $\langle E \rangle_{\max}$ can be done for Coulomb excitation where the interaction is separable [cf. Eq.(4)] but are in general, for nuclear interactions, rather complicated. However, the identity of (11) and (12), which holds for any field χ which is a function of the coordinates of the nucleons only, provides a relation which is useful in many other contexts. By appropriately choosing the field χ one obtains a relation which because (11) is a sum of positive numbers, sets limits (sum rules) on the magnitude of the matrix elements of χ.

We first assume that χ is the external field V_{aA} acting on the target A. We write it as the sum of single-particle potentials generated by the projectile, i.e.

$$V_{aA}(\mathbf{r}_{aA}) = \sum_k U_{1a}(|\,\mathbf{r}_{kA} - \mathbf{r}_{aA}\,|) \; . \tag{15}$$

The energy weighted sum rule corresponding to this interaction is

$$(\bar{S})_{\mathbf{r}_{aA}} \equiv \sum_n (E_n^A - E_0^A)\, |\, \langle n|\, V_{aA}\, |\, 0\rangle\, |^2$$

$$= \tfrac{1}{2} \langle 0|\, [V_{aA}, [H_A, V_{aA}]]|\, 0\rangle$$

$$= \frac{\hbar^2}{2M} \langle 0 | \sum_k (\nabla_k U_{1a}(| \mathbf{r}_{kA} - \mathbf{r}_{aA} |))^2 | 0 \rangle$$

$$= \frac{\hbar^2}{2M} \langle 0 | \sum_k [U'_{1a}(| \mathbf{r}_{kA} - \mathbf{r}_{aA} |)]^2 | 0 \rangle \ . \qquad (16)$$

A similar sum rule can be obtained for the different multipole components of the field $V_{aA}(\mathbf{r}_{aA})$. For this purpose we define

$$U_{1a}(| \mathbf{r}_{kA} - \mathbf{r}_{aA} |) = \sum_{\lambda\mu} g_\lambda(r_k, r_{aA}) Y_{\lambda\mu}(\hat{r}_{kA}) Y^*_{\lambda\mu}(\hat{r}_{aA}) \ , \qquad (17)$$

where

$$g_\lambda(r_k, r_{aA}) = 2\pi \int_{-1}^{1} dx \ U_{1a}(\sqrt{r_{aA}^2 + r_k^2 - 2r_{aA} r_k x}) P_\lambda(x) \ . \qquad (18)$$

We define the sum rule for mulipolarity λ by utilizing the multipole expansion of the matrix elements of V_{aA} given in Appendix A. For target excitation of even nuclei

$$\langle n | V_{aA} | 0 \rangle = \frac{(-1)^\lambda}{\sqrt{2\lambda + 1}} Y^*_{\lambda\mu}(\hat{r}_{aA}) f_\lambda^{(n)}(r_{aA}) \ . \qquad (19)$$

We define

$$(S(\lambda))_{r_{aA}} = \sum_n (s_n(\lambda))_{r_{aA}} \qquad (20)$$

with

$$s_n(\lambda) = (E_n^A - E_0^A) | f_\lambda^{(n)}(r_{aA}) |^2 \ , \qquad (21)$$

where the summation over n, at variance with (16), does not include summation over magnetic quantum numbers. Utilizing he expansion (17), we obtain [cf. Bohr and Mottelson (1975), p. 401]

$$(S(\lambda))_{r_{aA}} = \frac{2\lambda + 1}{4\pi} \frac{\hbar^2}{2M}$$

$$\times \langle 0 | \sum_k \left[\left(\frac{\partial g_\lambda}{\partial r_k} \right)^2 + \lambda(\lambda + 1) \left(\frac{g_\lambda}{r_k} \right)^2 \right] | 0 \rangle, \qquad (22)$$

where we used the relation (A.8) between f_λ and g_λ. We obtain furthermore the following relation:

$$(\bar{S})_{r_{aA}} = \frac{1}{4\pi} \sum_\lambda (S(\lambda))_{r_{aA}} \qquad (23)$$

between the quantities (20) and (16). More general relations are found in Appendix A.

As an application we consider first the case of the Coulomb interaction, where we find

$$(S(E\lambda))_{r_{aA}} = \left(\frac{4\pi Z_a e}{2\lambda + 1} \right)^2 r_{aA}^{-2\lambda-2} \sum_n (E_n^A - E_0^A) B(E\lambda; 0 \to n)$$

$$= \left(\frac{4\pi Z_a e}{2\lambda + 1} \right)^2 S(E\lambda) r_{aA}^{-2\lambda-2}, \qquad (24)$$

where $S(E\lambda)$ is defined in (II.4.16). Thus for the Coulomb field, the r_{aA}-dependence of the sum rule can be factorized and one can more conveniently use the sum rule discussed in Sec. II.4. We may also formulate this result by stating that the oscillator strength

$$f_n(E\lambda) = \frac{(s_n(E\lambda))_{r_{aA}}}{(S(E\lambda))_{r_{aA}}} = \frac{s_n(E\lambda)}{S(E\lambda)} \qquad (25)$$

is independent of the distance [cf. Eq. (II.4.17)].

In the following we shall often use the sum rules associated with a fictitious external field which has the same radial dependence as the

Coulomb field but acts equally on protons and neutrons. We define in analogy to the Coulomb case

$$s_n(\lambda; \tau = 0) = (E_n^A - E_0^A) |\langle n \| \mathscr{M}(\lambda) \| 0 \rangle|^2 , \qquad (26)$$

where $\mathscr{M}(\lambda)$ is the mass multipole moment defined by

$$\mathscr{M}(\lambda\mu) = \int \rho(\mathbf{r}')r'^{\lambda} Y_{\lambda\mu}(\hat{r}') \, d^3r' . \qquad (27)$$

The index $\tau = 0$ indicates the isoscalar character of the field, ρ being the total density. With this definition one finds the sum rule

$$S(\lambda; \tau = 0) = \sum_n s_n(\lambda; \tau = 0)$$

$$= \frac{3(2\lambda + 1)\lambda}{4\pi} \frac{\hbar^2 A}{2M} (R^{(0)})^{2\lambda - 2} \qquad (28)$$

in analogy to (II.4.16).

In a collective (macroscopic) model, where for the isoscalar modes, neutrons and protons move in phase, the multipole matrix element in (26) would be A/Ze times the electric multipole moment. On the other hand, the electric field can excite both isoscalar and isovector modes, while the operator $\mathscr{M}(\lambda\mu)$ has the selection rule $\tau = 0$. For nuclei with $N = Z$, where the excitations carry a well-defined isospin, the isoscalar multipole matrix elements would be

$$\langle n \| \mathscr{M}(\lambda) \| 0 \rangle = \frac{2}{e} \langle n \| \mathscr{M}(E\lambda) \| 0 \rangle . \qquad (29)$$

In the summation on the left-hand side of (28) one would therefore obtain a sum which is $4/e^2$ times the electric sum rule if it were not for the fact that only half of the states (i.e. those of $\tau = 0$) contribute. Thus the actual sum is $2/e^2$ times the electric sum rule, in agreement with (28) ($A = 2Z$). For $N > Z$ one might in a similar way preserve the macroscopic relation

$$\langle n \,||\, \mathscr{M}(\lambda) \,||\, 0 \rangle = \begin{cases} \dfrac{A}{Ze} \langle n \,||\, \mathscr{M}(E\lambda) \,||\, 0 \rangle & \text{for} \quad \tau = 0, \\[2em] 0 & \text{for} \quad \tau = 1, \end{cases} \tag{30}$$

if the contribution to the electric sum rule from the excitation of isovector modes $\Sigma_{n\,(\tau \approx 1)} S_n(E\lambda)$ would be N/A times the full sum rule, i.e.

$$\sum_{\substack{n \\ (\tau \approx 1)}} s_n(E\lambda) = \sum_n (E_n - E_0) \; \Big| \langle n \,(\tau \approx 1) \,||\, \mathscr{M}(E\lambda) \,||\, 0 \rangle \; \Big|^2$$

$$= \frac{N}{A} S(E\lambda). \tag{31}$$

The symbol $\langle n\,(\tau \approx 1) |$ indicates a mode of approximate isovector character.

The relations (30) and (31) are only approximately fulfilled in actual nuclei. For the case of quadrupole excitations in ^{208}Pb the results of a microscopic calculation based on the random-phase approximation are given in Fig. 12.

The factorization mentioned above for the Coulomb field does not apply to the nuclear part of the ion-ion interaction. One may still factorize the part of the sum $(S_{\text{coll}}(\lambda))_r = \Sigma_{n \in \text{coll}}(s_n(\lambda))_r$ which is associated with excitation of collective states. According to (3.15) we thus find

$$(S_{\text{coll}}(\lambda))_{r_{aA}} = \frac{2\lambda + 1}{8\pi} \hbar^2 (R_A^{(0)})^2 \left(\frac{\partial U_{aA}^N}{\partial r} \right)^2 \sum_n (D_{n\lambda})^{-1}$$

$$= \frac{\lambda(2\lambda + 1)\hbar^2}{6Z_A M} \left(\sum_n f_n(E\lambda) \right) \left(\frac{\partial U_{aA}^N}{\partial r} \right)^2 , \tag{32}$$

where we have expressed $1/D_{n\lambda}$ in terms of the electric oscillator strength [cf. Eq. (II.4.17)]. The sum in (32) can be estimated from the fact that only isoscalar states contribute, and we find

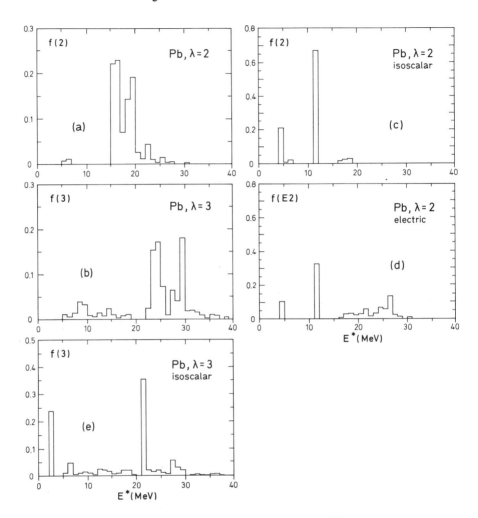

Fig. 12 Quadrupole and octupole response functions in ^{208}Pb. In (a) and (b) are shown the unperturbed strength distributions. The quantity $f_n(\lambda)$ $= s_n(\lambda, \tau = 0)/S(\lambda, \tau = 0)$ [cf. (6.26) and (6.28)], has been evaluated for $\lambda = 2$ and $\lambda = 3$ for a set of pure particle-hole configurations resulting from a Hartree-Fock calculation utilizing a Skyrme III force. The histogram indicates the isoscalar oscillator strength for energy bins of 1 MeV. In (c), (d) and (e) are given the results of an RPA calculation based on the above unperturbed single-particle states. In (c) is given the isoscalar ($\lambda = 2$) strength distribution while (d) shows the ($\lambda = 2$) electric strength distribution (6.25). In (e) is given the isoscalar oscillator

$$(S_{\text{coll}}(\lambda))_r \approx \frac{\lambda(2\lambda+1)\hbar^2}{6\,A_A M}\left(\frac{\partial U_{aA}^N}{\partial r}\right)^2. \tag{33}$$

This radial dependence is, however, different from the radial dependence of the total sum (22). In fact for large distances (33) decreases as $\exp\left[-r_{aA}/(a/2)\right]$, while the matrix element in (22), which receives its main contribution from the surface of the projectile, decreases as $\exp(-r_{aA}/a_d)$.

We may thus conclude that inelastic scattering to the collective states becomes progressively less important outside the sum of the nuclear radii. High-lying noncollective states will become important at large distances because the associated form factors decay more slowly than the collective form factors. Thus for a particle-hole state where the particle is unbound, the form factor decays approximately like $\exp(-\kappa r_{aA})$, proportional to the wavefunction of the hole outside the nucleus.

One might expect that such transitions would be hindered because of the adiabatic cutoff (cf. Sec. 1) and thus be of no consequence for inelastic scattering. However, the actual Q-value for the reaction may be quite different from the energy difference appearing in (16) or (21), because the particle may occupy a low-lying state in the projectile, thus leading to a transfer reaction. Such reactions are the subject of Chap. V.

---

strength for the $\lambda=3$ modes. A comparison of (a) and (c) (note the change of scale) shows how the residual interactions leads to a concentration of the oscillator strength mainly on two states. They correspond to the low-lying quadrupole vibration at ≈ 4 MeV and to the giant quadrupole resonance at ≈ 11 MeV. The iso-properties of the states can be seen from the comparison of (d) with (c). The isoscalar strengths of the 4-MeV and 11-MeV states are about a factor of two larger than the electric strength, which is somewhat smaller than the value expected from (30) ($A/Z \approx 2.5$). The rest of the electric oscillator strength is found in a broad region around 22 MeV. Some isoscalar strength is found close to 20 MeV, associated with iso-impurities in the broad isovector resonances around 22 MeV seen in (d). The comparison between the unperturbed particle-hole isoscalar response function for $\lambda=3$ in (b) and the correlated isoscalar response function in (e), shows a similar concentration of strength into a low-lying octupole vibration at ≈ 2.6 MeV and a giant resonance at ≈ 21 MeV. For more details on the RPA calculation see caption to Fig. 5. This figure is due to Pier Francesco Bortignon and Carlos Dasso.

7 OSCILLATOR MODEL AND ABSORPTION

In calculating the cross sections for inelastic scattering on the basis of the form factors derived in the previous sections, one faces the problem that multiple excitation will occur even for grazing collisions. One should thus solve the coupled equations (1.1) including also transfer reactions. One tries to circumvent this problem by introducing an absorptive potential in the coupled equations, which is supposed to take into account the depopulation into channels not considered, and solving a truncated set of coupled equations, e.g. by perturbation theory. We shall here show how the concept of a complex potential arises and can be calculated using a schematic model which includes only inelastic scattering of pure harmonic oscillators.

(a) Classical treatment

We assume that the total oscillator strength for inelastic scattering is exhausted by the collective modes described as pure harmonic oscillators (cf. Secs. II.4 and IV.3). In this model we can solve the coupled equations (1.1) explicitly and find the amplitude

$$a_{\{N_{\lambda\mu}^{(i)}(n)\}} = a_0 \prod_{i\lambda\mu n} (N_{\lambda\mu}^{(i)}(n)!)^{-1/2} [i(-1)^{\mu} \chi_{\lambda-\mu}^{(i)}(n)]^{N_{\lambda\mu}^{(i)}(n)} \tag{1}$$

with

$$\chi_{\lambda\mu}^{(i)}(n) = -\sqrt{\frac{\hbar\,\omega_{\lambda}^{(i)}(n)}{2C_{\lambda}^{(i)}(n)}} R_i \bar{q}_{\lambda\mu}(\omega_{\lambda}^{(i)}(n)) \tag{2}$$

and

$$a_0 = \exp\left\{ -\frac{1}{2}\sum_{i\lambda\mu n} |\chi_{\lambda\mu}^{(i)}(n)|^2 \right\}$$

$$= \exp\left\{ -\frac{1}{2}\sum_{in\lambda\mu} \frac{\hbar\,\omega_{\lambda}^{(i)}(n)}{2C_{\lambda}^{(i)}(n)} R_i^2 |\bar{q}_{\lambda\mu}(\omega_{\lambda}^{(i)}(n))|^2 \right\}. \tag{3}$$

The notation is the same as used in Sec. II.4, except that n may also indicate volume modes and that we have included by the index i, which can be equal to a or A, the possibility of target as well as projectile excitation. The quantity $\bar{q}$ is defined by

$$\bar{q}_{\lambda\mu}(\omega) = -\frac{1}{\hbar} \int_{-\infty}^{\infty} \frac{\partial U_{aA}^N}{\partial r} Y_{\lambda\mu}(\hat{r}) e^{i\omega t} \, dt . \tag{4}$$

In (1) we have neglected the Coulomb part of the interaction and we have used the macroscopic nuclear form factor.

The factor a_0 indicates the amplitude for staying in the ground state of both target and projectile, while the excitation amplitude, e.g. for exciting one phonon of type $n\lambda\mu$ in the target nucleus, would be given by

$$a_{1\lambda\mu}^{(A)(n)} = -i \sqrt{\frac{\hbar \, \omega_\lambda^{(A)}(n)}{2 C_\lambda^{(A)}(n)}} R_A (-1)^\mu \bar{q}_{\lambda -\mu}(\omega_\lambda^{(A)}(n)) a_0 . \tag{5}$$

We shall evaluate (4) by utilizing the fact that the main contribution to the integral arises from the part of the trajectory close to the classical turning point. We may here use the expansion

$$r(t) = r_0 + \tfrac{1}{2}\ddot{r}_0 t^2 , \tag{6}$$

$$\phi(t) = \dot{\phi}_0 t , \tag{7}$$

where $\phi(t)$ is the azimuthal angle in the scattering plane and r_0 is the distance of closest approach. The acceleration and the angular velocity at the turning point are denoted by $\ddot{r}_0$ and $\dot{\phi}_0$ respectively. Utilizing the empirical potential (III.1.36), we can evaluate (4) to give

$$\bar{q}_{\lambda\mu}(\omega) = \frac{1}{\hbar} \sqrt{\frac{2\pi}{a\ddot{r}_0}} U_{aA}^N(r_0) Y_{\lambda\mu}(\tfrac{1}{2}\pi,0) \exp\left\{ -\frac{(\mu\dot{\phi}_0 + \omega)^2 a}{2\ddot{r}_0} \right\} . \tag{8}$$

We have here used the coordinate system A of Fig. II.1, where the z-axis is perpendicular to the scattering plane.

According to (3) the probability for the system to be in the entrance channel after the collision is given by

$$P_0 = |a_0|^2 = \exp\left\{-\frac{2\pi a}{\ddot{r}_0 \hbar^2}\left(\frac{\partial U_{aA}^N(r_0)}{\partial r}\right)^2 \sigma^2\right\},\qquad (9)$$

where

$$\sigma^2 = \sum_{in\lambda} \frac{2\lambda+1}{4\pi}\frac{\hbar\,\omega_\lambda^{(i)}(n)}{2C_\lambda^{(i)}(n)}R_i^2 f(\omega_\lambda^{(i)}(n))\qquad (10)$$

with

$$f(\omega) = \frac{4\pi}{2\lambda+1}\sum_\mu |Y_{\lambda\mu}(\tfrac{1}{2}\pi,0)|^2 \exp\left\{-\frac{a}{\ddot{r}_0}(\mu\dot{\phi}_0 + \omega)^2\right\}.\qquad (11)$$

Apart from the factor $f(\omega)$, the quantity (10) signifies the fluctuation in the sum of the nuclear radii [cf. (3.18)]

$$\sigma = \langle (R_a + R_A - R_a^{(0)} - R_A^{(0)})^2 \rangle^{1/2}.\qquad (12)$$

The factor $f(\omega)$ is unity for the optimal Q-value and leads to an adiabatic cutoff when

$$\sqrt{\frac{a}{2\ddot{r}_0}}(\mu\dot{\phi}_0 + \omega) \gtrsim 1.\qquad (13)$$

Estimating roughly the force, $m_{aA}\ddot{r}_0$, at the turning point as the bombarding energy divided by r_0 [cf. (9.32) below], we find that (13) is equivalent to

$$\frac{\sqrt{(r_0)_{fm}}}{300v/c}(\Delta E)_{MeV} + \frac{\rho}{r_0}\frac{\mu}{\sqrt{2(r_0)_{fm}}} \gtrsim 1,\qquad (14)$$

where ρ is the impact parameter. For $v/c \approx 0.1$ and $r_0 \approx 10$ fm we find the adiabatic cutoff energy

$$(\Delta E)_{\mathrm{MeV}} + \mu \approx 10 \ . \tag{15}$$

Thus we may roughly conclude that collective modes with $\hbar \omega < 10$ MeV contribute with their full strength to the sum (11), while states above 10 MeV may contribute to the extent that the term with $\mu = -\lambda$ compensates for part of the energy loss.

From this discussion we conclude that the quantity σ in (9) is essentially the total zero-point fluctuation in the sum of the nuclear radii for all low- and medium-frequency modes. A typical estimate of σ is $\sigma \approx 0.4$ fm.

We notice now that the amplitude (5) has the form of the excitation amplitude as it would obtain in first-order perturbation theory, times the amplitude a_0 of remaining in the ground state (of all oscillators). The result (5) could therefore also be obtained by adding an imaginary potential i $W(r)$ to the interaction V_{aA} in (1.1) and solving these equations:

$$i\hbar \, \dot{a}'_n(t) = \sum_m \langle m \mid V_{aA} + iW(r) - U_{aA} \mid n \rangle$$

$$\times e^{(i/\hbar)(E_m - E_n)t} a'_m(t) \tag{16}$$

to first order in V_{aA}. In fact the general solution of (16) is

$$a'_n(t) = a_n(t) e^{(1/\hbar) \int_{-\infty}^{t} W(r(t')) \, dt'} , \tag{17}$$

where $a_n(t)$ is the solution of (1.1). The amplitude (3) for remaining in the ground state is reproduced by solving (16) to zeroth order, i.e. setting $a_n = \delta(n,0)$. In general one may reproduce the exact result (1) by solving (16) to lowest nonvanishing order, still including the effect of W according to (17).

These results do not provide a prescription for evaluating the quantity $W(r)$, except that the action integral in (17) should agree with the exponent in (3), i.e.

$$\frac{1}{\hbar} \int_{-\infty}^{\infty} W(r(t))\, dt = -\frac{1}{2} \sum_{in\lambda\mu} \frac{\hbar\, \omega_\lambda^{(i)}(n)}{2 C_\lambda^{(i)}(n)} R_i^2 \mid \bar{q}_{\lambda\mu}(\omega_\lambda^{(i)}(n)) \mid^2 . \tag{18}$$

If, in order to determine $W(r)$, one adheres to the concept of a depopulation local in time, one should enforce (18) at any time t making use of the time dependent solution of (1.1), which is also given by (3) with ∞ substituted by t in Eq. (4) [cf. Eq. (II.4.22)]. One then finds

$$\frac{1}{\hbar} \int_{-\infty}^{t} W(r(t'))\, dt' = \ln a_0(t) , \tag{19}$$

or

$$W(r(t)) = \hbar\, \frac{\dot{a}_0(t)}{a_0(t)} . \tag{20}$$

The resulting $W(r(t))$ is different for $t = \pm \mid t \mid$, although $r(t) = r(-t)$ [cf. Eq. (II.9.6)]. Compromising on the locality in time, one may enforce locality in r by using

$$W(r) = \frac{\hbar}{2} \left(\frac{\dot{a}_0(t)}{a_0(t)} + \frac{\dot{a}_0(-t)}{a_0(-t)} \right) . \tag{21}$$

which also satisfies (18). This quantity is strongly angular-momentum and energy dependent.

Since $W(r)$ only enters in the solution of the coupled equations through the integral (18), it seems more appealing to enforce the locality in r by interpreting (18) as an identity in the distance of closest approach r_0. We write the identity in the form

$$\sqrt{\frac{2r_0}{\ddot{r}_0}} \int_1^{\infty} \frac{W(r_0 x)}{\sqrt{x-1}} dx = -\frac{\pi\sigma^2 a}{\ddot{r}_0 \hbar} \left(\frac{\partial U_{aA}^N(r_0)}{\partial r} \right)^2 , \tag{22}$$

where we have used the approximation (6) for the left-hand side of (18) and the corresponding approximation [cf. (9)] for the right-hand side.

Using again the exponential form of U_{aA}^N, this integral equation for $W(r)$ is solved by the *Ansatz* $W(r) \sim e^{-2r/a}$, leading to

$$W(r) = - \sigma^2 \sqrt{\frac{\pi a}{\ddot{r}_0 \hbar^2}} \left(\frac{\partial U_{aA}^N(r)}{\partial r} \right)^2 . \tag{23}$$

Qualitatively one may understand (23) by noting that $\sigma^2 (U'_{aA})^2$ is the sum of the squares of the inelastic form factors associated with states below the adiabatic cutoff. The absorptive potential is thus essentially given by this quantity times the collision time and divided by $\hbar$, i.e.

$$\frac{1}{\hbar} W\tau = - \frac{1}{\hbar^2} \tau^2 \sum_{ni\lambda\mu} | \langle 1_{\lambda\mu}^{(i)}(n) | V_{aA} | 0 \rangle |^2 . \tag{24}$$

The expression (23) for $W(r)$ is singular for $\ddot{r}_0 = 0$, which happens when the angular momentum of relative motion is equal to the grazing value l_g of (III.3.28). This is however only a fictitious problem, since the mere existence of an imaginary potential means that the classical trajectories should be complex. The associated turning point determined from

$$E = U_{\text{eff}}(r_0) + iW(r_0) \tag{25}$$

is also complex, and the acceleration

$$m_{aA}\ddot{r}_0 = - \frac{\partial}{\partial r}(U_{\text{eff}} + iW)_{r=r_0} \tag{26}$$

is in general nonvanishing. In fact the use of the expression (23) for $W(r)$ implies a self-consistency of the strength of W and of the value of the acceleration $\ddot{r}_0$. Numerical estimates given in Sec. 9 below indicates that $\ddot{r}_0$ is fairly constant and equal to

$$\ddot{r}_0 \approx \frac{E}{m_{aA}r_0} , \tag{27}$$

as would be guessed from a dimensional argument.

(b)   Quantal treatment

The above classical model suffers from the deficiency that the trajectory of relative motion is not influenced by the excitation of the nuclei. While this is a rather good approximation for Coulomb excitation, it is not expected to be appropriate for the nuclear interaction. This is because the ion-ion potential which determines the relative trajectory is such a sensitive function of the position of the nuclear surfaces. Thus the zero-point oscillations of these surfaces as well as their evolution in time are expected to play an important role in determining the actual deflection function.

In a completely classical treatment where the surface degrees of freedom are treated on a par with the trajectory, one finds strong nonlinear effects due to the coupling between the two types of degrees of freedom. While such a treatment gives an accurate description of the average properties, it is not suited to describe the angular distribution associated with the excitation of single-quantum states. Some of the nonlinear effects can be included in a rather simple way by noticing that the imaginary potential is likely to be related to the distance between the surfaces of the two nuclei and therefore to depend on their deformation in a similar way to the ion-ion potential. Such an imaginary potential will have nondiagonal matrix elements between vibrational states and will therefore cause excitations together with the real potential.

In the following we illustrate these features by solving a simple model, where the only degree of freedom of the target nucleus is assumed to be the amplitude α of the monopole breathing mode. The nuclear radius is thus given by

$$R = R^{(0)}\left(1 + \left(\frac{1}{4\pi} \right)^{1/2} \alpha \right) . \qquad (28)$$

We furthermore assume that the frequency of this mode is so low that we may use the sudden approximation [cf. Austern and Blair (1965)]. In this case the elastic-scattering amplitude is given by

$$f_0(\vartheta) = \langle\, 0\,|\,f(\vartheta,\alpha)\,|\,0\,\rangle, \qquad (29)$$

while the scattering amplitude for exciting the one-phonon state $|1\rangle$ is equal to

$$f_1(\vartheta) = \langle 1 | f(\vartheta,\alpha) | 0 \rangle \,, \qquad (30)$$

where $f(\vartheta,\alpha)$ is the scattering amplitude for fixed value of α. For the spherically symmetric deformation (28) one may use (III.4.1) to find

$$f(\vartheta,\alpha) = \frac{i}{2k} \sum_l (2l + 1)(1 - e^{2i\beta_l(\alpha)}) P_l(\cos \vartheta) \,, \qquad (31)$$

where the phase shift depends on the nuclear deformation parameter α.
  The elastic-scattering amplitude may thus be written

$$f_0(\vartheta) = \frac{i}{2k} \sum_l (2l + 1)(1 - e^{2i\bar{\beta}_l}) P_l(\cos \vartheta) \,, \qquad (32)$$

where

$$e^{2i\bar{\beta}_l} = \langle 0 | e^{2i\beta_l(\alpha)} | 0 \rangle. \qquad (33)$$

  We may estimate the (complex) phase shift $\bar{\beta}_l$ by expanding the phase shift $\beta_l(\alpha)$ in powers of α, i.e.

$$\beta_l(\alpha) = \beta_l + \alpha \frac{\partial \beta_l}{\partial \alpha} + \tfrac{1}{2}\alpha^2 \frac{\partial^2 \beta_l}{\partial \alpha^2} \,. \qquad (34)$$

The expectation value can then be evaluated, leading to

$$\bar{\beta}_l = \beta_l + \tfrac{1}{2} \langle \alpha^2 \rangle \frac{\partial^2 \beta_l}{\partial \alpha^2} + i \left(\langle \alpha \rangle \frac{\partial \beta_l}{\partial \alpha} \right)^2 \,, \qquad (35)$$

where we have kept only terms of first order in $\langle \alpha \rangle^2 \, \partial^2 \beta_l / \partial \alpha^2$,

$$\langle \alpha \rangle = \sqrt{\frac{\hbar \omega}{2C}} \tag{36}$$

being the zero-point amplitude of the mode. The complex phase shift (35) can be reproduced by introducing a complex ion-ion potential, i.e. by making the substitution

$$U^N(r) \rightarrow U^N(r) + \Delta U^N(r) + iW(r) . \tag{37}$$

In the WKB approximation the phase shift can be evaluated according to (III.6.15), while the depopulation in the ground state is given by (3) and (4), both quantities evaluated along the complex trajectory C.

The equations which determine ΔU^N and W are therefore

$$\mathrm{Im} \int_C \sqrt{\frac{2m_{aA}}{\hbar^2}(E - U_{\mathrm{eff}} - \Delta U^N - iW)}\, dr$$

$$= \left(\langle \alpha \rangle \frac{1}{2\hbar} \int_C \frac{\partial U^N}{\partial \alpha} \frac{dr}{\sqrt{(2/m_{aA})(E - U_{\mathrm{eff}})}} \right)^2 \tag{38}$$

and a corresponding expression for the real part of the phase shift $\bar{\beta}_l$. As in the classical description, we obtain an integral relation for W for each value of l. In the quantal description, however, the imaginary potential also influences the "trajectory" on the left-hand side, giving rise to a complex turning point. By the additional requirement that W be l-independent, one may obtain a unique function of r. For small values of $W(r)$ we may thus write (38) as

$$\frac{1}{2\hbar} \int_{-\infty}^{\infty} W(r(t))\, dt = - \frac{1}{4\hbar^2} \frac{\hbar \omega}{2C} \left(\int_{-\infty}^{\infty} \frac{\partial U^N}{\partial \alpha}\, dt \right)^2 , \tag{39}$$

which is identical to (18) in the sudden approximation. If we consider this equation as an identity in the turning point r_0, we are led to the expression (23) for $W(r)$.

The correction to the real phase shift in (35) is mainly due to a renormalization of U^N due to the averaging over the fluctuating position of the nuclear surface [cf. Eq. (3.19)].

We now evaluate the scattering amplitude for the excitation of the one-phonon state:

$$f_1(\vartheta) = -\frac{i}{2k}\sum_l (2l + 1)\langle 1 \mid e^{2i\beta_l(\alpha)} \mid 0 \rangle P_l(\cos\vartheta). \quad (40)$$

We can evaluate this quantity by the technique introduced in (3.12), i.e.

$$\langle 1 \mid e^{2i\beta_l(\alpha)} \mid 0 \rangle = \sqrt{\frac{\hbar\omega}{2C}}\, \langle 0 \mid \frac{\partial}{\partial\alpha}(e^{2i\beta_l(\alpha)}) \mid 0 \rangle. \quad (41)$$

Utilizing the expansion (34), we thus find

$$\langle 1 \mid e^{2i\beta_l(\alpha)} \mid 0 \rangle$$

$$= \sqrt{\frac{\hbar\omega}{2C}}\left(2i\frac{\partial\beta_l}{\partial\alpha}\langle 0 \mid e^{2i\beta_l(\alpha)} \mid 0 \rangle + 2i\frac{\partial^2\beta_l}{\partial\alpha^2}\langle 0 \mid e^{2i\beta_l(\alpha)} \mid 0 \rangle \right)$$

$$= \left[2i\left(\frac{\hbar\omega}{2C}\right)^{1/2}\frac{\partial\beta_l}{\partial\alpha} - 4\left(\frac{\hbar\omega}{2C}\right)^{3/2}\frac{\partial\beta_l}{\partial\alpha}\frac{\partial^2\beta_l}{\partial\alpha^2} \right]_{\alpha=0} e^{2i\bar{\beta}_l}. \quad (42)$$

Using the *WKB* phase shifts, we may write

$$\langle 1 \mid e^{2i\beta_l(\alpha)} \mid 0 \rangle = \frac{i}{\hbar}\sqrt{\frac{\hbar\omega}{2C}}\Bigg\{ - \int_{-\infty}^{\infty}\frac{\partial U^N}{\partial\alpha}dt$$

$$+ \frac{i}{2\hbar}\frac{\partial}{\partial\alpha}\left(\sqrt{\frac{\hbar\omega}{2C}}\int_{-\infty}^{\infty}\frac{\partial U^N}{\partial\alpha}dt \right)^2 \Bigg\}_{\alpha=0} e^{2i\bar{\beta}_l}$$

$$= -\frac{i}{\hbar} \int_{-\infty}^{\infty} \langle 1 | [U^N(\alpha) + iW(\alpha)] | 0 \rangle \, dt \, e^{2i\bar{\beta}_l} . \qquad (43)$$

In the last expression we have introduced the imaginary potential according to (23), considering it as a function of α through the radius parameter (28). The reaction matrix for inelastic scattering is thus equal to [cf. (9.13) below]

$$a_1 = -\frac{i}{\hbar} \int_{-\infty}^{\infty} \langle 1 | (U^N + iW) | 0 \rangle \, dt . \qquad (44)$$

This result is very similar to the classical result (5), except that (44), in addition to the modification of the orbit due to W, contains the quantum-mechanical effects of the excitation of the phonon through the nondiagonal matrix of W. To the order in which the zero-point fluctuations are taken into account, W only changes the phase and therefore the angular distribution of inelastic scattering. It should be noticed that the result (43) [like (5)] is valid not only in perturbation theory, but holds also if $|a_1| > 1$.

8 CROSS-SECTIONS FOR INELASTIC SCATTERING

One of the main implications of the present chapter is that the coupling between the relative motion of the colliding ions and the surface degrees of freedom is so strong that a separation between these degrees of freedom is not possible. In particular this observation sets strong limitations on the applicability of the coupled equations (1.1), where such a separation is used.*

A way to deal with this problem is by including the surface degrees of freedom as classical variables on a par with the trajectory of relative

---

*In a quantal coupled-channel calculation the strong coupling means that the form factors as well as the diagonal matrix elements are state dependent.

motion. In this description (cf. Chap. VI) it is essential to take into account the zero-point fluctuations in the nuclear shapes in the entrance channel.

Insofar as one is only interested in the cross section for elastic scattering or for the excitation of definite low-lying vibrational or rotational states, one may achieve a separation by introducing an imaginary part in the potential. An approximate expression for this potential $W_i(r)$ is given in (7.23), i.e.

$$W_i(r) = \sigma^2 \sqrt{\frac{\pi a}{\ddot{r}_0 \hbar^2}} \left(\frac{\partial U_{aA}^N}{\partial r} \right)^2.$$ (1)

This potential has an exponential shape outside the barrier with a diffuseness parameter of $a_i = \frac{1}{2}a \approx 0.3$ fm. In addition there will be an absorption due to particle transfer between the two ions. If we use for this potential $W_t(r)$ an expression analogous to (1) but with $U^N(r)$ replaced by the single-particle-transfer form factor (cf. Chap. V), this part of the imaginary potential has a diffuseness parameter $a_t \approx 0.6$ fm. There is experimental evidence [Baltz et al. (1975), Kahana and Baltz (1977), and references therein] that the absorptive potential indeed has the two components expected from the above discussion [$W(r) = W_i(r) + W_t(r)$]. In this section we shall however only use the imaginary potentials determined empirically from elastic scattering. The quantitative estimates of $W_t(r)$ will be taken up in Chap. V.

In Fig. 13 we show the differential cross-section for a number of cases where the results of the analysis of the corresponding elastic scattering were shown in Fig. III.19. The macroscopic expression (3.15) for the form factor was used with U^N replaced by $U^N + iW$ as indicated by (7.44).

The qualitative features of the cross-sections can be understood from the elastic deflection function [cf. Fig. I.9]. For large impact parameters ($l = l_C$) only Coulomb excitation contributes. However, because of the double-valuedness of the deflection function there is a smaller impact parameter ($l = l_N$), leading to the same angle where excitation of the same state takes place mainly through nuclear interactions. The two contri-

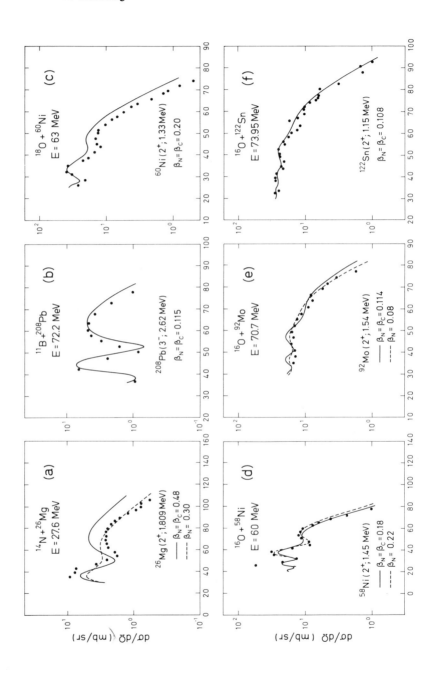

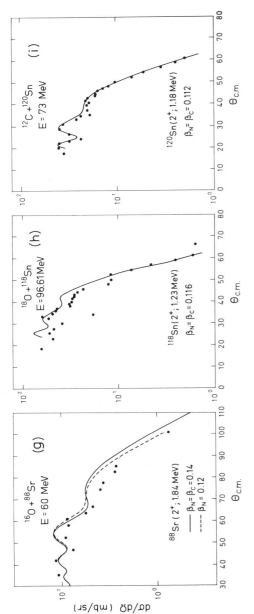

Fig. 13 Differential cross sections for inelastic scattering. The examples chosen are the same as the ones selected to illustrate angular distributions for elastic scattering in Fig. III.19. The experimental data are from the same references as were cited in the caption to this figure. The curves indicate the result of **DWBA** calculations where the optical potential used was the same as in Fig. III.19. The macroscopic form factors were used $[(3.15)$ with $U \rightarrow U + iW]$. The signature $\beta_N = \beta_C$ indicates that the standard prescription $\beta_C R^c = \beta_N R^{(0)}$ was used to determine the matrix element of $\alpha_{\lambda\mu}$ $(\beta = \sqrt{2\lambda + 1} \langle 1 | \alpha_{\lambda\mu} | 0 \rangle)$, the quantity β_C being determined from measured $B(E\lambda)$ values. This figure is due to Andrea Vitturi.

261

butions interfere and give rise to the oscillations observed in the angular distributions. For angles beyond the rainbow angle (ϑ_R) the cross-section drops exponentially into the classically forbidden region.

The overall agreement with the experimental data is striking in view of the fact that the optimal parameters utilized were the ones used to calculate the angular distribution for elastic scattering and the matrix elements of $\alpha_{\lambda\mu}$ were taken from the measured $B(E\lambda)$ value.

Some general features of the cross-section are noteworthy. Thus the oscillations in the angular distributions are small in situations where the energy of the excited state is low. In this case the inelastic process is dominated by Coulomb excitation, as was discussed in Sec. 1. One also notices that the oscillations in the inelastic cross-section are 180° out of phase with the oscillations in the corresponding elastic cross section. These two features of the Coulomb-nuclear interference will be discussed in more detail in the following section.

All examples given in Fig. 13 pertain to spherical nuclei. In Fig. 14 we show an example of elastic and inelastic scattering for a deformed nucleus. The effective ion-ion potential and the form factors for inelastic scattering were calculated from the expression (4.7)–(4.8), using for $U^N(r)$ the standard potential (4.15) plus an empirical imaginary potential.

9 COULOMB-NUCLEAR INTERFERENCE

The results displayed in Figs. 13 and 14 were calculated utilizing the standard distorted-wave Born approximation (DWBA), i.e.

$$\left(\frac{d\sigma}{d\Omega}\right)_{\alpha \to \alpha'} = \frac{k_{\alpha'}}{k_\alpha} \, | f_{\alpha' \to \alpha}(\vartheta,\varphi) |^2, \qquad (1)$$

where the scattering amplitude $f_{\alpha' \to \alpha}$ is given by

$$f_{\alpha \to \alpha'}(\vartheta,\varphi) = - \frac{m_{aA}}{2\pi\hbar^2} \int d^3 r_{aA} \, \chi^{(-)*}(\mathbf{k}'_\alpha, \mathbf{r}_{aA})$$

$$\times \langle \alpha' | V_{aA} | \alpha \rangle \, \chi^{(+)}(\mathbf{k}_\alpha, \mathbf{r}_{aA}). \qquad (2)$$

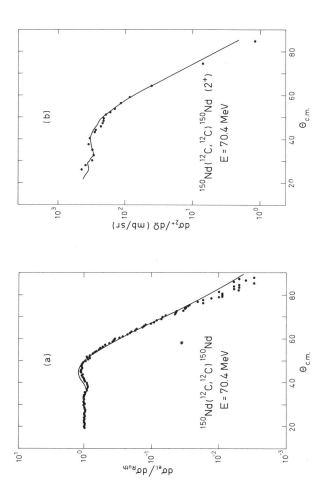

Fig. 14 Differential cross sections for elastic scattering and inelastic scattering in ^{12}C bombardment of ^{150}Nd at 70.4 MeV. The experimental data are from Hillis *et al.* (1977). The real part of the optical potential and the form factor were calculated according to (4.7) with the standard potential (4.15). The imaginary parts were calculated with the same formula from a spherical W with parameters $W_0 = 41.09$ MeV. $r_w = 1.31$ fm and $a_w = 0.433$ fm. The figure is due to Andrea Vitturi.

The functions $\chi^{(-)}$ and $\chi^{(+)}$ are the distorted plane waves in entrance and exit channel which are solutions of the wave equation for scattering in the complex potential $U^N + iW$. The expression (2) is evaluated using the partial-wave expansion [cf. (III.5.15)]

$$\chi^{(+)}(\mathbf{k},r) = (\chi^{(-)}(-\mathbf{k},r)\,)^{*}$$

$$= \frac{4\pi}{kr} \sum_{lm} i^l e^{i\beta_l} \chi_l(r) Y_{lm}^{*}(\hat{k}) Y_{lm}(\hat{r}). \tag{3}$$

For target excitation of an even nucleus one thus finds

$$f_{0 \to \lambda -\mu}(\vartheta,\varphi) = - \frac{2m_{aA}}{\hbar^2 k_\alpha k_{\alpha'}} \sum_{l_\alpha l_{\alpha'}} \langle\, l_{\alpha'}\, \mu\, \lambda - \mu \mid l_\alpha 0 \,\rangle$$

$$\times \langle\, l_{\alpha'} 0 \lambda 0 \mid l_\alpha 0 \,\rangle \sqrt{2 l_{\alpha'} + 1}$$

$$\times i^{l_{\alpha'} - l_\alpha} \exp\{i(\beta_{l_\alpha} + \beta_{l_{\alpha'}})\}\, I_{\alpha'\alpha} Y_{l_{\alpha'}\mu}(\hat{k}_{\alpha'}), \tag{4}$$

where

$$I_{\alpha'\alpha} = \int_0^\infty \chi_{l_{\alpha'}}(r) f_\lambda(r) \chi_{l_\alpha}(r)\, dr, \tag{5}$$

and where f_λ is given by (1.6) and (1.9) with U^N replaced by $U^N + iW$. We have here used a coordinate system with the z-axis along the incoming beam and have evaluated the amplitude for exciting the state of spin $\lambda, -\mu$.

In order to estimate the radial matrix element one may use the WKB approximation for the radial wave functions which were discussed in Sec. III.6 [cf. Landowne et al. (1976)]. To this end one first determines the outermost turning point in the complex r-plane and the associated regions of analyticity of the radial WKB wavefunctions. A typical situation is depicted in Fig. 15. In order to evaluate $I_{\alpha'\alpha}$ we deform the path of integration in (5) from the real axis to a path P that goes through the two turning points. The two integrals (from 0 to ∞) are equal, since the region between the two paths does not contain any singularity of the integrand.

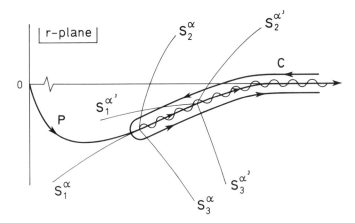

Fig. 15 Turning points and their associated Stokes lines for entrance and exit channels. The path P is used to evaluate the DWBA radial integral. The path C may be used to evaluate the WKB approximation to the radial integral, in which case the wavy line indicates the cuts associated with the turning points. The figure is taken from Landowne *et al.* (1976).

The sequence in which the path P goes through the turning points $r_{0\alpha}$ and $r_{0\alpha'}$ is such that it first encounters the point whose domain $\mathscr{D}$ (cf. Fig. III.11) contains the other turning point.

Having deformed the path of integration to P, we now substitute the WKB approximation (III.6.12) for the radial wavefunctions in (5), i.e.

$$I_{\alpha'\alpha} = \int \chi_{l_{\alpha'}}(r)f_\lambda(r)\chi_{l_\alpha}(r)\,dr$$

$$= \tfrac{1}{4}\int_{r_{0\alpha}}^{r_{0\alpha'}} \psi_{l_{\alpha'}}^{\text{in}}(r)f_\lambda(r)(\psi_{l_\alpha}^{\text{in}}(r) + \psi_{l_\alpha}^{\text{out}}(r))\,dr$$

$$+ \tfrac{1}{4}\int_{r_{0\alpha'}}^{\infty} (\psi_{l_{\alpha'}}^{\text{in}}(r) + \psi_{l\alpha'}^{\text{out}}(r))f_\lambda(r)(\psi_{l_\alpha}^{\text{in}}(r) + \psi_{l_\alpha}^{\text{out}}(r))\,dr, \quad (6)$$

where we have neglected the contribution from $r = 0$ to the innermost turning point $r_{0\alpha}$, which contains two exponentially decreasing functions. We can furthermore neglect the rapidly oscillating contributions containing the products $\psi_{l_{\alpha'}}^{in}$, $\psi_{l_\alpha}^{in}$ and $\psi_{l\alpha}^{out}$ in the second integral. In fact these contributions are of the same order of magnitude as the term neglected above, as can be shown by deforming the path of integration to follow the Stokes line $S_3^{\alpha'}$ or $S_2^{\alpha'}$ respectively and closing it at infinity, where the form factor vanishes.

The four remaining terms can be written as a single-loop integral along the path C in Fig. 15, i.e.

$$I_{\alpha'\alpha} = -\tfrac{1}{4} \int_C \psi_{l_{\alpha'}}^{in}(r) f_\lambda(r) \psi_{l\alpha}^{out}(r)\, dr$$

$$= -\tfrac{1}{4}\sqrt{k_\alpha k_{\alpha'}} \int_C dr\, f_\lambda(r) \frac{\exp\left\{ i\left[\int_{r_{0\alpha'}}^{r} k_{\alpha'}(r')dr' - \int_{r_{0\alpha}}^{r} k_\alpha(r')dr' \right]\right\}}{\sqrt{k_\alpha(r) k_{\alpha'}(r)}}.$$

$$(7)$$

This expression is very accurate for heavy-ion reactions, even for large Q-values, as was shown in Landowne et al. (1976).

If the two turning points are not too far apart, we may approximate the exponential in (7) by introducing the average turning point r_0, i.e.

$$\int_{r_{0\alpha}}^{r} k_\alpha(r')\, dr' - \int_{r_{0\alpha'}}^{r} k_{\alpha'}(r')\, dr'$$

$$\approx \int_{r_0}^{r} \left\{ (E_{\alpha'} - E_\alpha)\frac{\partial k}{\partial E} + (l_\alpha - l_{\alpha'})\frac{\partial k}{\partial l} \right\} dr', \qquad (8)$$

where k is the average wavenumber. The derivatives of k can be estimated by

$$\frac{\partial k}{\partial E} = \frac{m_{aA}}{\hbar} \frac{\partial v(r)}{\partial E} = \frac{1}{\hbar}\frac{1}{v(r)} \qquad (9)$$

and

$$\frac{\partial k}{\partial l} = \frac{m_{aA}}{\hbar} \frac{\partial v(r)}{\partial l} = -\dot{\phi}(r) \frac{1}{v(r)}, \tag{10}$$

where $v(r)$ is the average radial velocity, while $\dot{\phi} = L/(m_{aA} r^2)$ is the angular velocity.

Inserting (9) and (10) in (8), we find that the radial integral can be written

$$I_{\alpha' \alpha} = -\frac{v}{4} \int_C \frac{dr}{v(r)} f_\lambda(r)$$

$$\times \exp\left\{ i \int_{r_0}^r \left[(E_{\alpha'} - E_\alpha) \frac{dr'}{\hbar \, v(r')} - (l_\alpha - l_{\alpha'}) \dot{\phi}(r') \frac{dr'}{v(r')} \right] \right\}. \tag{11}$$

This integral is proportional to the classical first-order excitation amplitude [cf. (7.5) with $a_0 = 1$]

$$a_{l_{\lambda \mu'}}^{(A)}(\infty) = -\frac{i}{\hbar} \frac{(-1)^{\lambda + \mu'}}{\sqrt{2\lambda + 1}} \frac{4}{v} Y_{\lambda - \mu'}(\tfrac{1}{2}\pi, 0) I_{\alpha' \alpha}, \tag{12}$$

with

$$\mu' = l_\alpha - l_{\alpha'}$$

calculated along a complex trajectory in coordinate system A (cf. Fig. II.1).

The scattering amplitude (4) is more conveniently written in terms of the amplitudes in coordinate system C [cf. Fig. III.21(b).], i.e.

$$f_{0 \to \lambda \mu}(\vartheta, \varphi) = \frac{\sqrt{\pi}}{ik} \sum_l \sqrt{2l + 1} \, i^{-\mu} a_{l_{\lambda \mu}}^{(C)}(\infty) e^{2i\beta_l} Y_{l-\mu}(\vartheta, \varphi). \tag{14}$$

In deriving this formula we have used the asymptotic expression

$$\langle l_\beta m_\beta \lambda \mu | l_\alpha m_\alpha \rangle = D^\lambda_{l_\beta - l_{\alpha'} - \mu}(0, \alpha, 0), \tag{15}$$

with

$$\cos \alpha = \frac{m_\beta}{l_\beta + \frac{1}{2}} \quad (0 \le \alpha \le \pi), \tag{16}$$

which holds for l_α and $l_\beta \gg 1$. Furthermore we have used

$$\beta_{l_\alpha} = \beta_{l_{\alpha'}} + (l_\alpha - l_{\alpha'})\frac{\partial \beta_l}{\partial l} + (E_{\alpha'} - E_\alpha)\frac{\partial \beta_l}{\partial E}, \tag{17}$$

with

$$\frac{\partial \beta_l}{\partial l} = \tfrac{1}{2}\Theta(l) + i\frac{\partial(\text{Im}\,\beta_l)}{\partial l}, \tag{18}$$

where $\Theta(l)$ is the classical deflection angle [cf.(III.4.7)]. In the expression (14) we have neglected the last term in (17) and the imaginary part of (18). The amplitude $a^{(C)}$ is the first-order amplitude

$$a^{(C)}_{1_{\lambda\mu}}(\infty) = -\frac{i}{\hbar} \int_{-\infty}^{\infty} \langle 1_{\lambda\mu} | U + iW | 0 \rangle e^{i\omega t}\, dt$$

$$= \sum_{\mu'} D^\lambda_{\mu'\mu}\left(\frac{\pi + \Theta(l)}{2}, \frac{\pi}{2}, -\frac{\pi}{2}\right) a^{(A)}_{1_{\lambda\mu'}}, \tag{19}$$

evaluated in the coordinate system C [cf. Fig. III,21(b)], i.e., the same system in which the scattering amplitude is given ($\Theta(l) > 0$).

   Utilizing the expression (14) with the WKB approximation for the excitation amplitude and for the phase shift leads to an accurate description for the inelastic scattering amplitude in heavy-ion collisions. For more qualitative considerations we may use the stationary-phase method to evaluate the sum over l. Following the proceduce discussed in Sec. III.4, one obtains

$$f_{0 \to \lambda\mu}(\vartheta,\varphi) = \frac{1}{k\sqrt{\sin\vartheta}} \sum_{\bar{l}} \frac{(\bar{l} + \tfrac{1}{2})^{1/2}}{\sqrt{|\Theta'(\bar{l})|}} \, a_{\lambda\mu}^{(C)}(\infty)$$

$$\times \, e^{i[2\beta_{\bar{l}} - (l + \frac{1}{2})\Theta(l) + S(l)\frac{1}{2}\pi]} e^{-i\mu\varphi}, \qquad (20)$$

where

$$S(\bar{l}) = \begin{cases} -1 & \text{for} \quad \bar{l} = l_C, \\ 0 & \text{for} \quad \bar{l} = l_N. \end{cases} \qquad (21)$$

In deriving (20) we have used the asymptotic expression

$$Y_{lm}(\vartheta,\varphi) = \frac{1}{\pi} \frac{e^{im\varphi}}{\sqrt{\sin\vartheta}} \cos[(l + \tfrac{1}{2})\vartheta - \tfrac{1}{4}\pi + \tfrac{1}{2}m\pi]. \qquad (22)$$

The cross-section can be written

$$\frac{d\sigma}{d\Omega} = \sum_{\mu} |f_{0 \to \lambda\mu}(\vartheta,\varphi)|^2$$

$$= \sum_{\mu} \left(\frac{d\sigma_C}{d\Omega}\right)_{\mu} \left[1 + \left(\frac{d\sigma_{in}}{d\sigma_C}\right)_{\mu}\right.$$

$$\left. - 2\sqrt{\left(\frac{d\sigma_{in}}{d\sigma_C}\right)_{\mu}} \sin(\delta_c - \delta_N)\right] \qquad (23)$$

where $(d\sigma_C/d\Omega)_{\mu}$ is the cross-section for Coulomb excitation of the state $|\lambda\mu\rangle$ on the trajectory with $l = l_C$, while $(d\sigma_{in}/d\Omega)_{\mu}$ is the differential cross-section for inelastic scattering of the same state on the trajectory with $l = l_N$ [cf. Fig. I.9]. The phases are independent of μ and given by

$$\delta_C = 2\beta_{l_C} - (l_C + \tfrac{1}{2})\vartheta,$$

$$\delta_N = 2\beta_{l_N} - (l_N + \tfrac{1}{2})\vartheta. \qquad (24)$$

269

In deriving (23) we have assumed that the phases of the amplitudes are those obtained in first-order pertubation theory on real trajectories. The minus sign on the interference term comes about because the form factor for nuclear excitation generated by the attractive nuclear force has the opposite sign to the form factor for Coulomb-excitation.

Comparing (23) with the cross-section (III.4.10) for elastic scat-

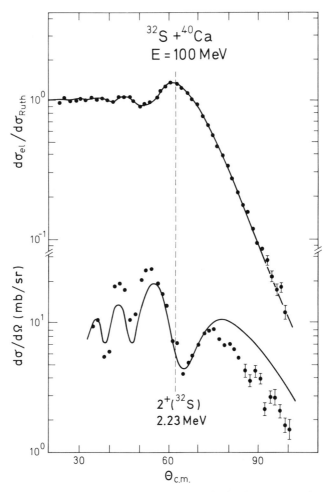

Fig. 16 Ratio of elastic to Rutherford and inelastic cross-sections for the excitation of the 2.23-MeV 2^+ state of ^{32}S in the scattering on ^{40}Ca at 100 MeV [Rehm *et al.* (1975)]. The solid curve is a DWBA fit from this reference.

tering, one observes that the oscillations in the two angular distributions are 180° out of phase [Malfliet *et al.* (1973)]. This rule is independent of the multipolarity of the excitation. A typical experimental result which shows the accuracy of the rule is shown in Fig. 16. Further examples can be obtained by comparing Fig. 13 with Fig. III.19. The deviations from the 180° rule can be understood as due to the phases we neglected, that is, (a) the last term in (17), (b) the phase associated with the fact that the scattering for $l = l_N$ takes place along a complex trajectory and (c) the phase due to higher-order effects in the excitation process.

In Fig. 17 is shown an example where target and projectile were both excited in the same experiment. In this case the first two effects mentioned above should be rather similar for the two inelastic angular distributions. It has however proved difficult to account for the observed shift between them in terms of higher-order effects (reorientation) [cf. Videbæk *et al.* (1976)]. [cf. also Vigezzi and Winther (1989)].

The magnitude of the Coulomb nuclear interference is determined by the ratio

$$\sqrt{\left(\frac{d\sigma_{\text{in}}}{d\sigma_C}\right)_\mu} = \sqrt{\left(\frac{d\sigma_N}{d\sigma_R}\right)} \left| \frac{(a_{\lambda\mu})_{\text{inel}}}{(a_{\lambda\mu})_{\text{Coul}}} \right| , \qquad (25)$$

where $d\sigma_N$ and $d\sigma_R$ are the elastic cross-sections associated with $l = l_N$ and $l = l_C$ respectively, and are discussed in Chap. III. The amplitude for Coulomb excitation is given in first-order perturbation theory by

$$(a_{\lambda\mu}(l = l_C))_{\text{Coul}} = -\frac{i}{\hbar} \int_{-\infty}^{\infty} \langle 1_{\lambda\mu} \mid V_E(r) \mid 0 \rangle e^{i\omega t} \, dt$$

$$= -i \sqrt{\frac{\hbar \omega_\lambda}{2C_\lambda}} \frac{Z_a Z_A e^2}{\hbar v} \left(\frac{R_A}{r_0}\right)^\lambda \frac{3(\lambda - 1)! 2^\lambda}{(2\lambda + 1)!!} \qquad (26)$$

$$\times \sqrt{\frac{2\lambda + 1}{\pi}} (-1)^\mu \left(\frac{1 + \varepsilon}{2}\right)^\lambda R_{\lambda-\mu}(\vartheta, \xi),$$

where we have used the results of Chap. II [e.g. (II.4.24)].

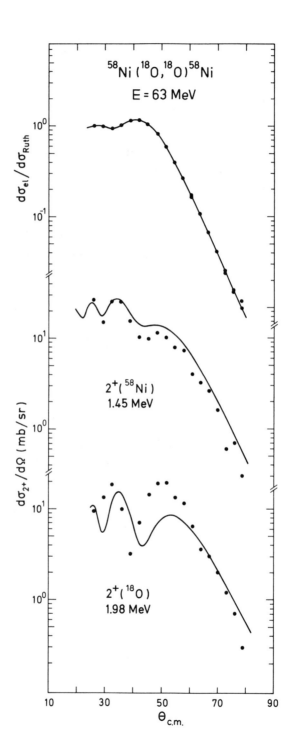

$^{58}\text{Ni}\,(^{18}\text{O},^{18}\text{O})\,^{58}\text{Ni}$

E = 63 MeV

$d\sigma_{el}/d\sigma_{Ruth}$

$d\sigma_{2^+}/d\Omega$ (mb/sr)

$2^+(^{58}\text{Ni})$
1.45 MeV

$2^+(^{18}\text{O})$
1.98 MeV

$\Theta_{c.m.}$

272

The excitation amplitude $(a_{\lambda\mu})_{\text{inel}}$ is defined by

$$(a_{\lambda\mu})_{\text{inel}} = -\frac{i}{\hbar} \int_{-\infty}^{\infty} \langle 1_{\lambda\mu} \mid V_E + V_N + iW \mid 0 \rangle e^{i\omega t}\, dt$$

$$= (a_{\lambda\mu}(l = l_N))_{\text{Coul}} + (a_{\lambda\mu}(l = l_N))_{\text{nucl}}, \qquad (27)$$

with

$$(a_{\lambda\mu}(l = l_N))_{\text{nucl}} = -\frac{i}{\hbar} \frac{(-1)^{\lambda}}{\sqrt{2\lambda + 1}} \int_{-\infty}^{\infty} f_{\lambda}(r) Y_{\lambda\mu}^*(\hat{r}) e^{i\omega t}\, dt, \qquad (28)$$

where we have used (1.5). The Coulomb-excitation amplitude in (27) is given by the same expression as (26) except that it is evaluated for a complex trajectory with $l = l_N$. In the coordinate system A the nuclear-excitation amplitude (28) is identical to the result (12). The orbital integral (11) can be estimated by making a quadratic approximation to the trajectory in the neighborhood of the turning point. As was shown in (7.5)–(7.9), this leads to the result

$$(a_{\lambda\mu}^{(A)}(l = l_N))_{\text{nucl}} = \frac{i}{\hbar} \sqrt{\frac{\hbar \omega_{\lambda}}{2C_{\lambda}}} (-1)^{\mu} R_A \frac{\partial (U + iW)}{\partial r} \bigg|_{r_0}$$

$$\times \sqrt{\frac{2\pi a}{\ddot{r}_0}}\, e^{-d_{\mu}^2/2}\, Y_{\lambda\,-\mu}(\tfrac{1}{2}\pi, 0), \qquad (29)$$

Fig. 17 Angular distributions for elastic and inelastic reactions to the lowest 2^+ states in target and projectile in the scattering of ^{18}O on ^{58}Ni at 63 MeV. The experimental data are taken from Rehm et al. (1975). The curves are results of DWBA calculations with optical parameters $V_0 = 55.44$ MeV, $a = 0.63$ fm, $r_0 = 1.18$ (fm), $W_0 = 41.66$ MeV, $a_w = 0.604$ fm and $r_w = 1.18$ fm. The macroscopic form factors were used with the strength obtained from the electromagnetic $B(E\lambda)$ value.

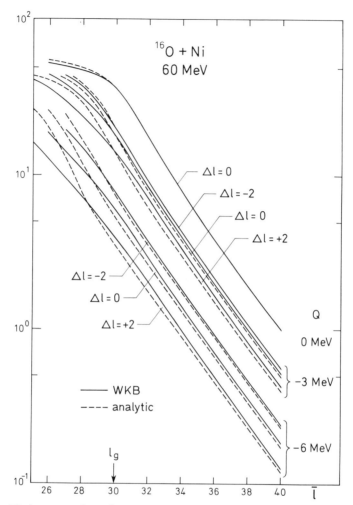

Fig. 18 A comparison between the WKB matrix element (9.7) and the quadratic approximation leading to (9.29). The modulus of the complex path integral (9.7) is given for the reaction ^{16}O on ^{58}Ni at 60 MeV for different Q-values and for different angular-momentum transfers $\Delta l = l'_\alpha - l_\alpha$ as a function of the average angular momentum $\bar{l}$ (cf. Fig. III.15). Also plotted is the modulus of the approximation $I_{appr} = \frac{1}{4} v \sqrt{2\pi a / \ddot{r}_0} f_\lambda(r_0) e^{-d^2/2}$, where d is given by (9.30), r_0 being the average complex turning point and $\ddot{r}_0$ the average (complex) acceleration at this point. The grazing angular momentum is indicated by an arrow (cf. also Figs. III.8–9). The average turning point and acceleration have been calculated as the turning point and acceleration for a trajectory with the average energy

where

$$d_\mu^2 = \frac{a}{\ddot{r}_0}(\omega - \mu\dot{\phi}_0)^2. \qquad (30)$$

The expression (29) gives rather accurate estimate of the amplitude and thus also of the WKB matrix element (7) and of the DWBA matrix element (5). A comparison of the integral (7) with the corresponding expression (29) is given in Fig. 18 for different values of l, μ, and ω in a specific case corresponding to the collision of ^{16}O on ^{58}Ni discussed in Figs. III.8–9 and III.15–16. It is seen that the accuracy is of the order of 10% or better for l-values larger than $l_g - 2$ and for $|Q| < 6$ MeV.

The expression (29) provides a decomposition of the nuclear excitation amplitude into a product of the form factor at the turning point times a collision time times an adiabatic cutoff factor, as was discussed in Sec. 7 [cf. e.g. Eq. (7.14)] in connection with real trajectories. In the present context, where the trajectory is complex, the quantity $\ddot{r}_0$ is always nonzero. We may interpret the quantity

$$\tau_{\text{char}} = \sqrt{\left|\frac{a}{\ddot{r}_0}\right|} \qquad (31)$$

as the collision time. The simple approximation

$$\ddot{r}_0 = -m_{aA}^{-1}\left(\frac{\partial U(r_0)}{\partial r} + \frac{l^2\hbar^2}{m_{aA}r_0^3}\right)$$

$$\approx \frac{2E - E_B}{m_{aA}|r_0|} \qquad (32)$$

is useful even for bombarding energies E above the barrier, as is illustrated in Fig. 19.

$E = (E_\alpha + E_{\alpha'})/2$ and average angular momentum $\bar{l} = (l_\alpha + l'_\alpha)/2$. The phase of I_{appr} is sometimes in error by about 20°. After an alternative averaging procedure where $r_0 = \frac{1}{2}(r_{0\alpha} + r_{0\alpha'})$ and $\ddot{r}_0 = \frac{1}{2}(\ddot{r}_{0\alpha} + \ddot{r}_{0\alpha'})$, the modulus of I_{appr} is very much the same as given in the figure, while the phase is more accurate and reproduces the phase of the WKB matrix element to better than 5° for the case illustrated.

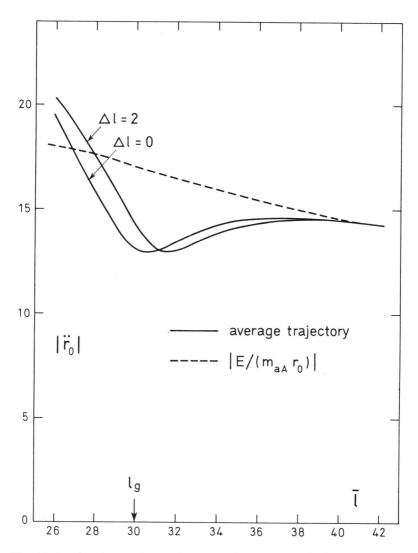

Fig. 19 Acceleration at the turning point. The modulus $|\ddot{r}_0|$ of the average acceleration is plotted in units of $\mathrm{MeV}^2\,\mathrm{fm}/\hbar^2$ as a function of the average angular momentum for the case of $Q = -6\,\mathrm{MeV}$ described in Fig. 18. The dashed curve is the result of the simple approximation $|E/(m_{aA}r_0)|$, where E is the average energy and r_0 the average turning point.

While the qualitative discussion of the relative magnitude of the two terms in (27) was given in Sec. 1, we can now estimate the magnitude of the Coulomb-nuclear interference in the angular distribution (23), making use of the fact that the Coulomb excitation amplitude (26) does not depend very strongly on l. We may thus estimate the magnitude of the interference (25) by

$$\sqrt{\left(\frac{d\sigma_{in}}{d\sigma_C}\right)_\mu} \approx \sqrt{\left(\frac{d\sigma_N}{d\sigma_R}\right)}$$

$$\times \left[1 - \tfrac{2}{3}(\lambda + \tfrac{1}{4})^{3/2} \sqrt{\frac{a}{r_0}} \left(\frac{r_0}{R_A}\right)^{\lambda-1} \frac{e^{-d_\mu^2/2}}{J_\lambda(\tilde{\xi}_\mu(\vartheta))} \right], \qquad (33)$$

where the function J_λ is given in (II.3.6). It is noted that $\tilde{\xi}_\mu(\vartheta) \approx \sqrt{r_0/(2a)}d_\mu$. The slow variation of the Coulomb form factor along the trajectory, as compared to the nuclear form factor, implies that the function J_λ depends more sensitively on the Q-value mismatch.

The μ-dependence of the cross-sections can be measured by coincidence experiments between the scattered particle and the subsequently emitted gamma-quantum. An example of this is shown in Fig. 20. It is seen that the cross-section for the substate with $\mu = -2$, where the Q-value mismatch is largest, shows the most violent interference effects. The total inelastic cross-section does not show large oscillations, since it is dominated by the cross-section associated with $\mu = +2$.

For low-lying excited states where the sudden approximation applies, both nuclear excitation and Coulomb excitation will lead to a final nuclear state which has the angular momentum isotropically distributed in a plane perpendicular to the symmetry axis of the trajectory, i.e. has $\mu = 0$ in the coordinate system B of Fig. II.2.

APPENDIX A   GENERAL DEFINITION OF RADIAL
FORM FACTORS AND SUM RULES

The matrix element appearing in (1.1) is specified by the spins and magnetic quantum numbers $I_A M_A(I'_A M'_A)$ and $I_a M_a(I'_a M'_a)$ of the target and projectile respectively. The form factor is then defined by

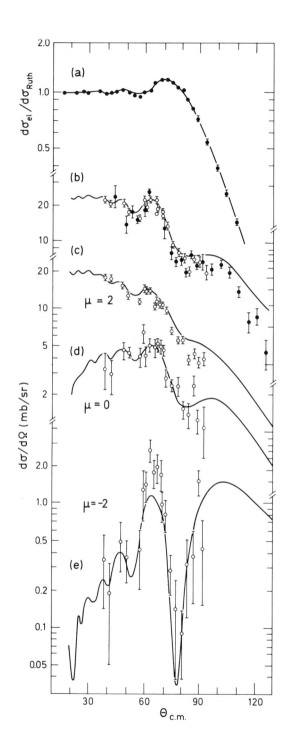

278

$$\langle \psi^a_{I_a M_a} \psi^A_{I'_A M'_A} \mid V_{aA} - U_{aA} \mid \psi^a_{I_a M_a} \psi^A_{I_A M_A} \rangle = f_{\alpha'\alpha}(\mathbf{r}_{aA})$$

$$= \sum_{\substack{JJ' \\ MM'\lambda\mu}} \langle I_A M_A JM \mid I'_A M'_A \rangle \langle I'_a M'_a J'M' \mid I_a M_a \rangle$$

$$\times \langle \lambda\mu JM \mid J'M' \rangle f^{(I_A J)I'_A,(I'_a J')I_a}_{\lambda\mu} (\mathbf{r}_{aA}) \tag{1}$$

or the reciprocal relation

$$f^{JJ'}_{\lambda\mu}(\mathbf{r}_{aA}) \equiv f^{(I_A J)I'_A,(I'_a J')I_a}_{\lambda\mu}(\mathbf{r}_{aA})$$

$$= \frac{(2J+1)(2\lambda+1)}{(2I'_A+1)(2I_a+1)} \sum_{M_A M'_A M_a M'_a MM'} \langle I_A M_A JM \mid I'_A M'_A \rangle$$

$$\times \langle I'_a M'_a J'M' \mid I_a M_a \rangle \langle \lambda\mu JM \mid J'M' \rangle f_{\alpha'\alpha}(\mathbf{r}_{aA}). \tag{2}$$

The quantities J, M and J', M' are the angular momenta transferred to target and from projectile respectively, while λ, μ is the angular momentum transferred to the relative (orbital) from the intrinsic motion. Since the function $f^{JJ'}_{\lambda\mu}(\mathbf{r}_{aA})$ is a tensor of rank λ, one can obtain the explicit dependence on two of its variables, utilizing an intrinsic system whose z-axis is along the relative position vector $\mathbf{r}_{aA}$, i.e.

$$f^{JJ'}_{\lambda\mu}(\mathbf{r}_{aA}) = \sum_{\mu'} D^\lambda_{\mu\mu'}(\varphi,\theta,0)[f^{JJ'}_{\lambda\mu'}(r_{aA})]_{\text{intr}}. \tag{3}$$

The form factor $[f^{JJ'}_{\lambda\mu}(r)]_{\text{intr}}$ is given by the same expression (2) except that it is evaluated in a coordinate system where $\mathbf{r}_{aA}$ is the quantization axis. Because of the axial symmetry around this vector, only the term with $\mu' = 0$ in (3) contributes, and we thus define the radial form factor by the equation

---

Fig. 20 Elastic and inelastic angular distributions for the scattering of ^{16}O on ^{56}Fe at 43 MeV. In (a) is shown the ratio of the elastic-scattering cross section to the Rutherford cross section. In (b) is shown the differential cross section for the excitation of the 0.856-keV 2^+ state in ^{56}Fe. The differential cross sections for the excitation of the different magnetic substates of $\mu = 2, 0$ and -2 in the coordinate system (A) with z-axis along $\mathbf{k}_f \times \mathbf{k}_i$ are given in (c)–(e). [Steadman et al. (1974).]

$$f_{\lambda\mu}^{JJ'}(\mathbf{r}_{aA}) = f_{\lambda}^{JJ'}(r_{aA})Y_{\lambda\mu}(\hat{r}_{aA}), \qquad (4)$$

where

$$f_{\lambda}^{JJ'}(r_{aA}) = \frac{4\pi}{\sqrt{2\lambda+1}}[f_{\lambda 0}^{JJ'}(\mathbf{r}_{aA})]_{\text{intr}}. \qquad (5)$$

For target excitation the energy weighted sum rule (6.16) takes the form

$$(\bar{S})_{r_{aA}} = \sum_{I'_A M'_A} (E_{I'_A} - E_{I_A})|\langle I'_A M'_A \mid V_{aA} - U_{aA}\mid I_A M_A\rangle|^2$$

$$= \frac{\hbar^2}{2M}\langle I_A M_A \mid \sum_k [U'_{ka}(\mid \mathbf{r}_{ka} - \mathbf{r}_{aA}\mid)]^2 \mid I_A M_A\rangle, \qquad (6)$$

which in general depends on M_A and thus on the direction of $\mathbf{r}_{aA}$.

The sum (6.20) is in general defined by

$$(S(\lambda))_{r_{aA}} = \frac{1}{(2\lambda+1)(2I_A+1)}\sum_{I'_A M'_A}(E_{I'_A}^A - E_{I_A}^A)|f_{\lambda}(r_{aA})|^2. \qquad (7)$$

Utilizing the expression

$$f_{\lambda}(r_{aA}) = (-1)^{\lambda}\left(\frac{2\lambda+1}{2I'_A+1}\right)^{1/2}$$

$$\langle I'_A \parallel \sum_k g_{\lambda}(r_{kA},r_{aA})Y_{\lambda}(\hat{r}_{kA}) \parallel I_A\rangle, \qquad (8)$$

for the form factor in terms of the function g_{λ} defined in (6.18), we find

$$(S(\lambda))_{r_{aA}} = \sum_{I'_A M'_A \mu}(E_{I'_A} - E_{I_A})$$

$$\times |\langle I'_A M'_A \mid \sum_k g_{\lambda}(r_{kA},r_{aA})Y_{\lambda\mu}(\hat{r}_{ka}) \mid I_A M_A\rangle|^2$$

$$= \frac{2\lambda + 1}{4\pi} \frac{\hbar^2}{2M}$$

$$\times \langle I_A M_A \mid \sum_k \left[\left(\frac{\partial g_\lambda}{\partial r} \right)^2 + \lambda(\lambda + 1) \left(\frac{g_\lambda}{r} \right)^2 \right] \mid I_A M_A \rangle \mid^2, \quad (9)$$

which is independent of M_A.

The connection between (6) and (7) is

$$\sum_\lambda S_\lambda(r_{aA}) = \frac{4\pi}{2I_A + 1} \sum_{M_A} \bar{S}(\mathbf{r}_{aA}). \quad (10)$$

APPENDIX B   INTERACTION BETWEEN TWO DEFORMED NUCLEI BY YUKAWA FOLDING

The Yukawa folding technique described in Appendix III.A can be used advantageously also for deformed systems [cf. Davies and Nix (1976), Arnould and Howard (1976)]. We thus consider the interaction

$$V_{aA} = \int \rho^a(\mathbf{r}_1)\rho^A(\mathbf{r}_2) \, V_{12}(\mid \mathbf{r} + \mathbf{r}_1 - \mathbf{r}_2 \mid) \, d^3r_1 \, d^3r_2,$$

$$(1)$$

where the two densities ρ^a and ρ^A are not necessarily spherically symmetric. Assuming

$$V_{12}(r) = S\frac{e^{-\kappa r}}{r}, \quad (2)$$

Eq. (1) represents the nuclear interaction between the ions if ρ^a and ρ^A are the mass densities and V_{12} the effective two-body interaction. The Coulomb interaction between the two systems is obtained for $\kappa = 0$ and $S = 1$ if ρ^a and ρ^A are the charge densities.

In order to evaluate (1) we use the Fourier representation

$$\frac{e^{-\kappa|r+r_1-r_2|}}{|r+r_1 - r_2|} = \int d^3k \frac{e^{ik\cdot(r+r_1-r_2)}}{2\pi^2(\kappa^2 + k^2)} \tag{3}$$

of the interaction (2). The double integral (1) thus separates into an integral over r_1 and an integral over r_2. Utilizing

$$e^{ik\cdot x} = \sum_{\lambda\mu} 4\pi i^\lambda j_\lambda(kx) Y_{\lambda\mu}(\hat{x}) Y_{\lambda -\mu}(\hat{k})(-1)^\mu \tag{4}$$

and integrating over the direction of **k** we find the result

$$V_{aA}(\mathbf{r}) = 16\sqrt{\pi}S \sum_{\substack{\lambda_1\lambda_2 \\ \mu\mu_1\mu_2}} (-1)^{(\lambda+\lambda_1-\lambda_2)/2}$$

$$\times \frac{\sqrt{(2\lambda + 1)(2\lambda_1 + 1)(2\lambda_2 + 1)}}{(2\lambda_1 + 1)!!(2\lambda_2 + 1)!!}$$

$$\times Y_{\lambda\mu}(\hat{r}) \begin{pmatrix} \lambda_1 & \lambda_2 & \lambda \\ \mu_1 & \mu_2 & \mu \end{pmatrix} \begin{pmatrix} \lambda_1 & \lambda_2 & \lambda \\ 0 & 0 & 0 \end{pmatrix}$$

$$\times \int dk \frac{k^{\lambda_1+\lambda_2+2}}{\kappa^2 + k^2} j_\lambda(kr) \mathcal{M}_a(k, \lambda_1\mu_1) \mathcal{M}_A(k, \lambda_2\mu_2). \tag{5}$$

The k-dependent multipole moments are defined by

$$\mathcal{M}(k, \lambda\mu) = \frac{(2\lambda + 1)!!}{k^\lambda} \int \rho(\mathbf{r}) j_\lambda(kr) Y_{\lambda\mu}(\hat{r}) \, d^3r \tag{6}$$

in such a way that they agree with the multipole moments (II.1.9) for $k = 0$, i.e.

$$\mathcal{M}(0, \lambda\mu) = \mathcal{M}(\lambda\mu) = \int \rho(\mathbf{r}) r^\lambda Y_{\lambda\mu}(\hat{r}) \, d^3r.$$

Note that for $\kappa = 0$ one may perform the integration over k. When $r > r_1 + r_2$ one may use

$$\int dk \, j_\lambda(kr) j_{\lambda_1}(kr_1) j_{\lambda_2}(kr_2)$$

$$= \frac{\pi^{3/2}}{8} \frac{r_1^{\lambda_1} r_2^{\lambda_2}}{r^{\lambda_1 + \lambda_2 + 1}} \frac{\Gamma\left(\dfrac{\lambda + \lambda_1 + \lambda_2 + 1}{2}\right)}{\Gamma\left(\dfrac{2\lambda_1 + 3}{2}\right) \Gamma\left(\dfrac{2\lambda_2 + 3}{2}\right)} \delta(\lambda, \lambda_1 + \lambda_2), \tag{7}$$

to obtain the familiar result (Alder and Winther 1975) for the interaction between two deformed charge distributions.

While the result (5) holds for any charge or mass distribution $\rho(r)$, a further simplification is achieved if the density can be written as the folding of a Yukawa function with a sharp uniform distribution $\rho_0(\mathbf{r}) = \rho_0 \vartheta(R(\mathbf{r}) - r)$, where R is a function of the direction of $\mathbf{r}$, i.e.

$$\rho(\mathbf{r}) = \rho_0 * \frac{1}{4\pi a_d^2} Y\left(\frac{1}{a_d}\right)$$

$$= \frac{\rho_0}{4\pi a_d^2} \int \vartheta(R(\hat{r}) - r') \frac{\exp(|\mathbf{r} - \mathbf{r}'|/a_d)}{|\mathbf{r} - \mathbf{r}'|} d^3 r'. \tag{8}$$

Inserting (8) in (6) one finds

$$\mathcal{M}(k, \lambda\mu) = \frac{1}{1 + a_d^2 k^2} \mathcal{M}_0(k, \lambda\mu), \tag{9}$$

where $\mathcal{M}_0$ is the k-dependent multipole moment corresponding to a sharp uniform distribution

$$\mathcal{M}_0(k, \lambda\mu) = \frac{(2\lambda + 1)!!}{k^\lambda} \int \rho_0(\mathbf{r}) j_\lambda(kr) Y_{\lambda\mu}(\hat{r}) \, d^3 r$$

$$= \rho_0 \frac{(2\lambda + 1)!!}{k^{\lambda+3}} \int d\Omega \, F_\lambda(kR(\hat{r})) Y_{\lambda\mu}(\hat{r}) \tag{10}$$

with

$$F_\lambda(z) = \int_0^z x^2 j_\lambda(x)\, dx.$$ (11)

It is noted that in the limit of $k \to 0$

$$\mathcal{M}(\lambda\mu) = \mathcal{M}_0(\lambda\mu).$$ (12)

For the Yukawa folded densities the usual multipole moments for a diffuse surface are therefore the same as the ones corresponding to a sharp surface.

Part II:
Transfer Reactions

CHAPTER V

TRANSFER REACTIONS

Processes in which nucleons are transferred between target and projectile play a dominant role in heavy-ion collisions. Such transfer reactions take place when the surfaces of the two nuclei come to distances where the tail of the wavefunctions of the nucleons in one of the systems starts to overlap with the attractive nuclear field of the other system.

In fact, these processes take place at distances larger than those relevant for nuclear inelastic scattering. This is because the associated form factors being proportional to the single-particle wavefunctions have a longer range than the form factors for inelastic scattering, which are proportional to the product of two single-particle wavefunctions (cf. Fig. 1).

In grazing collisions where the two nuclei only have a small overlap, transfer reactions are relatively weak, and one can make detailed spectroscopic studies of discrete states. The information which can be obtained this way about the tail of the form factors differs in an important way from similar information obtained with light-ion reactions in that the transfer processes almost always take place together with inelastic processes, especially Coulomb excitation (cf. Fig. II.4 (Vol. I)).

In closer collisions where the transfer reactions become very prolific, the properties of the individual states lose their relevance and phase space considerations become more appropriate. Such reactions play an important role in the damping of the relative motion leading to deep inelastic and fusion reactions. The transfer reactions determine the frictional forces acting between the nuclear surfaces. They also usually control the depopulation of elastic channels and thus provide the main component of the absorptive potential to be used in the description of grazing reactions.

Transfer reactions with heavy ions are very sensitive to the tail of the nuclear wavefunctions, and thus to the radius and diffusivity of the shell model potential. Standard shell model potentials, used for nuclear structure calculations, tend to overestimate this tail. Therefore, there is need for a systematic study of the absolute cross section of transfer

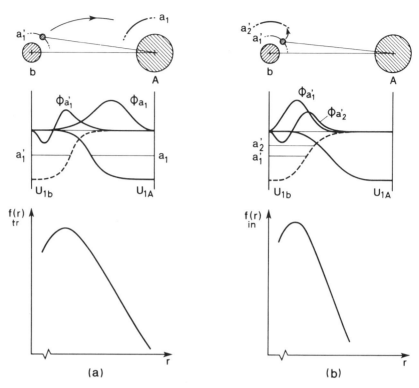

Fig. 1 Schematic representation of the radial dependence of the one-particle transfer and inelastic form factors. In (a) a nucleon moving in the orbital with quantum numbers a_1' in the projectile a is transferred under the action of the shell model potential U_{1A} to the target nucleus A into an orbital a_1. The dependence of the form factor on the distance between the two nuclei is determined by the overlap of the product of the single-particle wavefunctions $\phi_{a_1'}$ and ϕ_{a_1} with the potential U_{1A}. A schematic representation of this dependence is given at the bottom of (a). In (b) a nucleon in the projectile a is excited under the influence of the target field U_{1A} from the single-particle orbital with quantum numbers a_1' to the orbital with quantum numbers a_2'. The dependence of the form factor on the distance between the cores is here determined by the overlap of the product of the functions $\phi_{a_1'}$ and $\phi_{a_2'}$ with the potential U_{1A}. A representation of this dependence is shown at the bottom of (b).

reactions, which goes beyond the examples given in this chapter.

In the present chapter, as in previous chapters, the main emphasis is put on reactions at bombarding energies of the order of, or less than, 10 MeV per nucleon. Semiclassical description are also appropriate at higher energies where one may use further simplifications (cf., e.g., Hasan and Brink 1978, Lo Monaco and Brink 1985, and Bonaccorso et al. 1987).

1 SEMICLASSICAL THEORY OF TRANSFER

Transfer processes are already included in the description provided by the semiclassical coupled equation (IV.1.1), when one includes among the states n and n' also continuum states.

To illustrate this point, we write down these coupled equations for the simple model where the nucleus A is described by the motion of a single nucleon in a shell-model potential U_{1A}, while the projectile is represented by a potential which moves along a prescribed classical trajectory $\vec{R}(t)$, that is

$$
\begin{aligned}
i\hbar\dot{a}_n(t) = &\sum_m \langle \psi_n^A \mid U_{1a} \mid \psi_m^A \rangle \, e^{i(E_n - E_m)t/\hbar} a_m(t) \\
&+ \int d^3k \langle \psi_n^A \mid U_{1a} \mid \psi^A(\vec{k}) \rangle e^{i(E_n - E(k))t/\hbar} a(\vec{k}; t),
\end{aligned}
\tag{1a}
$$

and

$$
\begin{aligned}
i\hbar\dot{a}(\vec{k}; t) = &\sum_m \langle \psi^A(\vec{k}) \mid U_{1a} \mid \psi_m^A \rangle e^{i\left(E(k) - E_m\right)t/\hbar} a_m(t) \\
&+ \int d^3k' \langle \psi^A(\vec{k}) \mid U_{1a} \mid \psi^A(\vec{k}') \rangle e^{i\left(E(k) - E(k')\right)t/\hbar} a(\vec{k}'; t).
\end{aligned}
\tag{1b}
$$

The wavefunction $\psi^A(\vec{k})$ describes an eigenstate in the continuum in the shell-model potential $U_{1A}(r_{1A})$ with energy $E(k) = \hbar^2 k^2/2M$, while ψ_n^A is the wavefunction of a bound state with energy $E_n < 0$. The nuclear field generated by the projectile a depends on the position of the coordinate $\vec{r}_1$ through $U_{1a} = U_{1a}(\vec{r}_{1A} - \vec{R}(t))$. Similar models were discussed in Breit and Ebel (1956), Bertsch and Schaeffer (1977), and Esbensen et al. (1983).

The first term in (1a) describes inelastic processes among bound states while the first term in (1b) describes the excitation of continuum states. The feedback from these states to the bound states is described by the second term in (1a), while the second term in (1b) takes into account the redistribution among continuum states. Although these equations are a simple generalization of the coupled equations used in previous chapters, they are fundamentally different in structure through the fact that the continuum-continuum matrix elements do not necessarily vanish at $t = +\infty$ since the continuum wavefunctions have non-vanishing amplitudes at large distances from A. Coherent superpositions of the continuum wavefunctions can lead to localized wavepackets which follow the potential U_{1a}. This coherent state can be maintained up to $t = +\infty$ if the potential is attractive and displays single-particle bound states.

We can eliminate this difficulty by extracting from the states $\psi^A(\vec{k})$, the components which asymptotically coincide with the bound states ψ_n^a in a. These components $\widetilde{\psi_n^a}$ can be constructed by solving the coupled equations (1), integrating them from $t = +\infty$ backwards in time with initial conditions $G(\infty)\psi_n^a$, where G is the Galilean transformation corresponding to the asymptotic velocity of the potential U_{1a}.

Making use of the complete set of states formed by ψ_m^A, $\widetilde{\psi_n^a}$ and $\left(1 - \sum | \widetilde{\psi_n^a} \rangle \langle \widetilde{\psi_n^a} | \right) \psi^A(\vec{k})$, one obtains coupled equations where the coupling only acts for a short period of time around $t = 0$. Since these equations are rather cumbersome, we shall instead of $\widetilde{\psi_n^a}$ use the states $G(t)\psi_n^a$, where $G(t)$ is the local Galilean transformation which transforms the eigenstates of a into the frame of the potential U_{1A}, i.e.

$$G(t) = \exp \frac{i}{\hbar} \left\{ M\vec{v}(t) \cdot \vec{r}_{1A} - \int_0^t \frac{1}{2} M \left(\vec{v}(t') \right)^2 dt' \right\}$$

$$\times \exp \left\{ -\frac{i}{\hbar} \vec{R}(t) \cdot \vec{p} \right\}, \tag{2}$$

where $\vec{p} = \sum \vec{p}_i$ is the total momentum operator and M is the mass of the nucleon. The problems associated with non-orthogonality between ψ_n^A and $G(t)\psi_n^a$ at $t \approx 0$ is simpler to deal with than those associated with the construction of the states $\widetilde{\psi^a}$.

In many cases, one expects that making use of the basis $\{\psi_n^A, G(t) \psi_m^a\}$ one would be able to describe the solutions of the Schrödinger equation with sufficient accuracy. The truncation of the basis states to the bound states in the two potentials may under circumstances be too restrictive. The set may be supplemented with continuum states $\psi_{\lambda\mu}(k)$

of any finite angular momentum λ since such states are orthogonal to the states $G(t)\psi_n^a$ for $t \to \infty$, implying that all couplings vanish asymptotically.

To obtain a realistic description of transfer processes, one needs to include the reaction of the transfer on the relative motion. Thus, the model discussed above is insufficient. In the remaining part of this section, we shall therefore use the scheme discussed above but try to derive improved equations which include effects of transfer of mass, energy, and angular momentum on the trajectory. Because of the quantal nature of this process, we shall study the quantal description in the limit of small wavelength of relative motion.

We thus consider the general situation in which the nucleus a impinges on the nucleus A and where the two nuclei in the exit channel b and B may differ from the nuclei in the entrance channel by the transfer of one or several nucleons.

In the center-of-mass system, the total Hamiltonian may be written as

$$H = T_{aA} + H_a + H_A + V_{aA}, \tag{3}$$

where T_{aA} is the kinetic energy of relative motion

$$T_{aA} = -\frac{\hbar^2}{2m_{aA}} \nabla_{aA}^2, \tag{3a}$$

m_{aA} being the reduced mass. The Hamiltonians describing the intrinsic motion in nucleus a and A are denoted by H_a and H_A, respectively while V_{aA} is the effective interaction between the nucleons in a and the nucleons in A.

The Hamiltonian may also be written as

$$H = T_{bB} + H_b + H_B + V_{bB} \tag{4}$$

where the quantities have similar significance for the nuclei b and B.

We shall solve the time-dependent Schrödinger equation

$$i\hbar \frac{\partial \psi}{\partial t} = H\psi \tag{5}$$

with the initial condition that the nuclei a and A are in their ground states, and where the relative motion is described by a narrow wavepacket of rather well-defined impact parameter and velocity.

V   Transfer Reactions

We expand ψ on the channel wavefunctions

$$\psi_\beta(t) = \psi_m^b(\zeta_b)\psi_n^B(\zeta_B)\exp(i\delta_\beta) \qquad (6)$$

where ψ^b and ψ^B are the intrinsic wavefunctions of the two nuclei satisfying the equations

$$H_b\psi_m^b(\zeta_b) = E_m^b\,\psi_m^b(\zeta_b) \qquad (7)$$

and

$$H_B\psi_n^B(\zeta_B) = E_n^B\,\psi_n^B(\zeta_B) \qquad (8)$$

while ζ_b and ζ_B denote the intrinsic coordinates of the two systems (cf. Fig. 2).

The phase δ_β is defined by

$$\delta_\beta = \frac{1}{\hbar}\left\{m_\beta\vec{v}_\beta(t)\cdot\left(\vec{r}_\beta - \vec{R}_\beta(t)\right)\right.$$

$$\left. - \int_o^t \left(U_\beta\left(R_\beta(t')\right) - \frac{1}{2}m_\beta\left(\vec{v}_\beta(t')\right)^2\right)dt'\right\}. \qquad (9)$$

The index β labels both the partition of nucleons into b and B, as well as the quantal states of the two nuclei. The quantity U_β is the ion-ion potential in this channel. It is equal to the expectation value of $V_\beta = V_{bB}$ in the channel β. The distance between the centers of mass of the two systems is denoted by

$$\vec{r}_\beta \equiv \vec{r}_{bB} = \vec{r}_b - \vec{r}_B. \qquad (10)$$

The quantity $\vec{R}_\beta$ and its derivative $\vec{v}_\beta = \dot{\vec{R}}_\beta$ are supposed to describe the motion of the centers of the wavepackets, and satisfy the classical equation of motion

$$m_\beta\dot{\vec{v}}_\beta = -\vec{\nabla}_\beta U_\beta(\vec{R}_\beta). \qquad (11)$$

The initial conditions are discussed below.

The phase factor $\exp(i\delta_\beta)$ in the channel wavefunction is essentially a Galilean transformation where an additional phase has been added to eliminate, as far as possible, the diagonal matrix elements in the coupled equations.

Using the notation $E_\beta = E_m^b + E_n^B$ and inserting the ansatz

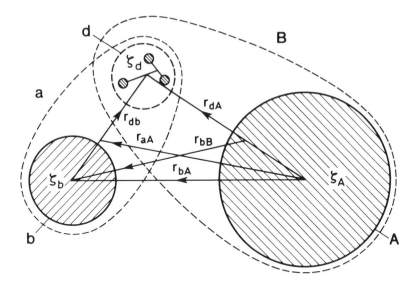

Fig. 2 The connection between the core-core coordinate $\vec{r}_{bA}$, the relative center-of-mass coordinates $\vec{r}_{aA}$ and $\vec{r}_{bB}$, and the center-of-mass of the transferred nucleons $\vec{r}_d$. For the case of a single nucleon transfer $\vec{r}_{db}$ and $\vec{r}_{dA}$ coincide with the corresponding coordinate of the transferred nucleon. In the calculations of the form factors describing the transfer of many nucleons all the degrees of freedom of the transferred nucleons are explicitly taken into account. The intrinsic coordinates ζ_d stand for the relative coordinate of the transferred nucleons as well as for their spin coordinates. The same applies to ζ_b and ζ_A, which describe the intrinsic coordinates of the nuclei b and A.

$$\psi = \sum_\beta c_\beta\big((r_\beta - R_\beta), t\big)\psi_\beta(t)e^{-iE_\beta t/\hbar}, \qquad (12)$$

in the time-dependent Schrödinger equation one obtains the following set of coupled equations for the coefficients c:

$$\sum_\beta \psi_\beta \left\{ \frac{\hbar^2}{2m_\beta}\nabla^2 + i\hbar\left(\frac{\partial}{\partial t} + \vec{v}_\beta \cdot \vec{\nabla}\right)\right\} c_\beta e^{-iE_\beta t/\hbar}$$

$$= \sum_\beta \left\{ \big(V_\beta - U_\beta(\vec{R}_\beta(t))\big) + m_\beta \dot{\vec{v}}_\beta \cdot \big(\vec{r}_\beta - \vec{R}_\beta(t)\big)\right\} c_\beta \psi_\beta e^{-iE_\beta t/\hbar}. \quad (13)$$

These equations should be solved with the initial condition that $c_\alpha(t = -\infty)$ represents a wave packet in the entrance channel concentrated around $\vec{r}_\alpha = \vec{R}_\alpha(t)$ impinging on the target nucleus and at the same time that $c_\beta(t = -\infty) = 0$ for $\beta \neq \alpha$. At time $t \approx 0$, when the interaction energy in (13) becomes non-vanishing over the volume of the wavepacket, new wavepackets develop in other channels.

We shall express the function

$$c_\beta = a_\beta(t)\chi_\beta(\vec{r}_\beta - \vec{R}_\beta(t), t), \tag{14}$$

as a product of an amplitude a_β and a shape function χ_β which is normalized according to

$$\int d^3r \mid \chi_\beta(\vec{r} - \vec{R}_\beta(t), t) \mid^2 = 1. \tag{15}$$

We assume that the shape functions develop in time as free wavepackets i.e.

$$i\hbar\left(\frac{\partial}{\partial t} + \vec{v}_\beta \cdot \vec{\nabla}_\beta\right)\chi_\beta = -\frac{\hbar^2}{2m_\beta}\nabla^2\chi_\beta. \tag{16}$$

The elastic trajectory $\vec{R}_\beta(t)$ is only in part determined by the energy and angular momentum. For the rest we see from Eq. (13) that χ_β is created around $t \simeq 0$, and that it here has a perfect overlap with the wavepackets associated with those channels that mainly feed the channel β.

In order to determined the coefficients a_β, we multiply Eqs. (13) by $\psi_\gamma^* \chi_\gamma^*(\vec{r}_\gamma - \vec{R}_\gamma(t), t)$ and integrate over the set of coordinates $\{\zeta_f, \zeta_F, \vec{r}_\gamma\}$ associated with the channel γ. We thus find

$$i\hbar \sum_\beta \int d^3r_\gamma d\zeta_f d\zeta_F \psi_\gamma^*(\zeta_f \zeta_F \vec{r}_\gamma t) e^{-iE_\beta t/\hbar}$$

$$\times \chi_\gamma^*(\vec{r}_\gamma - \vec{R}_\gamma(t), t)\chi_\beta(\vec{r} - \vec{R}_\beta(t), t)\dot{a}_\beta(t)\psi_\beta(\zeta_b\zeta_B\vec{r}_\beta t)$$

$$= \sum_\beta \int d^3r_\gamma d\zeta_f d\zeta_F \psi_\gamma^*(\zeta_f \zeta_F \vec{r}_\gamma t)\chi_\gamma^*(\vec{r}_\gamma - \vec{R}_\gamma(t), t)$$

$$\times (V_\beta - U_\beta(r_\beta))\psi_\beta(\zeta_b\zeta_B\vec{r}_B t)\chi_\beta(\vec{r}_\beta - \vec{R}_\beta(t), t)a_\beta(t)e^{-iE_\beta t/\hbar} \tag{17}$$

where the index γ runs over all channels included in (12). In deriving (17), use was made of (16) even for times $t \approx 0$ and of the relation

$$U_\gamma(R_\gamma) - m_\gamma \dot{\vec{v}}_\gamma \cdot (\vec{r}_\gamma - \vec{R}_\gamma(t)) \approx U_\gamma(\vec{r}_\gamma), \tag{18}$$

which follows from (11) and from the assumption that the wavepackets are narrow in the interaction region.

In carrying out the integration, the difference $\vec{r}_\gamma - \vec{r}_\beta$ which is proportional to the center-of-mass coordinate $\vec{r}_d$ of the transferred particles (cf. Eqs. (4.9)–(4.10)) is kept fixed. For Gaussian wavepackets, we have

$$\chi_\beta \chi_\gamma \sim \exp\left\{-(\vec{r}_\beta - \vec{R}_\beta)^2/2\sigma^2\right\} \exp\left\{-(\vec{r}_\gamma - \vec{R}_\gamma)^2/2\sigma^2\right\}$$

$$= \exp\left\{-\frac{[(\vec{r}_\beta + \vec{r}_\gamma)/2 - (\vec{R}_\beta + \vec{R}_\gamma)/2]^2}{\sigma^2}\right\}$$

$$\exp\left\{-\frac{[(\vec{r}_\beta - \vec{r}_\gamma) - (\vec{R}_\beta - \vec{R}_\gamma)]^2}{4\sigma^2}\right\} \tag{18a}$$

where the width σ is a function of time. For narrow wavepackets, we find that the integral over $\vec{r}_\gamma$ in (17) receives the main contribution when $(\vec{r}_\beta + \vec{r}_\gamma)/2 = (\vec{R}_\beta + \vec{R}_\gamma)/2$, while the second factor is approximately unity. The dependence on $\vec{r}_\beta - \vec{r}_\gamma \approx 0$ is not significant when $\vec{R}_\gamma \approx \vec{R}_\beta$ because it only depends slowly on $\vec{r}_d$. We thus obtain the result (Broglia and Winther 1972)

$$i\hbar \sum_\beta \dot{a}_\beta(t) \langle \psi_\xi \mid \psi_\beta \rangle_{\vec{R}_{\xi\gamma}} e^{-iE_\beta t/\hbar}$$

$$= \sum_\gamma \langle \psi_\xi \mid V_\gamma - U_\gamma(r_\gamma) \mid \psi_\gamma \rangle_{\vec{R}_{\xi\gamma}} a_\gamma(t) e^{-iE_\gamma t/\hbar}, \tag{19}$$

where the sub-index on the matrix elements indicate that in the integration over the degrees of freedom of the two nuclei, the average center-of-mass coordinate $\vec{r}_{\beta\gamma} = \frac{1}{2}(\vec{r}_\beta + \vec{r}_\gamma)$ should be identified with the average classical coordinate, i.e.

$$\vec{r}_{\beta\gamma} \to \vec{R}_{\beta\gamma} = \frac{1}{2}(\vec{R}_\beta + \vec{R}_\gamma). \tag{20}$$

The Jacobian for the transformation from $d^3 r_\gamma$ to $d^3 r_{\beta\gamma}$ will be included in the definition of the matrix elements.

Equations (19), which are generalizations of the semiclassical equation used in Vol. I, form the basis for the discussions in the following chapters. They can also be derived from the Feynman path integral method.

2   THE COUPLED EQUATIONS

To solve Eqs. (1.19), we need to know the classical trajectories $\vec{R}_\beta(t)$, $\vec{R}_\xi(t)$,.... They have to satisfy the Newtonian equation ((1.11)). The initial conditions for the solution of these equations are in part known from the energy and angular momentum conservation. Still each trajectory can be rotated in the plane perpendicular to the angular momentum. Furthermore, the time at which the classical particle is at a given point in its trajectory is still undetermined. The choice of these two parameters should be done so as to maximize the overlap of the trajectories in the interaction region. In a situation of very weak interactions (grazing collisions), one would choose the trajectories to have common symmetry axis and the ions to move such that they are at the distance of closest approach at $t = 0$ (cf. Fig. 3(a)). For very strongly coupled situations, the number of steps through which the process evolves may become large. The trajectory corresponding to a definite channel should then be adjusted to approximately osculate the trajectories of the channels through which it is most strongly populated, at the time where this population takes place (cf. Fig. 3(b)). It is in this limit of multistep processes that macroscopic dissipation functions like friction have a well-defined meaning. This subject is discussed in more detail in Chapter VI.

The coupled equations (1.19) are of first order in time and can be solved knowing the initial condition at time $t = -\infty$,

$$a_\gamma(-\infty) = \delta(\gamma, \alpha), \tag{1}$$

where α labels the entrance channel, that is, the nuclei a and A in their ground state. The quantities $a_\beta(+\infty)$ describe the amplitudes on the various exit channels. The cross section for the reaction $\alpha \to \beta$ is determined by the geometrical relationship between the trajectory $\vec{R}_\alpha$ of a given impact parameter ρ and the trajectory $\vec{R}_\beta$ which at time $t = +\infty$ defines a scattering angle ϑ_β with respect to the beam in the center-of-mass system (cf. Fig. 3 and Fig. III.4). The functional relation $\Theta_\beta = \theta_\beta(\rho)$ is called the deflection function associated with the exit channel (cf. Sect. III.2). The classical expression for the cross section is thus given by (cf. Eq. (III.2.7))

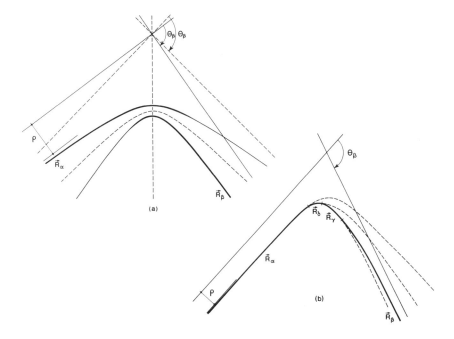

Fig. 3 Classical trajectories of relative motion associated with a grazing collision (a) and with a strongly coupled situation (b). In (a) the two trajectories
of relative motion $\vec{R}_\alpha(t)$ and $\vec{R}_\beta(t)$ associated with the entrance and exit channels of a one-step reaction are shown where the transition takes place in an
interval of time around $t = 0$. Also indicated with a dashed curve is the aver-
age trajectory $\vec{R}_{\alpha\beta}(t) = \left(\vec{R}_\alpha(t) + \vec{R}_\beta(t)\right)/2$. The angle between the directions
$\vec{R}_\alpha(t = -\infty)$ and $\vec{R}_\beta(t = +\infty)$ is the scattering angle θ_β. Since the angle
between the directions $\vec{R}_{\alpha\beta}(t = -\infty)$ and $\vec{R}_{\alpha\beta}(t = +\infty)$ is the same, the
scattering angle is correctly described by the average trajectory. In (b) the
trajectory associated with a strongly coupled situation is shown. In this case
there are so many elementary processes taking place that the average energy
and angular momentum loss is a well-defined function of time and the relative
motion evolves from one trajectory to the next continuously. The resulting
average trajectory is indicated by a continuous thick curve.

$$\left(\frac{d\sigma}{d\Omega}\right)_{\alpha\to\beta} = \mid a_\beta(t = +\infty)\mid^2 \left|\frac{\rho}{\sin\Theta_\beta}\frac{d\rho}{d\,\Theta_\beta}\right|. \qquad (2)$$

A characteristic feature of the coupled equations (1.19) is the presence of the overlap $\langle \psi_\xi \mid \psi_\beta \rangle$ on the lefthand side. If ξ and β describe two channels of the same partition, e.g., $\xi = \beta'$, the prime indicating excited states of the nuclei b and B, the overlap matrix is

$$\bar{g}(\vec{R}) = \langle \psi_{\beta'} \mid \psi_\beta \rangle_{\vec{R}} = \delta(\beta', \beta). \tag{3}$$

If ξ and β describe different partitions, the overlap $\langle \psi_\xi \mid \psi_\beta \rangle$ is different from zero in the region where the densities in the two nuclei overlap. The range over which the overlaps and the form factors

$$\bar{f}(\vec{R}) = \langle \psi_\xi \mid V_\gamma - U_\gamma(\vec{r}_\gamma) \mid \psi_\gamma \rangle_{\vec{R}}, \tag{4}$$

are non-vanishing depends on the binding energy of the nucleons which are transferred between the two systems. In the coupled equations (1.19) we shall only include bound states. This implies that the functions $\bar{g}$ and $\bar{f}$ decay exponentially as functions of R. It is noted that both the overlap and the form factor matrices $\bar{g}_{\xi\beta}$ and $\bar{f}_{\xi\beta}$ are Hermitean.

The coupled equations (1.19) can be written in a more compact way by introducing the adjoint channel wavefunction

$$\omega_\xi = \sum_\gamma g_{\xi\gamma}^{-1} \psi_\gamma \tag{5}$$

where g^{-1} is reciprocal of the overlap matrix

$$g_{\xi\gamma} = \langle \psi_\xi \mid \psi_\gamma \rangle, \tag{6}$$

that is

$$\sum_\xi g_{\gamma\xi} g_{\xi\beta}^{-1} = \sum_\xi g_{\gamma\xi}^{-1} g_{\xi\beta} = \delta(\gamma, \beta). \tag{7}$$

With this definition, we find

$$(\omega_\xi, \psi_\beta) = \delta(\xi, \beta). \tag{8}$$

The adjoint wavefunction ω should be constructed at each instant of time. Outside the interaction region where $g_{\gamma\xi} = \delta(\gamma, \xi)$, one finds $\psi_\gamma = \omega_\gamma$. In the construction of the state ω one must include the same states as are included in the coupled equations. This condition which follows directly from Eq. (1.19) is not explicit in Eq. (5).

Making use of Eqs. (5) and (7), we may write Eqs. (1.19) in the form

$$i\hbar \dot{a}_\beta(t) = \sum_\gamma \langle \omega_\beta \mid V_\gamma - U_\gamma \mid \psi_\gamma \rangle_{\vec{R}_{\beta\gamma}} e^{i\left(E_\beta - E_\gamma\right)t/\hbar} a_\gamma(t). \qquad (9)$$

These equations are formally identical to the coupled equations for inelastic scattering (IV.1.1) to which they reduce for $\gamma = \beta'$. However, in contrast to the solutions of Eq. (IV.1.1), the quantity $\sum \mid a_\beta(t) \mid^2$ is only equal to unity for $t = \pm \infty$. The conservation of probability takes in this case a somewhat more involved form i.e.

$$\frac{d}{dt} \langle \psi \mid \psi \rangle = \frac{d}{dt} \sum_{\beta\gamma} \int d^3 r_\beta a_\beta^*(t) \chi^*(\vec{r}_\beta - \vec{R}_\beta(t)) \langle \psi_\beta \mid \psi_\gamma \rangle$$

$$\times a_\gamma(t) \chi(\vec{r}_\gamma - \vec{R}_\gamma(t), t) e^{i\left(E_\beta - E_\gamma\right)t/\hbar} \qquad (10)$$

$$\approx \frac{d}{dt} \sum_{\beta\gamma} a_\beta^*(t) a_\gamma(t) \langle \psi_\beta \mid \psi_\gamma \rangle_{\vec{R}_{\beta\gamma}} e^{i\left(E_\beta - E_\gamma\right)t/\hbar} = 0.$$

Utilizing the coupled equations (1.19), Eq. (10) can be written as

$$\sum_{\beta\gamma} \Big\{ \langle \psi_\beta \mid V_\gamma - U_\gamma \mid \psi_\gamma \rangle_{\vec{R}_{\beta\gamma}} - \langle \psi_\beta \mid V_\beta - U_\beta \mid \psi_\gamma \rangle_{\vec{R}_{\beta\gamma}}$$

$$+ i\hbar \frac{d}{dt} \langle \psi_\beta \mid \psi_\gamma \rangle_{\vec{R}_{\beta\gamma}} - (E_\beta - E_\gamma) \langle \psi_\beta \mid \psi_\gamma \rangle_{\vec{R}_{\beta\gamma}} \Big\} \qquad (11)$$

$$\times a_\beta^*(t) a_\gamma(t) e^{i\left(E_\beta - E_\gamma\right)t/\hbar} = 0.$$

Since this relation has to be fulfilled for all possible initial conditions, we may conclude that

$$\langle \psi_\beta \mid V_\gamma - U_\gamma \mid \psi_\gamma \rangle_{\vec{R}_{\beta\gamma}} - \langle \psi_\beta \mid V_\beta - U_\beta \mid \psi_\gamma \rangle_{\vec{R}_{\beta\gamma}}$$

$$= -i\hbar \dot{g}_{\beta\gamma}(\vec{R}_{\beta\gamma}) + (E_\beta - E_\gamma) g_{\beta\gamma}(\vec{R}_{\beta\gamma}), \qquad (12)$$

which is known as the post-prior symmetry relation.

In view of the relation (10), it may be natural to define adjoint amplitudes (cf. Dietrich and Hara, 1973) through the relation

$$\bar{a}_\beta = \sum_\gamma a_\gamma(t) \langle \psi_\beta \mid \psi_\gamma \rangle_{\vec{R}_{\beta\gamma}} e^{i\left(E_\beta - E_\gamma\right)t/\hbar}. \tag{13}$$

These amplitudes satisfy the equation

$$i\hbar\dot{\bar{a}}_\beta(t) = \sum_\gamma \langle \psi_\beta \mid V_\beta - U_\beta \mid \psi_\gamma \rangle a_\gamma(t) e^{i\left(E_\beta - E_\gamma\right)t/\hbar}, \tag{14}$$

as can be seen making use of (9) and (12). Another form of these equations expressed solely in terms of the adjoint amplitudes $\bar{a}_\beta$ is

$$i\hbar\dot{\bar{a}}_\beta(t) = \sum_\gamma \langle \psi_\beta \mid V_\beta - U_\beta \mid \omega_\gamma \rangle_{\vec{R}_{\beta\gamma}} \bar{a}_\gamma(t) e^{i\left(E_\beta - E_\gamma\right)t/\hbar}. \tag{15}$$

Equations (9) and (15) are equivalent. The amplitudes a and $\bar{a}$ are identical outside the interaction region but are at intermediate times different. Making use of a and $\bar{a}$, one can define a quantity which can be interpreted as the probability $P_\beta(t)$ of the system to be in channel β at any instant of time, that is,

$$P_\beta(t) = Re\left(\bar{a}_\beta(t)a_\beta^*(t)\right). \tag{16}$$

Making use of (10), it is seen that

$$\frac{d}{dt} \sum_\beta P_\beta(t) = 0. \tag{17}$$

In most cases Eq. (16) provides a useful definition of probability although, exceptionally, it can become negative over short intervals of time.

The coupled equations derived above describe both inelastic and transfer processes and the interplay between them. The evaluation of the transfer form factors and overlaps will be the main subject of this chapter. The inelastic scattering form factors have been discussed in Chapter IV.

For a discussion of the validity of the approximation of only including bound states in the solution of the coupled equations ((1.19)), we

refer to Esbensen *et al.*, 1983. In practice, an even more drastic truncation has to be carried out. To compensate for the loss of probability for
states not included, an imaginary potential is introduced. As we shall
see, the main part of the imaginary potential in grazing collisions is due
to transfer reactions. This quantity is calculated in Section 7 below.

Through the ansatz (1.6) and (1.12), we have neglected the antisymmetrization between the wavefunction describing the two nuclei in
all channels.

It can be shown (cf. Appendix B) that Eqs. (1.19) describe the
channel amplitudes on totally antisymmetric wavefunctions provided the
full matrix elements

$$
\langle \psi_\gamma \mid V_\beta - U_\beta \mid \psi_\beta \rangle_a =
$$
$$
= \frac{1}{\sqrt{N_{fF} N_{bB}}} \sum_{P_\beta P_\gamma} (-1)^{P_\beta + P_\gamma} \langle \psi_\gamma \mid V_\beta - U_\beta \mid \psi_\beta \rangle , \qquad (18)
$$

are used in the coupled equations. Here P_β indicates the N_{bB} permutations between the nucleons in b and in B.

Equation (18) expresses the fully antisymmetricized transfer matrix element in terms of a linear combination of form factors for simple
transfer with additional exchange of nucleons (cf. also §4 below).

The identity of the nucleons also implies that the transfer of the
cluster $d = b - a$ in a collision with scattering angle ϑ cannot be distinguished from a pick-up reaction of the cluster $e = B - a$ in a collision with
scattering angle $\pi - \vartheta$. The condition of antisymmetrization determines
the phase of the interference between these processes (cf. Appendix D)
which is similar to the interference observed in the elastic scattering of
identical nuclei.

3 PERTURBATION EXPANSION

The set of coupled equations (1.19) describe all the reactions that can take place in grazing collisions between heavy ions, since they contain both the inelastic processes as well as the transfer of particles. In this description are not included the situations where the two nuclei move around each other in a bound state (cf. Appendix F). These trajectories are important in a detailed description of fusion.

A great simplification in the description can be achieved if the interaction is so weak that a perturbation expansion can be made. However, for the Coulomb part of the interaction, this condition is for most target-projectile combinations not satisfied (cf. Fig. II.4). For the nuclear part of the interaction, the conditions for the validity of a perturbative expansion are often satisfied for grazing collisions (cf. Section 8). For the inelastic channels it was shown in Section IV.1 that the strength parameter (cf. IV.1.14)

$$\chi_N^{(\gamma)} = \frac{1}{\hbar} f_\lambda^N (r_o) \tau_{\text{char}}, \tag{1}$$

is, in general, smaller than unity for $r_o > r_g$, at least for non-deformed nuclei. The strength of the transfer reaction is measured by the same quantity (1) inserting for f^N the quantity (2.4) and for the characteristic collision time (cf. Eq. (IV.1.11))

$$\tau_{\text{char}} \approx \sqrt{\frac{a_{tr}}{\ddot{r}_o}} \approx \frac{\sqrt{2(R_a + R_A)a_{tr}}}{v}, \tag{2}$$

where $a_{tr} = 1.2$ fm is the inverse slope of the exponential decay of the transfer form factor (cf. Section 4 below), v the relative velocity at large distances, while $\ddot{r}_o$ is the acceleration at the turning point. Since the transfer form factor at $r_o > r_g$ is of the order of or smaller than 1 MeV and τ_{char} is $\approx 0.3\hbar/\text{MeV}$, the strength parameter (1) is relatively small.

In this chapter we shall limit the discussion to situations where Coulomb excitation is weak (hatched area in Fig. II.4).

(a) First-order perturbation theory

In first-order perturbation theory the amplitude for the transition $\alpha \rightarrow \beta$ is given by

$$\left(a_\beta(t)\right)_{(1)} = \frac{1}{i\hbar} \int_{-\infty}^{t} \langle \omega_\beta \mid V_\alpha - U_\alpha \mid \psi_\alpha \rangle_{\vec{R}_{\beta\alpha}(t)} e^{i(E_\beta - E_\alpha)t'/\hbar} dt', \quad (3)$$

which is obtained from (2.9) by inserting the zeroth-order amplitude

$$\left(a_\gamma(t)\right)_{(o)} = \delta(\gamma, \alpha), \quad (4)$$

α denoting the entrance channel. Because of the presence of the $\mid \omega_\beta \rangle$ state (cf. Eq. (2.5)), this expression differs from the corresponding expression for inelastic scattering. Since, in grazing collisions, the overlap between the two nuclei is small, it is convenient to write the quantity (2.6) in the form

$$g_{\alpha\beta}(\vec{R}) = \delta(\alpha, \beta) + \varepsilon_{\alpha\beta}(\vec{R}), \quad (5)$$

where the matrix elements of ε are non-vanishing only if β indicates a rearrangement channel, that is, situations in which transfer or exchange of nucleons have taken place. The inverse matrix to g is given by the expansion

$$g^{-1} = 1 - \varepsilon + \varepsilon^2 - \dots \quad (6)$$

For inelastic scattering $(\beta = \alpha')$, $\omega_{\alpha'}$ differs from $\psi_{\alpha'}$ only in second order in ε (i.e. $\sim \sum (\psi_\alpha \mid \psi_\gamma)(\psi_\gamma \mid \psi_{\alpha'})$), and should be neglected (cf. Eq. (2.5) and subsequent discussion). One thus obtains

$$\left(a_{\alpha'}(t)\right)_{(1)} = \frac{1}{i\hbar} \int_{-\infty}^{t} \langle \psi_{\alpha'} \mid V_\alpha - U_\alpha \mid \psi_\alpha \rangle_{\vec{R}_{\alpha\alpha'}} e^{i(E_{\alpha'} - E_\alpha)t'/\hbar} dt' \quad (7)$$

which coincides with the results of Chapter IV.

For transfer reactions $(\beta \neq \alpha)$

$$\omega_\beta = \psi_\beta - \langle \psi_\alpha \mid \psi_\beta \rangle_{\vec{R}_{\alpha\beta}} \psi_\alpha, \quad (8)$$

leading to

$$\left(a_\beta(t)\right)_{(1)} = \frac{1}{i\hbar} \int_{-\infty}^{t} \langle \psi_\beta \mid V_\alpha - U_\alpha \mid \psi_\alpha \rangle_{\vec{R}_{\alpha\beta}} e^{i(E_\beta - E_\alpha)t'/\hbar} dt', \quad (9)$$

303

where we have used that

$$\langle \psi_\alpha \mid V_\alpha - U_\alpha \mid \psi_\alpha \rangle = 0, \tag{10}$$

i.e., we have defined the ion-ion potential U_α to be the expectation value of the interaction V_α in the entrance channel. If this had not been the case, the second term in (8) leads to a contribution

$$\delta a_\beta(t) = -\frac{1}{i\hbar} \int_{-\infty}^{t} \langle \psi_\beta \mid \langle \psi_\alpha \mid V_\alpha - U_\alpha \mid \psi_\alpha \rangle \mid \psi_\alpha \rangle \, e^{i(E_\beta - E_\alpha)t'/\hbar} dt'. \tag{11}$$

The total amplitude is therefore given by

$$\left(a_\beta(t) \right)_{(1)} = \frac{1}{i\hbar} \int_{-\infty}^{t} \langle \psi_\beta \mid V_\alpha - \langle V_\alpha \rangle \mid \psi_\alpha \rangle_{\vec{R}_{\alpha\beta}} \, e^{i(E_\beta - E_\alpha)t'/\hbar} dt', \tag{12}$$

which coincides with (9) if $\langle V_\alpha \rangle_\alpha = \langle \psi_\alpha \mid V_\alpha \mid \psi_\alpha \rangle = U_\alpha$. The interaction inducing the transfer is therefore $V_\alpha - \langle V_\alpha \rangle_\alpha$ irrespective of whether the elastic scattering is determined by $\langle V_\alpha \rangle_\alpha$ or not.

For the elastic scattering amplitude ($\beta = \alpha$) we find up to first order in overlaps and matrix elements

$$a_\alpha(t) = 1 + \frac{1}{i\hbar} \int_{-\infty}^{t} \langle \psi_\alpha \mid V_\alpha - U_\alpha \mid \psi_\alpha \rangle_{\vec{R}_\alpha} \, dt' \tag{13}$$

which is equal to unity only if $\langle V_\alpha \rangle_\alpha = U_\alpha$.

If instead of solving Eq. (2.9), we solve (2.14) or (2.15), we find the adjoint amplitude

$$\left(\bar{a}_\beta(t) \right)_{(1)} = \frac{1}{i\hbar} \int_{-\infty}^{t} \langle \psi_\beta \mid V_\beta - U_\beta \mid \psi_\alpha \rangle_{\vec{R}_{\alpha\beta}} \, e^{i(E_\beta - E_\alpha)t'/\hbar} dt'. \tag{14}$$

The two form factors appearing in Eqs. (9) and (14) are very different, as the interaction acts in one case in the entrance channel, while in the second case it acts in the exit channel. Since the two equations have the same physical content, they must lead to the same final amplitude, i.e.,

$$\left(a_\beta(+\infty) \right)_{(1)} = \left(\bar{a}_\beta(+\infty) \right)_{(1)}. \tag{15}$$

They provide the so-called prior- and post-representations of the first-order transfer process, respectively. During the collision the two amplitude are different. From Eq. (2.12), one can calculate the difference to be

$$\left(\bar{a}_\beta(t)\right)_{(1)} - \left(a_\beta(t)\right)_{(1)} = \langle \psi_\beta \mid \psi_\alpha \rangle_{\vec{R}_{\alpha\beta}} e^{i(E_\beta - E_\alpha)t/\hbar}. \tag{16}$$

The result (15) follows from this equation since at $t = +\infty$ the overlap $\langle \psi_\beta \mid \psi_\alpha \rangle_{\vec{R}_{\alpha\beta}}$ vanishes.

The product of the two amplitudes (9) and (14)

$$P_\beta(t) = Re\left((a(t))_{(1)}(\bar{a}(t))^*_{(1)}\right), \tag{17}$$

may be used as a measure of the transfer probability at time t (cf. Eq. (2.16)).

(b) Second-order perturbation theory

By further iteration of the semiclassical coupled equations, one can obtain the amplitude to second order

$$a_\beta(t) = \left(a_\beta(t)\right)_{(0)} + \left(a_\beta(t)\right)_{(1)} + \left(a_\beta(t)\right)_{(2)}, \tag{18}$$

where $(a_\beta)_{(0)}$ and $(a_\beta)_{(1)}$ are given by (4) and (3), respectively, while

$$\left(a_\beta(t)\right)_{(2)} = \left(\frac{1}{i\hbar}\right)^2 \sum_\gamma \int_{-\infty}^t dt' \, \langle \omega_\beta \mid V_\gamma - U_\gamma \mid \psi_\gamma \rangle_{\vec{R}_{\beta\gamma}(t')} \, e^{i(E_\beta - E_\gamma)t'/\hbar}$$

$$\times \int_{-\infty}^{t'} dt'' \, \langle \omega_\gamma \mid V_\alpha - U_\alpha \mid \psi_\alpha \rangle_{\vec{R}_{\gamma\alpha}(t'')} \, e^{i(E_\gamma - E_\alpha)t''/\hbar}. \tag{19}$$

The state vectors $\mid \omega \rangle$ have to include now the non-orthogonality effects between the channels β, γ and α. To second order, one finds (cf. (6))

$$\omega_\beta = \psi_\beta - \sum_{\gamma \neq \beta} \langle \psi_\gamma \mid \psi_\beta \rangle_{\vec{R}_{\gamma\beta}} \psi_\gamma +$$

$$+ \sum_{\gamma \neq \beta, \alpha} \langle \psi_\alpha \mid \psi_\gamma \rangle_{\vec{R}_{\alpha\gamma}} \langle \psi_\gamma \mid \psi_\beta \rangle_{\vec{R}_{\gamma\beta}} \psi_\alpha. \tag{20}$$

V Transfer Reactions

Out of the very many second-order processes, we shall here only discuss the second-order effects in a two-nucleon transfer reaction $A + a$ ($= b + 2$ nucleons) $\rightarrow B$ ($= A + 2$ nucleons) $+ b$, where the intermediate channel γ corresponds to the one-nucleon transfer channel $F(= A + 1$ nucleon$) + f(= b + 1$ nucleon$)$. The second-order process thus describes the successive transfer of two nucleons.

Inserting ω_β in (3) and (19), one finds that the two-nucleon transfer amplitudes can be written as

$$a(\infty) = (a_\beta)_{(1)} + (a_\beta)_{\text{orth}} + (a_\beta)_{\text{succ}}, \tag{21}$$

up to second order of perturbation theory. The different quantities appearing in this equation are

$$(a_\beta)_{(1)} = \frac{1}{i\hbar} \int_{-\infty}^{\infty} dt \, \langle \psi_\beta \mid V_\alpha - \langle V_\alpha \rangle \mid \psi_\alpha \rangle_{\vec{R}_{\beta\alpha}} \, e^{i(E_\beta - E_\alpha)t/\hbar}, \tag{22}$$

$$(a_\beta)_{\text{succ}} = \left(\frac{1}{i\hbar}\right)^2 \sum_{\gamma \neq \beta} \int_{-\infty}^{\infty} dt \, \langle \psi_\beta \mid V_\gamma - \langle V_\gamma \rangle \mid \psi_\gamma \rangle_{\vec{R}_{\beta\gamma}(t)} \, e^{i(E_\beta - E_\gamma)t/\hbar} \tag{23}$$

$$\times \int_{-\infty}^{t} dt' \, \langle \psi_\gamma \mid V_\alpha - \langle V_\alpha \rangle \mid \psi_\alpha \rangle_{\vec{R}_{\gamma\alpha}(t')} \, e^{i(E_\gamma - E_\alpha)t'/\hbar}$$

and

$$(a_\beta)_{\text{orth}} = -\frac{1}{i\hbar} \sum_{\gamma \neq (\beta,\alpha)} \int_{-\infty}^{\infty} dt \, \langle \psi_\beta \mid \psi_\gamma \rangle_{\vec{R}_{\beta\gamma}} \tag{24}$$

$$\times \langle \psi_\gamma \mid V_\alpha - \langle V_\alpha \rangle \mid \psi_\alpha \rangle_{\vec{R}_{\alpha\gamma}} \, e^{i(E_\beta - E_\alpha)t/\hbar}.$$

The first term $(a_\beta)_{(1)}$ which is identical to (12) describes the first-order simultaneous transfer of two nucleons. The two-step successive transfer is described by $(a_\beta)_{\text{succ}}$, while $(a_\beta)_{\text{orth}}$ is a second-order contribution arising from the non-orthogonality of the channels considered.

There are two limiting situations where (21) becomes especially simple. In the independent particle limit the non-orthogonality term cancels exactly the simultaneous transfer contribution as is shown in detail in Section 9 below. The transfer reaction is then described as a purely successive process by the amplitude (23) as was to be expected.

In the opposite limit of strong correlation between the two transferred particles, the non-orthogonality term (24) is expected to cancel the successive transfer contribution (23). This can most easily be seen by rewriting the second-order perturbation amplitudes in a mixed representation. The form (23) is the prior-prior expression of the second-order amplitude as the interaction $V - \langle V \rangle$ is the one associated with the initial channel in both steps. The mixed representation is obtained from (2.14) by inserting on the righthand side $a_\gamma(t) = \left(a_\gamma(t)\right)_{(0)} + \left(a_\gamma(t)\right)_{(1)}$ (cf. Eqs. (3) and (4)). One thus obtains

$$\bar{a}_\beta(\infty) = a_\beta(\infty) = (\bar{a}_\beta)_{(1)} + (\bar{a}_\beta)_{(2)} \tag{25}$$

where $(U_\beta = \langle V_\beta \rangle)$

$$(\bar{a}_\beta)_{(1)} = \frac{1}{i\hbar} \int_{-\infty}^{\infty} dt\, \langle \psi_\beta \mid V_\beta - \langle V_\beta \rangle \mid \psi_\alpha \rangle_{\vec{R}_{\beta\alpha}} e^{i(E_\beta - E_\alpha)t/\hbar}, \tag{26}$$

and

$$(\bar{a}_\beta)_{(2)} = \left(\frac{1}{i\hbar}\right)^2 \sum_\gamma \int_{-\infty}^{\infty} dt\, \langle \psi_\beta \mid V_\beta - \langle V_\beta \rangle \mid \psi_\gamma \rangle_{\vec{R}_{\beta\gamma}} e^{i(E_\beta - E_\gamma)t/\hbar}$$

$$\times \int_{-\infty}^{t} dt'\, \langle \psi_\gamma \mid V_\alpha - \langle V_\alpha \rangle \mid \psi_\alpha \rangle_{\vec{R}_{\gamma\alpha}} e^{i(E_\gamma - E_\alpha)t'/\hbar}. \tag{27}$$

The expression (26) is identical to (22) because of the post-prior symmetry, while (27) differs from (23) by having the interaction in the exit channel in the second step. This representation is known as the post-prior representation (cf. Tohyama (1972) and Robson (1973)). In this representation the non-orthogonality term does not appear.

We envisage now the situation where the two particles to be transferred have a very strong mutual interaction V_{12}. In the intermediate state γ one has to break this pair. The transition matrix is thus very small if the transfer potential does not contain this interaction. In Eq. (27) neither transfer potentials contain V_{12} and $(a_\beta)_{(2)}$ is therefore expected to be very small in this limit. The transfer of two particles thus happens mainly as a simultaneous process. A more detailed argument is given in Sect. 10 below.

From the above discussion, it is seen that the separation of the two-nucleon transfer amplitude into a first- and a second-order amplitude is a subtle question.

4   FORM FACTORS AND OVERLAPS

The coupled equations (1.19) depend on the nuclear structure through the overlaps

$$\bar{g}_{\beta\alpha}(\vec{R}) = \langle \psi_\beta \mid \psi_\alpha \rangle_{\vec{R}} , \qquad (1)$$

and the form factors

$$\bar{f}^{(\alpha)}_{\beta\alpha}(\vec{R}) = \langle \psi_\beta \mid V_\alpha - U_\alpha \mid \psi_\alpha \rangle_{\vec{R}} , \qquad (2)$$

$$\bar{f}^{(\beta)}_{\beta\alpha}(\vec{R}) = \langle \psi_\beta \mid V_\beta - U_\beta \mid \psi_\alpha \rangle_{\vec{R}} , \qquad (3)$$

which we have written in both post- and prior-representations.† The channel wavefunctions ψ_α and ψ_β are given in terms of the intrinsic wavefunctions of the two colliding nuclei through (1.6)–(1.9).

Inserting (1.6) and the corresponding expression for ψ_α in (1), one may write the overlap in the form

$$\bar{g}_{\beta\alpha}\big(\vec{R}(t)\big) = g_{\beta\alpha}\big(\vec{k}, \vec{R}(t)\big)\exp\left(\frac{i}{\hbar}\gamma_{\beta\alpha}(t)\right), \qquad (4)$$

where

$$g_{\beta\alpha}(\vec{k}, \vec{R}) = \big\langle \psi^b \psi^B , \exp(i\sigma_{\beta\alpha})\psi^a \psi^A \big\rangle . \qquad (5)$$

We have here separated the difference between the phases δ_α and δ_β into a part $\gamma_{\beta\alpha}$ which only depends on time and a phase $\sigma_{\beta\alpha}$ which also depends on the center-of-mass coordinate of the transferred particles. We thus define

$$\gamma_{\beta\alpha}(t) = \int_o^t dt' \left\{ U_\beta\big(\vec{R}_\beta(t')\big) - \frac{1}{2}m_\beta v_\beta^2(t') - U_\alpha\big(\vec{R}_\alpha(t')\big) + \frac{1}{2}m_\alpha v_\alpha^2(t')\right\}$$

$$+ \vec{k}_{\beta\alpha}(t) \cdot \big(\vec{R}_\beta(t) - \vec{R}_\alpha(t)\big)\hbar, \qquad (6)$$

---

† Note that the definition of form factors and overlaps here include the Jacobian (28), in contrast to the definition in Broglia et al. (1977).

where $\vec{k}_{\beta\alpha}$ is the average wave vector

$$\vec{k}_{\beta\alpha}(t) = \frac{1}{2\hbar}\left(m_\alpha \vec{v}_\alpha(t) + m_\beta \vec{v}_\beta(t)\right). \tag{7}$$

Similarly,

$$\sigma_{\beta\alpha} = \vec{k}_{\beta\alpha}(t) \cdot \left(\vec{r}_\alpha - \vec{r}_\beta\right). \tag{8}$$

In deriving (6)–(8), we have used the relation $\vec{r}_\alpha + \vec{r}_\beta = \vec{R}_\alpha + \vec{R}_\beta$ (cf. Appendix A) and have neglected terms in second order in the mass of the transfer particles. The phase σ is characteristic for transfer processes since the dynamical variables $\vec{r}_\alpha$ and $\vec{r}_\beta$ are identical for inelastic scattering. It arises from the change in the center-of-mass coordinate taking place when mass is transferred from one system to the other. It gives rise to what is known as recoil effect. The magnitude of the phase σ is somewhat arbitrary. For example, one could add a numerical constant times $\vec{R}_{\alpha\beta}(t)$ correspondingly changing the definition of γ. As we shall see later, such a change corresponds to changing the scaling of the distored waves in a quantal distorted-wave Born approximation (DWBA) treatment. In the semiclassical treatment we shall deal with recoil to lowest order in the transferred mass.

In the following we consider a stripping reaction $a + A \rightarrow b + B$ where the projectile is stripped of a number of particles collectively denoted by d, while the residual nucleus is equal to the sum of the target A and the "cluster" d (cf. Fig. 2). The symbol d as well as the word "cluster" are used here for economy. Actually the degrees of freedom of each transferred particle are treated explicitly. In this case the coordinates of relative motion in entrance and exit channels are (cf. Fig. 4 and Appendix A).

$$\vec{r}_\alpha \equiv \vec{r}_{aA} = \vec{R}_{\alpha\beta} + \frac{m_d}{2m_{\alpha\beta}}\vec{r}_{d\delta}, \tag{9}$$

and

$$\vec{r}_\beta \equiv \vec{r}_{bB} = \vec{R}_{\alpha\beta} - \frac{m_d}{2\,m_{\alpha\beta}}\vec{r}_{d\delta} \tag{10}$$

where

$$m_{\alpha\beta} = \frac{(m_a + m_b)(m_A + m_B)}{4(m_a + m_A)}, \tag{11}$$

is the average reduced mass of the system. The point δ lies on the vector $\vec{R}_{\alpha\beta}$ dividing it in the ratio $(m_a + m_b):(m_A + m_B)$; i.e.,

$$\vec{r}_\delta = \frac{(m_a + m_b)}{4(m_a + m_A)}(\vec{r}_A + \vec{r}_B) + \frac{m_A + m_B}{4\,(m_a + m_A)}(\vec{r}_a + \vec{r}_b). \tag{12}$$

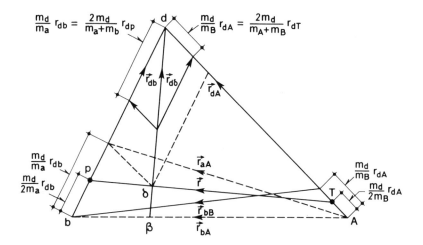

Fig. 4 Geometrical relationship between the position vectors used in describing the stripping reaction $a + A \to b + B$. The relative distance vectors in entrance and exit channels are labeled $\vec{r}_{aA} = \vec{r}_a - \vec{r}_A$ and $\vec{r}_{bB} = \vec{r}_b - \vec{r}_B$. The average position vector is denoted by $\vec{r}$, P and T being the end points of this vector. The transferred particle is denoted by d, the relative distance between its center of mass and the center of mass of the cores b and A being $\vec{r}_{db} = \vec{r}_d - \vec{r}_b$ and $\vec{r}_{dA} = \vec{r}_d - \vec{r}_A$. Also indicated is the vector $\vec{r}_{d\delta}$ which is proportional to

$$\vec{r}_{aA} - \vec{r}_{bB} = \frac{m_d}{m_a}\vec{r}_{db} + \frac{m_d}{m_B}\vec{r}_{dA}.$$

It defines the position of the transferred particle with respect to a point δ on the average position vector $\vec{r}$. The relative distance between the two cores is $\vec{r}_{bA}$. The auxiliary dotted lines are used to evaluate the length of $\vec{r}_{d\delta}$, etc. in Appendix A.

The recoil phase can be written as

$$\sigma_{\beta\alpha} \approx \vec{k}(t) \cdot \vec{r}_{d\delta}, \tag{13}$$

where

$$\vec{k}(t) = \frac{m_d \vec{v}_{\alpha\beta}(t)}{\hbar}, \tag{14}$$

$\vec{v}_{\alpha\beta}$ being the average

$$\vec{v}_{\alpha\beta}(t) = \frac{1}{2}\left(\vec{v}_\alpha(t) + \vec{v}_\beta(t)\right), \tag{15}$$

of the velocities of relative motion in entrance and exit channels. In (13) terms of order $(m_d/m_a)^2$ have been neglected. To the same approximation the quantity (6) may be written in terms of the average position (1.20) and velocity (15) as

$$\gamma_{\beta\alpha}(t) = \int_o^t dt' \left\{ U_\beta\left(R_{\alpha\beta}(t')\right) - U_\alpha\left(R_{\alpha\beta}(t')\right) \right.$$

$$\left. -\frac{1}{2}(m_\beta - m_\alpha)\vec{v}_{\alpha\beta}^2(t') \right\}. \tag{16}$$

In order to evaluate the overlap (5), we make use of the fractional parentage expansions

$$\psi_{I_B M_B}^B\left(\zeta_A, \vec{r}_{dA}, \zeta_d\right)$$

$$= \sum_{JMI_A'M_A'A'} \langle I_A'M_A'JM \mid I_B M_B \rangle \, \psi_{I_A'M_A'}^{A'}\left(\zeta_A\right) \phi_{JM}^{B(A')}\left(\vec{r}_{dA},\zeta_d\right), \tag{17}$$

and

$$\psi_{I_a M_a}^a\left(\zeta_b, \vec{r}_{db}, \zeta_d\right)$$

$$= \sum_{J'M'I_b'M_b'b'} \langle I_b'M_b'J'M' \mid I_a M_a \rangle \, \psi_{I_b'M_b'}^{b'}\left(\zeta_b\right) \phi_{J'M'}^{a(b')}\left(\vec{r}_{db},\zeta_d\right), \tag{18}$$

where ζ_b, ζ_A, and ζ_d are the intrinsic coordinates of the nuclei b, A, and the cluster d (cf. Fig. 2). The angular momentum quantum numbers I and M associated with the quantal states of the four nuclei a, A, b, and B have been written explicitly. Through the definition of the overlaps and form factors (1)–(3), we have neglected the identity of nucleons in a with nucleons in A, as well as the identity of nucleons in b with those in B. This implies that the transferred nucleons are labeled and that the fractional parentage expansions, which violate antisymmetrization, are appropriate in this context. The summations are carried out over all the states A' and b' of nuclei A and b, which have parentage to the nuclei B and a, respectively.

The parentage function $\phi^{B(A)}$ describes the motion of the nucleons of d around the core A to form B; i.e.,

$$\phi_{JM}^{B(A)}\left(\vec{r}_{dA},\zeta_d\right) = \frac{2J+1}{2I_B+1} \sum_{M_A M_B} \langle I_A M_A JM \mid I_B M_B \rangle$$

$$\times \int d\zeta_A \psi_{I_A M_A}^{A*}\left(\zeta_A\right) \psi_{I_B M_B}^B\left(\zeta_A, \vec{r}_{dA}, \zeta_b\right). \tag{19}$$

Similarly, one can write

$$\phi_{J'M'}^{a(b)}\left(\vec{r}_{db},\zeta_d\right) = \frac{2J'+1}{2I_a+1}\sum_{M_bM_a}\langle I_bM_bJ'M' \mid I_aM_a\rangle$$

$$\times \int d\zeta_b\psi_{I_bM_l}^{b*}\left(\zeta_b\right)\psi_{I_aM_a}^a\left(\zeta_b,\vec{r}_{db},\zeta_d\right). \tag{20}$$

The parentage functions (19) and (20) are thus normalized according to

$$\int \rho_{I_a}^a\left(\vec{r}_{db},\zeta_d\right)d^3r_{db}d\zeta_d = 1, \tag{21}$$

where

$$\rho_{I_a}^a\left(\vec{r}_{db},\zeta_d\right) = \sum_{b'J'M'}\frac{1}{2J'+1}\mid\phi_{J'M'}^{a(b')}\left(\vec{r}_{db},\zeta_d\right)\mid^2 \tag{22}$$

describes the density distribution of the nucleons which formed d in the nucleus a. The quantity $\rho_{I_a}\left(\vec{r}_{db},\zeta_d\right)$ describes the density of all the N_a nucleons that were specified in the wavefunctions ψ^a and ψ^b and between which the wavefunctions were antisymmetrized. Thus for a single neutron transfer with completely antisymmetrized wavefunctions, ρ_{I_a} describes the total neutron single-particle density of nucleus a in the state I_a (averaged over M_a) divided by N_a

$$\rho_{I_a}^a\left(\vec{r}_{1b}\right) = \frac{1}{2I_a+1}\sum_{M_a}\frac{1}{N_a}\langle I_aM_a \mid \rho_N\left(\vec{r}_{1b}\right)\mid I_aM_a\rangle, \tag{23}$$

where

$$\rho_N\left(\vec{r}_{1b}\right) = \sum_{\text{neutrons}}\delta\left(\vec{r}_1-\vec{r}_i'\right). \tag{24}$$

For two-particle transfer, one finds, similarly, that $\rho_{I_a}^a\left(\vec{r}_{db}\right)$ is related to the two-neutron density divided by $N_a(N_a-1)/2$ when all neutrons are specified explicitly.

In practice, one specifies only a few active nucleons in the wavefunctions ψ^a and ψ^b. The quantity ρ_{I_a} then describes the nucleon density of the active particles divided by the number of specified particles (for one-particle transfer) or the number of pairs of specified particles (for two-nucleon transfer).

While the function $\rho_{I_a}^a$ is always normalized to unity according to (21), the parentage function $\phi^{a(b)}$ to a definite state of b depends on the number of nucleons which are specified. One may compensate for this by renormalizing $\rho^{a(b)}$. In fact, the function

$$\tilde{\phi}_{J'M'}^{a(b)}(\vec{r}_{db}, \zeta_d) = \left[\binom{N_s}{N_d} \binom{Z_s}{Z_d} \right]^{1/2} \phi_{J'M'}^{a(b)}(\vec{r}_{db}, \zeta_d), \qquad (25)$$

has a normalization which is independent of the number N_s and Z_s of specified neutrons and protons, respectively. It is the function $\tilde{\phi}$ that leads to the correct spectroscopic factors for transfer reactions when antisymmetrization is taken properly into account also between the nucleons in target and projectile (for more details, cf. App. B). With these definitions, one can write the overlap (5) as

$$g_{\beta\alpha}(\vec{k}, \vec{R})$$

$$= \sum_{JJ'MM'} \langle I_b M_b J' M' \mid I_a M_a \rangle \langle I_A M_A J M \mid I_B M_B \rangle g_{MM'}^{JJ'}(\vec{k}, \vec{R}), \qquad (26)$$

with

$$g_{MM'}^{JJ'}(\vec{k}, \vec{R})$$

$$= J \int d^3 r_{d\delta} d\zeta_d \big(\tilde{\phi}_{JM}^{B(A)*}(\vec{r}_{dA}, \zeta_d) e^{i\sigma_{\beta\alpha}} \, \tilde{\phi}_{J'M'}^{a(b)}(\vec{r}_{db}, \zeta_d) \big)_{\vec{r}_{\alpha\beta} = \vec{R}}. \qquad (27)$$

The quantity JM signifies the angular momentum transferred to the target nucleus while $J'M'$ is the angular momentum taken from the projectile. In the integral the average center-of-mass distance $\vec{r}_{\alpha\beta}$ is kept fixed and equal to the classical value $\vec{R}(t)$. The Jacobian J depends on this choice as well as on the coordinates to be used for the integration over $\vec{r}_d$ (cf. Appendix A). The Jacobian associated with the choice of the variables $\vec{r}_{d\delta}$ and $\vec{r}_{\alpha\beta}$ is

$$J = [4 m_a m_B / (m_a + m_b)(m_A + m_B)]^3. \qquad (28)$$

A similar procedure can be used to calculate the form factors (2). In this case, however, because of the presence of the interaction potential, there will appear terms which are non-diagonal in the quantum number of the cores A and b, and the form factor can therefore be written as

$$\bar{f}_{\beta\alpha}(\vec{k}, \vec{R}) = \bar{f}_{\beta\alpha}(\vec{k}, \vec{R})_{\text{diag}} + \bar{f}_{\beta\alpha}(\vec{k}, \vec{R})_{\text{non-diag}}. \qquad (29)$$

313

V Transfer Reactions

The non-diagonal part gives rise to corrections in the cross sections which are expected to be of the same order of magnitude as the contribution from processes where the nucleus is excited before or after the transfer takes place. A more detailed discussion of these terms will be given in Chapter VI. In the present section, we neglect these terms and the evaluation of the form factors then runs along the same lines as the derivation given above for the overlaps. We thus find

$$\bar{f}_{\beta\alpha}^{(\alpha)}(\vec{k},\vec{R}) = f_{\beta\alpha}^{(\alpha)}(\vec{k},\vec{R}) \exp\left(\frac{i}{\hbar}\gamma_{\beta\alpha}(t)\right) \tag{30}$$

where

$$f_{\beta\alpha}^{(\alpha)}(\vec{k},\vec{R})$$

$$= \sum_{JJ'MM'} \langle I_A M_A J M \mid I_B M_B \rangle \langle I_b M_b J'M' \mid I_a M_a \rangle f_{MM'}^{JJ'(\alpha)}(\vec{k},\vec{R}), \tag{31}$$

with

$$f_{MM'}^{JJ'(\alpha)}(\vec{k},\vec{R}) = J \int d\zeta_d \, d^3 r_d \left[\tilde{\phi}_{JM}^{B(A)\,*}\left(\vec{r}_{dA},\zeta_d\right)\left(U_{dA}\left(\vec{r}_{dA},\zeta_d\right) - \langle U_{dA}\rangle\right)\right.$$

$$\left. \times e^{i\sigma_{\beta\alpha}} \tilde{\phi}_{J'M'}^{a(b)}\left(\vec{r}_{db},\zeta_d\right)\right]_{\vec{r}_{\alpha\beta}=\vec{R}} \tag{32}$$

In this diagonal part of the form factor, the interaction potential is defined as

$$U_{dA} - \langle U_{dA}\rangle = \int d\zeta_b \, d\zeta_A \mid \psi^b(\zeta_b)\mid^2 \mid \psi^A(\zeta_A)\mid^2 \left(V_{dA} + V_{bA} - U_{aA}\right)$$

$$= \int d\zeta_A \left\{\mid \psi^A(\zeta_A)\mid^2 V_{dA}\right\} - \left(U_{aA}(r_{aA}) - U_{bA}(r_{bA})\right). \tag{33}$$

The quantity U_{dA} is thus equal to

$$U_{dA}\left(\vec{r}_{db},\zeta_d\right) \approx \frac{1}{2I_A+1} \sum_{M_A} \int \left|\psi_{I_A M_A}^A(\zeta_A)\right|^2 V_{dA}\left(\zeta_A,\vec{r}_{dA},\zeta_b\right) d\zeta_A$$

$$= U_{1A}\left(\vec{r}_{1A},\zeta_1\right) + U_{2A}\left(\vec{r}_{2A},\zeta_2\right) + \dots, \tag{34}$$

314

where the definition

$$V_{dA}\left(\zeta_A, \vec{r}_{dA}, \zeta_d\right) \equiv V_{1A}\left(\zeta_A, \vec{r}_{1A}, \zeta_1\right) + V_{2A}\left(\zeta_A, \vec{r}_{2A}, \zeta_2\right) + \cdots \qquad (35)$$

has been used. In the expression (34) we neglected the dependence of U_{dA} on the magnetic quantum numbers of the state ψ^A. The numbers 1, 2,... label the nucleons in d.

The "average" interaction $\langle U_{dA} \rangle$ is, according to (33), defined as

$$\langle U_{dA} \rangle = U_{aA}(r_{aA}) - U_{bA}(r_{bA})$$

$$\approx \int \rho_{Ia}^a\left(\vec{r}'_{db}, \zeta'_d\right) U_{dA}^N\left(r'_{dA}, \zeta'_d\right) d^3r'_{db} d\zeta'_d$$

$$+U_{bA}^N(r_{aA}) - U_{bA}^N(r_{bA}) + U_{aA}^C(r_{aA}) - U_{bA}^C(r_{bA}). \qquad (36)$$

In this expression the nuclear potential appears in a form, where the uncertainty in this quantity is unimportant. Thus, one may, to a good approximation, use

$$U_{bA}^N(r_{aA}) - U_{bA}^N(r_{bA}) \approx U_{aA}^N(r_{aA}) - U_{aA}^N(r_{bA}) \qquad (37)$$

in evaluating (36). The potential U_{bA} between the cores is calculated as a double folding neglecting the possible dependence on the magnetic quantum numbers. The present discussion is thus limited to spherical nuclei. Transfer reactions involving deformed nuclei are discussed in Chapter VI.

Because of the symmetry in the scattering plane which contains both vectors $\vec{R}$ and $\vec{k}$, the form factor satisfies the symmetry relation

$$f_{MM'}^{JJ'}(\vec{k}, \vec{R}) = \mathcal{R}_\perp(180°) \mathcal{P} f_{MM'}^{JJ'}(\vec{k}, \vec{R}) \qquad (38)$$

where $\mathcal{R}_\perp$ is a rotation of 180° around the normal to the plane, and $\mathcal{P}$ is the parity operator. The rule is especially simple in the coordinate system A (cf. Fig. 5) where one finds

$$\left(f_{MM'}^{JJ'}(\vec{k}, \vec{R}) \right)_A = (-1)^{M-M'+\pi} \left(f_{MM'}^{JJ'}(\vec{k}, \vec{R}) \right)_A. \qquad (39)$$

Here π is the parity change $\pi_a + \pi_A - \pi_b - \pi_B$. This rule implies that the form factors are nonvanishing only if $M - M' + \pi = M_b + M_B - M_a - M_A + \pi$ is even.

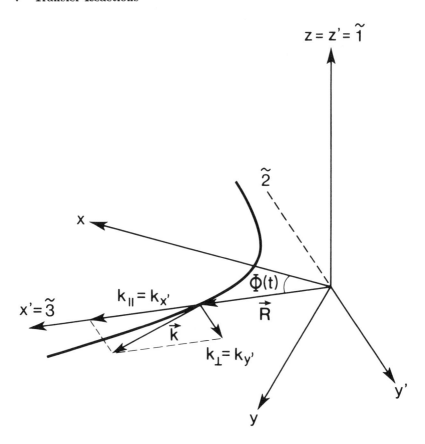

Fig. 5 The coordinate system $(A) \equiv (x, y, z)$ and the two intrinsic coordinate systems $S' \equiv (x', y', z')$ and $\tilde{S} \equiv (\tilde{1}, \tilde{2}, \tilde{3})$. In the coordinate system (A) the trajectory of relative motion lies in the x–y plane, and the x-axis is along the symmetry axis of the trajectory toward the projectile (cf. caption to fig. II.1). The z'– and $\tilde{1}$–axes of the intrinsic coordinate systems coincide with the z-axis of the (A)–system. The x'– and $\tilde{3}$–axes are along the distance of relative motion $\vec{R}$ implying that the y'–axis is chosen such that the y-component of the projectile velocity is positive. The momentum $\vec{k}$ of relative motion is also indicated. The two coordinate systems A and S' are connected by a rotation of angle $\phi(t)$ around the z-axis, while the systems S' and $\tilde{S}$ are connected by a rotation with Eulerian angles $(\alpha, \beta, \gamma) = (0, \pi/2, \pi)$.

For spherical nuclei, the parentage functions $\tilde{\phi}$ are usually expressed in the shell-model basis as

$$\tilde{\phi}_{JM}^{B(A)}(\vec{r}_{dA}, \zeta_d) = \sum_{\substack{a_1 a_2 \ldots \\ m_1 m_2 \ldots}} C^{(A)}(a_1 m_1, a_2 m_2, \ldots; JM)$$

$$\times \phi_{j_1 m_1}^{(A)}(a_1; \vec{r}_{1A}\zeta_1)\phi_{j_2 m_2}^{(A)}(a_2; \vec{r}_{2A}\zeta_2)\ldots, \qquad (40)$$

and similarly for $\tilde{\phi}^{a(b)}$. The label $a_1 \equiv (n_1, \ell_1, j_1)$ stands for the radial, orbital angular momentum and total angular quantum numbers of the normalized single-particle wave function

$$\phi_{j_1 m_1}^{(A)}(a_1; \vec{r}_{1A}\zeta_1) = R_{a_1}^{(A)}(r_{1A})\left[Y_{\ell_1}(\hat{r}_{1A})\chi(\zeta_1)\right]_{j_1 m_1}, \qquad (41)$$

describing the motion of a neutron or a proton around the nucleus A. The corresponding normalized radial wave function is denoted by $R_{a_1}^{(A)}(r_{1A})$, while $\chi(\zeta_1)$ is the spin function. The nuclear structure information, as well as the angular momentum coefficients and antisymmetrization phase are contained in the coefficients C.

These coefficients are more conveniently written in second quantization. We thus find

$$C^{(A)}(a_1 m_1, a_2 m_2, \ldots, JM)$$

$$= \sum_{M_A M_B} \sqrt{\binom{n_B}{n_A}}\frac{2J+1}{2I_B+1}\langle I_A M_A JM \mid I_B M_B \rangle \int \left(\phi_{j_1 m_1}^{(A)}(a_1\vec{r}_{1A}\zeta_1)\right)^*$$

$$\times \left(\phi_{j_2 m_2}^{(A)}(a_2\vec{r}_{2A}\zeta_2)\right)^* \cdots \left(\psi_{I_A M_A}^A(\zeta_A)\right)^* \psi_{I_B M_B}^B(\zeta_A\vec{r}_1\vec{r}_2\ldots\zeta_1\zeta_2\ldots)d\zeta_B$$

$$= \sum_{M_A M_B}\frac{2J+1}{2I_B+1}\frac{1}{\sqrt{(n_B-n_A)!}}$$

$$\times \langle I_A M_A JM \mid I_B M_B \rangle \langle I_B M_B \mid a_{a_1 m_1}^\dagger a_{a_2 m_2}^\dagger \cdots \mid I_A M_A \rangle, \qquad (42)$$

where the operator $a_{a_1 m_1}^\dagger$ creates a particle in the state (41).

For a single-particle, one finds

$$C^{(A)}(a_1 m_1, JM) = \frac{\langle I_B \|a_{a_1}^\dagger\| I_A \rangle^*}{\sqrt{2I_B+1}}\delta(j_1, J)\delta(m_1, M) \qquad (43)$$

$$\equiv C^{(A)}(I_A a_1; I_B),$$

317

which is the single-particle spectroscopic amplitude (cf. Appendix B). For the two-particle spectroscopic amplitude, we find similarly

$$C^{(A)}(a_1 m_1 a_2 m_2, JM)$$

$$= \langle j_1 m_1 j_2 m_2 \mid JM \rangle \frac{\langle I_B \| (a_{a_1}^\dagger a_{a_2}^\dagger)_J \| I_A \rangle}{\sqrt{2}}. \tag{43a}$$

The product representation of the wavefunction $\tilde{\phi}_{JM}^{B(A)}$ contains an ambiguity. While in (40) we referred the position of all particles to the core A, we could as well have assumed a product representation in which all particles were referred to the center of mass of the total system B. In evaluating the form factors and overlaps with the product wavefunctions, we shall assume that the transferred mass is so small compared to the masses of the cores that the result would be independent of this choice.

Making use of the fact that the recoil phase (13) can be written as $\sigma_{\beta\alpha} = \sigma_1 + \sigma_2 + \ldots$, where σ_i is associated with the transfer of nucleon i, one can write the overlap (27) in the form

$$g_{MM'}^{JJ'}(\vec{k}, \vec{R}) = \Sigma C^*(a_1 m_1 a_2 m_2 \ldots, JM) C(a_1' m_1' a_2' m_2' \ldots, J'M')$$

$$\times g_{m_1 m_1'}^{a_1 a_1'}(\vec{k}_1, \vec{R}) g_{m_2 m_2'}^{a_2 a_2'}(\vec{k}_2, \vec{R}_2) \ldots \tag{44}$$

where, for example

$$g_{m_1 m_1'}^{a_1 a_1'}(\vec{k}_1, \vec{R})$$

$$= J_1 \int d\zeta_1 d^3 r_{1\delta_1} \phi_{j_1 m_1}^{(A)*}(a_1; \vec{r}_{1A}\zeta_1) e^{i\sigma_1} \phi_{j_1' m_1'}^{(b)}(a_1'; \vec{r}_{1b}\zeta_1) \bigg|_{\vec{r}_{\alpha_1\beta_1} = \vec{R}}. \tag{45}$$

The recoil phase σ_1 is given by (M equal to nucleon mass)

$$\sigma_1 = \vec{k}_1 \cdot \vec{r}_{1\delta_1} = \frac{M \vec{v}_{\alpha_1\beta_1}}{\hbar} \cdot \vec{r}_{1\delta_1}. \tag{46}$$

The labels α_1 and β_1 refer to the channels connected through the transfer of particle 1, and the associated relative coordinate is then given by

$\vec{r}_{\alpha_1\beta_1} = (\vec{r}_{aA} + \vec{r}_{(a-1)(A+1)})/2$. The point δ_1 lies on this vector. The Jacobian is given by (28) by replacing b by $(a-1)$ and B by $A+1$.

The separation of the overlap in a product of single-particle overlaps (44) can also be achieved for the form factor (32). Utilizing the representation (34) and writing (36) in the form

$$\langle U_{dA} \rangle = U_{aA}(r_{aA}) - U_{bA}(r_{bA})$$

$$= U_{aA}(r_{aA}) - U_{(a-1)A}(r_{(a-1)A})$$

$$+ U_{(a-1)A}(r_{(a-1)A}) - U_{(a-2)A}(r_{(a-2)A}) \tag{47}$$

$$+ \ldots$$

$$+ U_{(b+1)A}(r_{(b+1)A}) - U_{bA}(r_{bA}),$$

one finds

$$f^{JJ'(\alpha)}_{MM'}(\vec{k}, \vec{R}) = \Sigma C^* (a_1 m_1 a_2 m_2, \ldots, JM) \, C \, (a'_1 m'_1, a'_a m'_2, \ldots, J'M')$$

$$\times \left\{ f^{a_1 a'_1(\alpha)}_{m_1 m'_1}(\vec{k}_1, \vec{R}(t)) g^{a_2 a'_2}_{m_2 m'_2}(\vec{k}_2, \vec{R}(t)) \ldots \right.$$

$$\left. + g^{a_1 a'_1}_{m_1 m'_1}(\vec{k}_1, \vec{R}(t)) f^{a_2 a'_2}_{m_2 m'_2}(\vec{k}_2, \vec{R}(t)) \cdots + \cdots \right\}. \tag{48}$$

The single-particle form factor is given by

$$f^{a_1 a'_1(\alpha)}_{m_1 m'_1}(\vec{k}, \vec{R}) = J_1 \int d\zeta_1 d^3 r_{1\delta} \phi^{(A)*}_{j_1 m_1}(a_1; \vec{r}_{1A}\zeta_1) e^{i\sigma_1}$$

$$\times \left(U_{1A}(r_{1A}, \zeta_1) - \langle U_{1A} \rangle \right) \phi^{(b)}_{j'_1 m'_1}(a'_1; \vec{r}_{1b}\zeta_1), \tag{49}$$

where

$$\langle U_{1A} \rangle = U_{(b+1)A}(r_{(b+1)A}) - U_{bA}(r_{bA}). \tag{50}$$

The separation of the form factor in a product of a single-particle form factors and overlaps is the basis for the cancellation of the non-orthogonality term (3.24) for independent particle motion in perturbation theory (cf. Section 9).

For the actual numerical evaluation of the form factor, it is advantageous to express the form factors in terms of the total angular momentum which is transferred from the two nuclei to the relative motion, i.e.

$$\vec{\lambda} = \vec{J}' - \vec{J}. \tag{51}$$

319

V   Transfer Reactions

Accordingly, we define the stripping form factor $f_{\lambda\mu}$ by

$$f_{MM'}^{JJ'}(\vec{k}, \vec{R}) = \sum_\lambda \langle \lambda\mu JM \mid J'M' \rangle f_{\lambda\mu}^{JJ'}(\vec{k}, \vec{R}), \qquad (52a)$$

or

$$f_{\lambda\mu}^{JJ'}(\vec{k}, \vec{R}) = \sum_{MM'} \langle \lambda\mu JM \mid J'M' \rangle \frac{2\lambda + 1}{2J' + 1} f_{MM'}^{JJ'}(\vec{k}, \vec{R}), \qquad (52b)$$

and similar for the overlap $g_{\lambda\mu}$. The angular momentum coupling which is implied by the definitions (31) and (52) is illustrated in Fig. 6. We have here introduced also the angular momenta of relative motion $\vec{\ell}_\alpha$ and $\vec{\ell}_\beta$ in entrance and exit channels as well as the channel spins $\vec{I}_\alpha$ and $\vec{I}_\beta$ and the total angular momentum $\vec{S}$.

Out of the six variables in (52), three can be extracted in terms of a D-function since the form factor is a tensor of rank λ. The remaining three integrals can be reduced to two if one chooses an intrinsic system $\tilde{S} = (\tilde{1}, \tilde{2}, \tilde{3})$ where the 3-axis points along the vector $\vec{R}$ while the $\tilde{1}$-axis points along the angular momentum. The connection between the systems A and $\tilde{S}$ is indicated in Fig. 5, and one thus finds

$$\left(f_{\lambda\mu}^{JJ'}(\vec{k}, \vec{R}) \right)_{(A)} = \sum_{\mu'} D_{\mu\mu'}^\lambda \left(\phi, \frac{\pi}{2}, \pi \right) \tilde{f}_{\lambda\mu'}^{JJ'}(k_\parallel k_\perp R) \qquad (53)$$

where we have explicitly indicated that the form factor in this system only depends on $k_{\tilde{3}} = k_\parallel, k_{\tilde{2}} = -k_\perp$ and R.

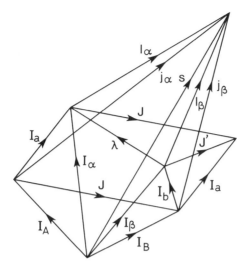

Fig. 6 Angular momentum coupling scheme for the reaction $a + A \to b + B$. The angular momenta of the nuclei involved are denoted by $\vec{I}_a, \vec{I}_A, \vec{I}_b,$ and $\vec{I}_B$. The angular momentum transferred to the target nucleus A is denoted by $\vec{J}$, while $\vec{J}'$ is the angular momentum transferred from the projectile. The total angular momentum transfer is $\vec{\lambda} = \vec{J}' - \vec{J} = \vec{\ell}_\beta - \vec{\ell}_\alpha$, where $\vec{\ell}_\alpha$ and $\vec{\ell}_\beta$ are the angular momenta of relative motion in entrance and exit channels, respectively. The total angular momenta in the relative motion of the projectile a, and of the outgoing nucleus b, are denoted by $\vec{j}_\alpha$ and $\vec{j}_\beta$, respectively. The total nuclear spins in entrance and exit channels are denoted by $\vec{I}_\alpha$ and $\vec{I}_\beta$, respectively. The total channel spin $\vec{S} = \vec{I}_\alpha + \vec{\ell}_\alpha = \vec{I}_\beta + \vec{\ell}_\beta$ is conserved in the reaction.

5 SINGLE-PARTICLE FORM FACTORS

When a particle is transferred in a heavy-ion collision, the center of mass of the two nuclei is changed. The associated single-particle form factor is thus a non-local quantity and depends on the coordinate of relative motion in entrance and exit channels. As was shown in the previous section, this non-locality in space can be translated into a momentum dependence which, in the semi-classical approximation, has a simple form. It was possible then to define single-particle transfer form factors that can be used as building blocks in the description of transfer reactions, similar to what was done in Chapter IV in the case of inelastic scattering processes. In the present section we discuss the evaluation of the single-particle transfer form factors in the case where recoil effects are moderate.

(a) Recoil effects

The overlaps and form factors are functions of six variables. They are the three components of the relative center-of-mass distance $\vec{r} = \vec{R}$ and the three components of the wavevector $\vec{k}$. The latter is associated with the linear momentum carried by the transferred particle and enters in the phase σ (cf. (4.13)).

It is convenient to evaluate the form factor (4.49) in the coordinate system A which was used in the previous chapters (cf. Fig. II.1). One finds

$$\left(f^{a_1 a_1'}_{m_1 m_1'}(\vec{k}, \vec{r}) \right)_{(A)} = e^{i\mu\phi} \left(f^{a_1 a_1'}_{m_1 m_1'}(k_{\|} k_{\perp}, r) \right)_{(S')} \tag{1}$$

where $\mu = m_1' - m_1$. The form factor on the right hand side is evaluated in the "intrinsic" frame S' which is related to A by a rotation through the angle ϕ around the z–axis such that the x–axis points along the vector $\vec{r}$ (cf. Fig. 5). In this system the form factor only depends on $r = |\vec{r}|$ and on the longitudinal $k_{x'} = k_{\|}$ and transverse $k_{y'} = k_{\perp}$ components of $\vec{k}$. The intrinsic form factor is explicitly given by

$$\left(f^{a_1 a_1'(\alpha)}_{m_1 m_1'}(k_{\|} k_{\perp}, r) \right)_{(S')}$$

$$= J \sum_{m_s, m_\ell, m_\ell'} \langle \ell_1 m_\ell \tfrac{1}{2} m_s \mid j_1 m_1 \rangle \langle \ell_1' m_\ell' \tfrac{1}{2} m_s \mid j_1' m_1' \rangle$$

$$\int dx'\, dy'\, dz\, R^{(A)}_{a_1}(r_{1A}) R^{(b)}_{a_1'}(r_{1b}) \left(U_{1A}(r_{1A}) - \langle U_{1A} \rangle \right)$$

$$\times Y^*_{\ell_1 m_\ell}(\theta_A', 0) Y_{\ell_1' m_\ell'}(\theta_b', 0)$$

$$\exp\{ -im_\ell \phi_A' + im_\ell' \phi_a' + ik_{\|}(x' - x_\delta) + ik_{\perp} y' \},$$

where the meaning of the symbols follows from Fig. 7.

From the expression (2), we may estimate the dependence of the form factor on the recoil momentum $\vec{k}$, which distinguishes the transfer

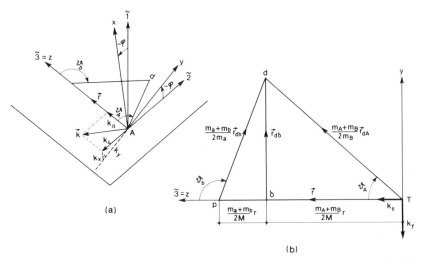

Fig. 7 The coordinates used for the evaluation of the form factors. In (a) is shown the intrinsic system $\tilde{S} \equiv (\tilde{1}, \tilde{2}, \tilde{3})$ (compare Fig. 5) where recoil momentum $\vec{k}$ is in the $(\tilde{2}, \tilde{3})$-plane, the transverse component $\vec{k}_\perp$ being along the negative $\tilde{2}$-axis. The coordinate system x, y, z which contains the vectors $\vec{r}, \vec{R}_{db}$ and $\vec{r}_{dA}$ in the (y,z) plane is obtained from this system by a rotation ϕ around the $\tilde{3}$ axis. The components of $\vec{k}$, along the $x-$ and $y-$axes are indicated. In (b) is shown the geometry in the z, y plane (compare Fig. 4). As can be seen, the variables entering in Eq. (5.6) are given in terms of y and z by

$$\cos\theta_A = \frac{z}{\sqrt{y^2 + z^2}}, \qquad \sin\theta_A = \frac{y}{\sqrt{y^2 + z^2}},$$

and

$$\cos\theta_b = \frac{z - r}{\sqrt{y^2 + (z - r)^2}}, \qquad \sin\theta_b = \frac{y}{\sqrt{y^2 + (z - r)^2}}$$

while

$$r_{1A} = \frac{2m_B}{m_A + m_B}\sqrt{y^2 + z^2}, \qquad r_{1b} = \frac{2m_a}{m_a + m_b}\sqrt{y^2 + (z - r)^2}.$$

from inelastic processes. We notice first, that because of the appearance of the potential $U_{1A}(r_{1A})$ the main contribution to the integral comes from the region of the surface of nucleus A facing a. While in

323

Broglia *et al.* 1977, the maximum of the integrand was found to be at the distance $r_{1A} \approx \bar{R}_A = 1.0A_A^{1/3}$ fm, more detailed investigation by Sørensen *et al.* 1988 show that the position of the maximum depends somewhat on the quantum numbers a_1 and that a better value is $\bar{R}_A = 1.25A_A^{1/3}$ fm. The width of the maximum in the x' direction is related to the exponential decay rate κ of the wavefunction $R^{(b)}$, i.e. $\Delta_x \approx \kappa^{-1}$ which for single-particle transfer is ≈ 1 fm. In the y' direction this exponential decay of the radial wavefunction gives rise to a width $\Delta_y \approx (2\kappa^{-1} \times (r - \bar{R}_A)\bar{R}_A/r)^{1/2} \approx 2.5$ fm for the single-particle case. Assuming the product $R_{a_1} R_{a'_1} U_{1A}$ to be proportional to $\exp(-x'^2/\Delta_x^2 - y'^2/\Delta_y^2)$, the integration over these variables can be carried out leading to

$$\left(f_{m_1 m'_1}^{a_1 a'_1(\alpha)}(k_\parallel k_\perp r) \right)_{(S')} \sim e^{i\bar{\sigma}(\alpha)}$$

$$\times \exp\left\{ -\left(\frac{\Delta_x}{2}k_\parallel\right)^2 - \left[\frac{\Delta_y}{2}\left(k_\perp - \frac{m_1}{\bar{R}_A} - \frac{m'_1}{r - \bar{R}_A}\right)\right]^2 \right\}, \quad (3)$$

where we have extracted the average recoil phase

$$\bar{\sigma}^{(\alpha)} = k_\parallel \left(\bar{R}_A - \frac{m_A + m_B}{2(m_a + m_A)} r \right). \quad (4)$$

In the derivation of (3) we used that $\phi'_A \approx y'/\bar{R}_A$ and $\phi'_a \approx \pi - y'/(r - \bar{R}_A)$.

From the expression (3), we see that through the recoil effect the form factors are significantly reduced when the recoil momenta become of the order of $1 - 2\text{fm}^{-1}$ which corresponds to an energy of relative motion of ≈ 20 MeV per nucleon. It is noticed, however, that the longitudinal component of k is zero at the classical turning point. This means that the longitudinal recoil effect essentially changes the slope of the effective form factor and not the value at the turning point. On the other hand, the transverse component of k is maximal at this point. In fact, $k_\perp$ is related to the classical angular momentum L of relative motion, i.e.

$$L = \hbar k_\perp r \frac{m_{aA}}{m_d} = \hbar k_\perp r \frac{A_a A_A}{A_a + A_A}. \quad (5)$$

The dependence on $k_\perp$ implies that the form factors tend to be largest if

$$\frac{m_1}{R_A} + \frac{m'_1}{R_b} \approx k_\perp. \quad (5a)$$

This rule can be understood in terms of a simple classical picture implying that in the region of transfer the linear momentum of the transferred
nucleon is conserved (cf. Brink (1972)). For small values of $k_\perp$ the angular momenta m_1 and m_1' are opposite and the total angular momentum
transfer $\mu = m_1' - m_1$ is large, cf. Fig. 8(a). This is even more so because
the spherical harmonics $Y_{\ell m}(\approx \pi/2, 0)$ appearing in eq. (2) are largest
if $\mid m \mid$ is close to ℓ. For larger values of $k_\perp$ (cf. Fig. 8(b)) this rule
is changed to favor parallel and positive values of m_1 and m_1' until for
large values of $k_\perp$ the form factors are strongly reduced.

(b) Evaluation of the form factors

For the more accurate numerical evaluation of the single-particle form
factors it is convenient to use the tensors $f_{\lambda\mu'}^{JJ'}(k_\parallel k_\perp r)$ in coordinate system $\tilde{S}$. For this purpose, we make use of (2) and of the relation (4.52b).
One finds in the cylindrical coordinates $x_{\tilde{1}} = -y\sin\varphi, x_{\tilde{2}} = y\cos\varphi$ and
$x_{\tilde{3}} = z$, that one can evaluate the integral over φ in terms of a Bessel
function, leading to

$$\tilde{f}_{\lambda\mu'}^{a_1 a_1'(\alpha)}(k_\parallel k_\perp r)$$

$$= 2\pi\left((2j_1 + 1)(2\lambda + 1)\right)^{1/2}\begin{Bmatrix} \ell_1 & \ell_1' & \lambda \\ j_1' & j_1 & \frac{1}{2} \end{Bmatrix}(-1)^{j_1' + \lambda + \frac{1}{2}}$$

$$\times \left(\frac{4m_a m_B}{(m_a + m_b)(m_A + m_B)}\right)^3$$

$$\times \int y\,dy\,dz R_{a_1}^{(A)}(r_{1A})R_{a_1'}^{(b)}(r_{1b})(U_{1A}(r_{1A}) - \langle U_{1A}\rangle)$$

$$\times \exp\left[ik_\parallel\left(z - \frac{m_A + m_B}{2(m_a + m_A)}r\right)\right]J_{\mu'}(k_\perp y)\sum_{m_1 m_1'}\langle\ell_1 m_1 \ell_1' m_1' \mid \lambda\mu'\rangle$$

$$\times Y_{\ell_1 m_1}(\theta_A, 0)Y_{\ell_1' m_1'}(\theta_b, 0). \tag{6}$$

The definition of the angles and distances r_{1A} and r_{1b} in terms of y and z
are given in Fig. 7. The overlap function $\tilde{g}_{\lambda\mu'}^{a_1 a_1}(k_\parallel k_\perp r)$ is defined by the
same formula except that the factor $U_{1A} - \langle U_{1A}\rangle$ is missing.

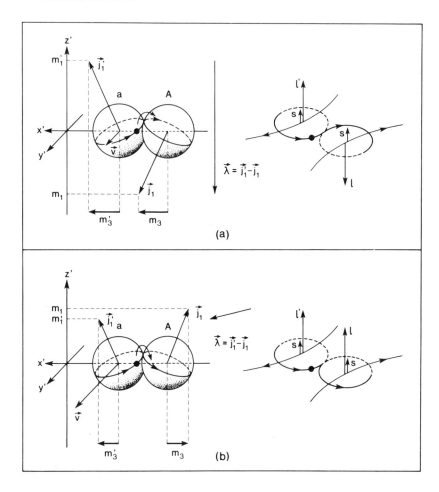

Fig. 8 Schematic representation of a single-particle transfer process. In (a) the velocity $\vec{v}$ of relative motion is assumed to be small. A nucleon which starts moving in a given orbit of the projectile is transferred smoothly into an orbit in the target nucleus almost without changing momentum. Because of this continuity, the projection of the angular momentum on the core-core relative motion vector is essentially the same before and after the transfer, and the x' (or $\tilde{3}$) axis is in this case a symmetry axis for the transferred nucleon motion. This is the reason why in this case the perpendicular recoil is negligible. The continuity in the nucleon motion has opposite consequence regarding the z' component of the angular momentum transferred. In fact,

The potential U_{1A} is the shell-model potential in nucleus A. As a standard potential, one may use (Bohr and Mottelson 1969, p. 239)

$$U(r) = V_o f(r) + V_{\ell s} (\vec{\ell} \cdot \vec{s}) r_o^2 \frac{1}{r} \frac{d}{dr} f(r), \qquad (7)$$

where $f(r)$ is the Fermi function

$$f(r) = \left(1 + \exp \frac{r - R}{a} \right)^{-1} \qquad (8)$$

with the parameters

$$R = r_o A^{1/3}, \qquad r_o = 1.27 \text{ fm}, \qquad a = 0.67 \text{ fm}, \qquad (8a)$$

while

$$V_o = -51 + 33\tau_z \frac{N - Z}{A} \text{MeV}, \qquad (8b)$$

and

$$V_{\ell s} = -0.44 V_o. \qquad (8c)$$

The quantity τ_z is equal to $+1$ for neutrons and -1 for protons, respectively. It contains besides the central potential a spin-orbit part. As long

---

when the nucleon goes from the target to the projectile, the reference point for its orbital motion changes from being the center of mass of the projectile to being the center of mass of the target A. The sign of the projection of the angular momentum along the z–axis is in this way changed leading to maximum angular momentum transfer. These features are independent of the orientation of the angular momenta. In (b) where the velocity of relative motion is large, a smooth transition can only be achieved if the angular momentum j_1' has a positive z'–component. Orbitals with negative j_1' components would only make transitions to continuum states of the target of high angular momenta. In the situation depicted the sign of the projection of the nucleon orbital angular momentum along the z–axis is not changed in the transfer process. On the other hand, the projection of the nucleon angular momentum on the x'–axis is inverted giving rise to a large component of the transferred angular momentum along this axis, and thus to a large perpendicular recoil effect. Because the standard ion-ion potential does not contain a spin-orbit term, the spin of the nucleon remains unchanged in the transfer process. The role of spin-flip and non-spin-flip transfer processes is altered in going from case (a) to case (b).

V   Transfer Reactions

as recoil effects are small, the spin-orbit force acts as a modification of the central field. Since

$$\vec{\ell} \cdot \vec{s} \, \phi(a_1; \vec{r} \, \zeta) = \frac{1}{2}(j_1(j_1 + 1) - \frac{3}{4} - \ell_1(\ell_1 + 1))\phi(a_1; \vec{r} \, \zeta), \qquad (9)$$

the effect of the spin-orbit force is to change the radius of the central potential by

$$\Delta R = 0.22 r_o A^{-1/3} \times \begin{cases} \ell_1 & \text{for } j_1 = \ell_1 + \frac{1}{2} \\ -(\ell_1 + 1) & \text{for } j_1 = \ell_1 - \frac{1}{2} \end{cases} \qquad (10)$$

The effect of the spin-orbit force on the recoil effect has not been investigated.

The potential $\langle U_{1A} \rangle$ is defined in terms of the total interaction between the two ions (cf. Eq. (4.36)) through

$$\langle U_{1A} \rangle = U_{aA}(r_{aA}) - U_{bA}(r_{bA}). \qquad (11)$$

The distances $\vec{r}_{aA}$ and $\vec{r}_{bA}$ are defined through

$$\vec{r}_{aA} = \vec{r} + \frac{m_d}{2m_{\alpha\beta}} \vec{r}_{1\delta}, \qquad (12)$$

$$\vec{r}_{bA} = \frac{m_a + m_B}{m_a + m_A} \left(\vec{r} - \frac{(m_A - m_b)m_d}{2m_a m_B} \vec{r}_{1\delta} \right), \qquad (13)$$

where the point δ lies on the vector $\vec{r}$ (the z–axis) at the distance

$$r_{\delta A} = \frac{m_a + m_B}{2(m_a + m_A)} r$$

from the origin (cf. Figs. 4 and 7 and Appendix A).

Except at small distances between the two ions, the main part of $\langle U_{1A} \rangle$ is due to the Coulomb field. One thus finds (cf. Fig. 4)

$$\langle U_{1A} \rangle \approx U_{aA}^C(r) - U_{bA}^C(r) + \frac{m_d}{m_a} \vec{r}_{1a} \cdot \vec{\nabla} U(r), \qquad (14)$$

to lowest order in m_d/m_a and Z_d/Z_a. For neutron transfer where the two first terms cancel, we find

$$U_{1A} - \langle U_{1A} \rangle \approx U_{1A}^N(r_{1A}) - \frac{m_d}{m_a} \vec{r}_{1a} \cdot \vec{\nabla} U_{aA}(r). \qquad (15)$$

The transfer is thus caused by a combination of the nuclear field and the acceleration field (compare with Eq. (1.13)). For proton transfer, the main part of $\langle U_{1A} \rangle$ cancels the Coulomb field in U_{1A}, and we find

$$U_{1A} - \langle U_{1A} \rangle \approx U_{1A}^N(r_{1A})$$

$$+ \left(\frac{Z_d}{Z_a} - \frac{m_d}{m_a} \right) \vec{r}_{1a} \cdot \vec{\nabla} U_{aA}(r) \tag{16}$$

Besides the nuclear field now occurs the difference between the acceleration and the dipole part of the Coulomb field acting on the transferred particle. If the charge to mass ration were the same for all nucleons, the two terms would cancel.

The acceleration term becomes especially important for transfer to loosely bound states in the target. The long tail of the wavefunction $R_{a1}(r_{1A})$ then gives rise to significant contributions at large distances between the ions.

The form factor (6) shows the symmetry relation

$$\tilde{f}_{\lambda-\mu'}^{a_1 a_1'}(k_\| k_\perp r) = (-1)^{\lambda+\pi} \tilde{f}_{\lambda\mu'}^{a_1 a_1'}(k_\| k_\perp r), \tag{17}$$

where π is the relative even or odd parity

$$\pi = \ell_1 + \ell_1', \tag{18}$$

of the two states a_1 and a_1'. This rule is a consequence of the reflection symmetry in the $\tilde{2}, \tilde{3}$ plane (cf. Eq. (4.39)).

As mentioned above, the longitudinal recoil effect associated with $k_\|$ can, to a large extent, be extracted by substituting z in the exponent by $\langle z \rangle \approx \bar{R}_A$ (cf. Eqs. (6) and (4)). If the transverse recoil effect associated with $k_\perp$ is small, only form factors with $\mu' = 0$ are non-vanishing. The transverse recoil effect is therefore associated with a transfer of angular momentum along the symmetry axis, μ' being the component of λ along this axis (cf. Fig. 8).

(c) Low recoil approximations

At low bombarding energy an expansion of the recoil effect in powers of $k_\perp$ and $k_\|$ is appropriate, i.e., for $\mu' \geq 0$

$$\tilde{f}_{\lambda\mu'}^{a_1 a_1'(\alpha)}(k_\| k_\perp r)$$

$$= \sum_{\nu n} (ik_\| r)^\nu (k_\perp r)^{\mu'+2n} e^{i\bar{\sigma}(\alpha)} \sqrt{\frac{2\lambda+1}{4\pi}} F_{\lambda\mu'}^{(\nu,2n)(\alpha)}(a_1 a_1' r). \tag{19a}$$

329

V Transfer Reactions

For $\mu' < 0$ one should use the symmetry relation given in Eq. (17). The quantity $F^{(\nu,2n)}_{\lambda\mu'}(\alpha)$ is given by†

$$F^{(\nu,2n)(\alpha)}_{\lambda\mu'}(a_1 a_1' r)$$

$$= \frac{4\pi^{3/2}(-1)^n}{2^{\mu'+2n} n! (\mu' + n)! \nu!} \sqrt{2j_1 + 1} (-1)^{j_1'+\lambda+\frac{1}{2}} \begin{Bmatrix} \ell_1 & \ell_1' & \lambda \\ j_1' & j_1 & \frac{1}{2} \end{Bmatrix}$$

$$\left(\frac{4m_a m_B}{(m_a + m_b)(m_A + m_B)}\right)^3 \int_{y>o} y \, dy \, dz \, R^{(A)}_{a_1}(r_{1A}) R^{(b)}_{a_1'}(r_{1b})$$

$$\times \left(U_{1A}(r_{1A}) - \langle U_{1A}\rangle\right) \left(\frac{y}{r}\right)^{\mu'+2n} \left(\frac{z - z_\beta}{r}\right)^\nu$$

$$\sum_{m_1 m_1'} \langle \ell_1 m_1 \ell_1' m_1' \mid \lambda\mu'\rangle \, Y_{\ell_1 m_1}(\theta_A, 0) Y_{\ell_1' m_1'}(\theta_b, 0). \tag{19b}$$

In deriving (19), we have, as in Eq. (4), extracted a factor $\exp(i\bar{\sigma}^{(\alpha)})$

$$\bar{\sigma}^{(\alpha)} = k_{\parallel} \left(z_\beta - \frac{m_A + m_B}{2(m_a m_A)} r\right), \tag{20}$$

before expanding the exponential function. By an appropriate choice of $z_\beta = z_\beta(r)$, we may in fact ensure that (19b) vanishes for $\nu = 1$. Detailed numerical studies show that the choice $z_\beta = \bar{R}_A = 1.25 \cdot A_A^{1/3}$ fm is more accurate for stripping reactions than the value $1.0 \, A^{1/3}$ fm suggested in Broglia et al., 1977. The scaling parameter β used in this reference is defined by $\beta = z_\beta(r)/r$.

Using this scaling, one finds to lowest order in the recoil (for given $\mu' \geq 0$)

$$\tilde{f}^{a_1 a_1'(\alpha)}_{\lambda\mu'}(k_{\parallel} k_\perp r) = e^{i\bar{\sigma}(\alpha)} \sqrt{\frac{2\lambda + 1}{4\pi}} (k_\perp r)^{\mu'}$$

$$\times F^{(0,0)\alpha}_{\lambda\mu'}(a_1 a_1' r). \tag{21}$$

---

† In Broglia et al. (1977) and Pollarolo et al. (1983) we used the notation

$$F^{(\nu,0)}_{\lambda\mu'}(a_1 a_1' r) = f^{a_1 a_1'}_{\lambda\mu'}(\nu, r).$$

In the no-recoil limit where $k_\parallel = k_\perp = 0$ the quantity

$$F_{\lambda 0}^{(0,0)\alpha}(a_1 a_1' r) \equiv f_\lambda^{a_1 a_1'(\alpha)} \tag{22}$$

plays the same role as the form factor for inelastic scattering.

For the overlap function $\tilde{g}$ we may use the same expansion (19a) and define functions $G_{\lambda \mu'}^{(\nu,2n)}(a_1 a_1' r)$ which are given by the expression (19b) without the factor $U_{1A} - \langle U_{1A} \rangle$. For the average phase, we may again use the expression (20) but with $z_\beta = 1/2(r + R_A - R_b)$ corresponding to the fact that in this case the maximum contribution to the integral comes from a region halfway between the nuclear surfaces. The term with $\mu' = 1$ is of special interest in the case where $\lambda + \pi$ is odd and where, according to (17), $\tilde{f}_{\lambda 0}^{a_1 a_1'}(k_\parallel k_\perp r) = 0$.

To second order in the recoil, we find for $\mu' = 0$

$$\tilde{f}_{\lambda 0}^{a_1 a_1'(\alpha)}(k_\parallel k_\perp r) = e^{i\bar{\sigma}(\alpha)} \left(\frac{2\lambda + 1}{4\pi} \right)^{1/2}$$

$$\times \left[f_\lambda^{a_1 a_1'(\alpha)}(r) + (k_\perp r)^2 F_{\lambda 0}^{(0,2)\alpha}(a_1 a_1' r) - (k_\parallel r)^2 F_{\lambda 0}^{(2,0)\alpha}(a_1 a_1' r) \right], \tag{23}$$

while to the same order the expressions (21) may be used for $\mu' = 1$ and 2. While the longitudinal recoil effect mainly affects the Q value dependence of the reaction (cf. §6 below) the transverse recoil effect is related to the angular momentum L through (5). The recoil effects are of the order of magnitude of $k_\perp \approx (2m_a^2(E - E_B)/\hbar^2 m_{aA})^{1/2}$ times the distance Δ_y over which the integrand in (6) is nonvanishing (cf. Eq. (3)), i.e.

$$k_\perp \Delta_y \approx \left(\frac{\varepsilon_{\text{MeV}}}{5} \right)^{1/2}, \tag{24}$$

where ε_{MeV} is the energy above the Coulomb barrier in MeV per nucleon.

In Fig. 9 we have illustrated the form factor given by Eq. (19) for a number of cases. In the region of interest outside the Coulomb barrier the form factors can be rather well approximated by exponential functions. Furthermore, the form factors with maximal $\lambda(j_1 + j_1'$ or $j_1 + j_1' - 1)$ are usually much larger than the form factors with smaller values of λ. This is a reflection of the rule (5a) according to which the angular momenta $\vec{\ell}_1$ and $\vec{\ell}_1'$ tend to be opposite. Finally, one notices that the form factor for $\mu' = 1$ is typically a factor of 10 smaller than the corresponding ones for $\mu' = 0$. This result corroborates the estimate (24) of the magnitude of the recoil effect.

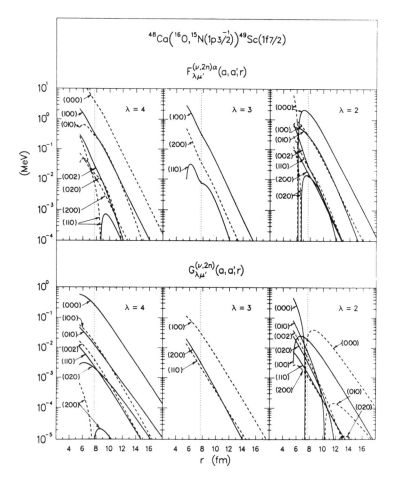

Fig. 9 The quantities (5.19b) used in the calculation of the local momentum approximation form factors in the prior representation (cf. Eq. (5.19a)) for the reactions $^{48}\text{Ca}(^{16}\text{O},^{15}\text{N}(1\text{p}_{3/2}^{-1}))^{49}\text{Sc}(1\text{f}_{7/2})$, $^{88}\text{Sr}(^{16}\text{O},^{15}\text{N}(1\text{p}_{1/2}^{-1}))^{89}\text{Y}(2\text{p}_{1/2})$, $^{208}\text{Pb}(^{16}\text{O},^{17}\text{O}(1\text{d}_{5/2}))^{207}\text{Pb}(1\text{i}_{13/2}^{-1})$, $^{208}\text{Pb}(^{16}\text{O},^{15}\text{N}(1\text{p}_{1/2}^{-1}))^{209}\text{Bi}(1\text{h}_{9/2})$.

The expressions given above and the numerical results are for form factors in the prior representation (4.2). The post representation (4.3)

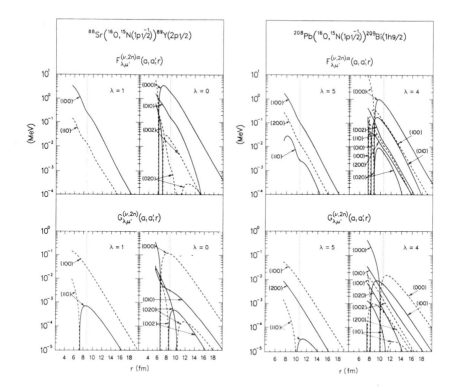

Fig. 9 (cont.) For each curve the values of $(\mu', \nu, 2n)$ are indicated. The sign of the form factor is positive for full-drawn and negative for dashed curves. The vertical dotted line indicates the distance $1.25\,(A_a^{1/3} + A_A^{1/3})$. For the potential U_{1A} the standard single-particle potential defined in Eq. (7) was adjusted to reproduce the binding energy of the different orbitals. The functions $R_{a_1}^{(A)}$ are eigenfunctions of this potential (continued p. 335).

is simply obtained by the substitution

$$U_{1A}(r_{1A}) - \langle U_{1A} \rangle \rightarrow U_{1b}(r_{1b}) - \langle U_{1b} \rangle , \qquad (25)$$

where

$$\langle U_{1b} \rangle = U_{bB}(r_{bB}) - U_{bA}(r_{bA}), \tag{26}$$

and

$$\vec{r}_{bB} = \vec{r} - \frac{m_d}{2m_{\alpha\beta}} \vec{r}_{1\delta}. \tag{27}$$

The form factors in the two representations are usually numerically rather different. Also the average recoil phase (4) will be different because the maximum of the integrand in the post representation is at the surface of the projectile, i.e.

$$e^{i\bar{\sigma}(\beta)} = \exp\left\{ ik_\parallel \left(r - \bar{R}_b - \frac{m_A + m_B}{2(m_a + m_A)} r \right) \right\}$$
$$= \exp\left\{ ik_\parallel \left(\frac{m_a + m_b}{2(m_a + m_A)} r - \bar{R}_b \right) \right\}, \tag{28}$$

with $\bar{R}_b \approx 1.25\, A_b^{1/3}$ fm. The overlaps (4.1) can also be evaluated from the same expressions by leaving out the potentials (25). In this case the recoil phase receives the main contribution halfway between the nuclear surfaces.

At low bombarding energies, where the recoil effects are unimportant, one may simplify the evaluation of the form factor (22) by using the momentum representation (Roberts (1972)) for single-particle wavefunctions, for instance,

$$\tilde{R}_{a_1}^{(A)}(k) = \frac{2}{\pi} \int_o^\infty j_{\ell_1}(kr) R_{a_1}^{(A)}(r) r^2 dr. \tag{29}$$

The quantity $j_{\ell_1}(kr)$ is the spherical Bessel function of order ℓ_1. We evaluate directly (4.49) by extracting the average phase $\exp(i\bar{\sigma})$ (compare with Eq. (4)) and use the Schrödinger equation

$$U_{1A}(r_{1A}\zeta_1)\phi_{j_1 m_1}^{(A)}(a_1; r_{1A}\zeta_1)$$
$$= \frac{\hbar^2}{2m_{1A}} \left[(\vec{\nabla})^2 - \kappa_{a_1}^2 \right] \phi_{j_1 m_1}^{(A)}(a_1; r_{1A}\zeta_1). \tag{30}$$

Here κ_{a_1} is related to the energy of the state a_1 in the target by

$$\kappa_{a_1} = \left(\frac{2m_{1A}(-E_{a_1})}{\hbar^2} \right)^{1/2} \approx \frac{m_A + m_B}{2m_B} \bar{\kappa}_{a_1}, \tag{31a}$$

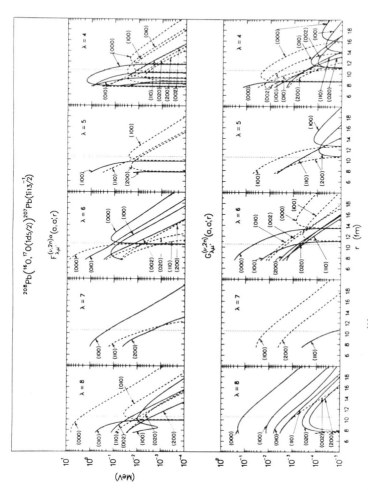

Fig. 9 (cont.) Similarly for $R_{a'}^{(b)}$. Also shown are the quantities $G_{\lambda'\mu'}^{(\nu,2n)}(a_1,a_1',r)$ needed in the calculation of the overlaps. Note that in this figure the average value z_β was set equal to $\bar{R}_A = 1.0\,A_A^{1/3}$. For the overlaps the value $z_\beta = 0.5\,r + 0.725\,(\bar{R}_A - \bar{R}_b)$ was used. This figure is due to J. Havskov Sørensen.

335

with

$$\bar{\kappa}_{a_1} = \sqrt{\frac{2m_d}{\hbar^2}(-E_{a_1})}. \tag{31b}$$

We find (cf. Appendix C)

$$f_{m_1 m_1'}^{a_1 a_1'(\alpha)}(\vec{k},\vec{r}) = \sum_{\lambda\mu} \langle \lambda\mu j_1 m_1 \mid j_1' m_1' \rangle \, e^{i\bar{\sigma}(\alpha)} f_\lambda^{a_1 a_1'(\alpha)}(r) Y_{\lambda\mu}(\hat{r}), \tag{32}$$

with

$$f_\lambda^{a_1 a_1'(\alpha)}(r) = F_{\lambda 0}^{(0,0)\alpha}(a_1 a_1'; r)$$

$$= \left(4\pi(2j_1+1)\right)^{1/2} \delta\left(\ell_1 + \ell_1' + \lambda, \text{even}\right) (-1)^{j_1' + \frac{1}{2}} i^{\ell_1' - \ell_1 + \lambda} \langle j_1 \tfrac{1}{2} j_1' - \tfrac{1}{2} \mid \lambda 0 \rangle$$

$$\times \int k^2 \, dk \, j_\lambda(kr) \tilde{R}_{a_1}^{(A)}\left(\frac{m_A+m_B}{2m_B}k\right) \tilde{R}_{a_1}^{(b)}\left(\frac{m_a+m_b}{2m_a}k\right)$$

$$\times \left\{\frac{\hbar^2}{2m_{1A}}\left[\left(\frac{m_A+m_B}{2m_B}k\right)^2 + \kappa_{a_1}^2\right] + \langle U_{1A}(r)\rangle\right\} \tag{33}$$

where we have assumed $\langle U_{1A}(r)\rangle$ is constant over the region of overlap. Making use of Eq. (4.52), the quantity $f_{\lambda\mu}^{JJ'}(\vec{k},\vec{R})$ can be written in a form similar to (IV.2.10). As indicated, this quantity is according to (4.53) and (21) identical to the quantity (19) with $\mu' = \nu = 0$.

The form factor in the post representation and the overlap are given by similar expressions. We note that they approximately fulfill the relation (cf. (4.36))

$$f_\lambda^{(\alpha)}(r) - f^{(\beta)}(r) = \left\{-\frac{\hbar^2}{2m_{1A}}\kappa_{a_1}^2 + \frac{\hbar^2}{2m_{1b}}\kappa_{a_1'}^2 - \langle U_{1A}\rangle + \langle U_{1b}\rangle\right\} g_\lambda(r)$$

$$= \left(E_A - E_b - U_{aA}(r) + U_{bB}(r)\right) g_\lambda(r) \tag{34}$$

in which we recognize in this no-recoil limit the post-prior relation (2.12).

(d) Simple parametrization

A simple parametrization of the form factor may be obtained in the tail region in situations where the acceleration term in (15) may be neglected.

For neutron transfer, this corresponds to the neglect of $\langle U_{1A} \rangle$. In the prior representation, the potential U_{1A} in (19) then ensures that only the tail of the wave function $R_{a_1}(r_{1b})$ contributes to the integral, and we may therefore use the approximation

$$R_{a_1'}^{(b)}(r_{1b}) = \bar{N}_{a_1'} k_{\ell_1'}(\kappa_{a_1'} r_{1b})/k_{\ell_1'}(\bar{\kappa}_{a_1'} R_b), \tag{35}$$

where $\bar{N}_{a_1'}$ is a number and κ_ℓ is a modified spherical Hankel function (cf. Appendix C). If one chooses the radius parameter in (35) as

$$R_b = 1.25\, A_b^{1/3} \text{fm}, \tag{36}$$

the normalization factor $\bar{N}_{a_1'} \approx R_{a_1'}^{(b)}(R_b)$ turns out to be quite independent of the binding energy, i.e., of the depth parameter V_o in the single-particle potential (7).

Inserting the momentum representation (C.9) of this tail in (33) and making use of (29) for $\tilde{R}_{a_1'}^{(b)}$, one may perform the integration over k (cf. (C.10)) to obtain (Buttle and Goldfarb (1966))

$$\left(f_\lambda^{a_1 a_1'(\alpha)}(r) \right)^{\nu S} = \left(4\pi(2j_1 + 1) \right)^{1/2} \delta(\ell_1 + \ell_1' + \lambda, \text{even})$$

$$\left(\frac{2m_a}{m_a + m_b} \right)^3 (-1)^{j_1 - \frac{1}{2} + \ell_1'} \bar{N}_{a_1'} \langle j_1 \frac{1}{2} j_1' - \frac{1}{2} \mid \lambda 0 \rangle \frac{k_\lambda(\bar{\kappa}_{a_1'} r)}{k_{\ell_1'}(\bar{\kappa}_{a_1'} R_b)}$$

$$\int r_{1A}^2 dr_{1A} R_{a_1}^{(A)}(r_{1A}) i_{\ell_1} \left(\frac{m_A + m_B}{2m_B} \bar{\kappa}_{a_1'} r_{1A} \right) \tag{37}$$

The label νS stands for neutron stripping.

If the binding energy of the neutron in the two nuclei is the same (i.e. $\bar{\kappa}_{a_1} = \bar{\kappa}_{a_1'}$), one may evaluate the integral in (37) explicitly by using that $R_{a_1}^{(A)}(r_{1A})$ satisfies the radial wave equation corresponding to (30). By partial integration, one may take advantage of the wave equation (C.14) for $i_\ell(\kappa r)$ and obtain (Buttle and Goldfarb, 1966)

$$\int r_{1A}^2 dr_{1A} R_{a_1}^{(A)}(r_{1A}) U_{1A}(r_{1A}) i_{\ell_1} \left(\frac{m_A + m_B}{2m_B} \bar{\kappa}_{a_1'} r_{1A} \right)$$

$$= -\frac{\hbar^2 \pi}{4\kappa_{a_1} m_{1A}} \frac{\bar{N}_{a_1}}{k_{\ell_1}(\bar{\kappa}_{a_1} R_A)}, \tag{38}$$

337

where we furthermore used that

$$R_{a_1}^{(A)}(r_1) \underset{r_1 \to \infty}{=} \bar{N}_{a_1} k_{\ell_1}(\kappa_{a_1} r_1)/k_{\ell_1}(\bar{\kappa}_{a_1} R_A).$$ (39)

Inserting this into Eq. (37), we find the result

$$\left(f_\lambda^{a'_1 a_1(\alpha)}(r) \right)^{\nu S} = J \pi^{3/2} \left((2j_1 + 1)(2\lambda + 1) \right)^{1/2} \delta(\ell_1 + \ell'_1 + \lambda, \text{even})$$ (40)

$$\times (-1)^{j'_1 + \ell'_1 + \frac{1}{2}} \begin{pmatrix} j_1 & j'_1 & \lambda \\ \frac{1}{2} & -\frac{1}{2} & 0 \end{pmatrix} \frac{\hbar^2}{2M\bar{\kappa}_{a'_1}} \bar{N}_{a_1} \bar{N}_{a'_1} \frac{k_\lambda(\bar{\kappa}_{a'_1} r)}{k_{\ell_1}(\bar{\kappa}_{a'_1} R_A) k_{\ell'_1}(\bar{\kappa}_{a'_1} R_a)}$$

Although the expression (38) was derived under the assumption that $\bar{\kappa}_{a_1} = \bar{\kappa}_{a'_1}$, it turns out to be very accurate also when the κ's are different. This is due to the fact that the value of the integral in (37) depends only weakly on the change in binding energy of the particle in A as is obtained by changing the depth V_o of the single-particle potential U_{1A}. One obtains formally the same expression for the form factor (38) as if $\bar{\kappa}_{a_1}$ were equal to $\bar{\kappa}_{a'_1}$. The accuracy of the approximation is discussed in detail in Quesada et al., 1985.

The situation for proton transfer is more complicated because in this case we cannot neglect the average potential $\langle U_{1A} \rangle$ in Eq. (19). As was discussed in connection with Eq. (16), the main effect of $\langle U_{1A} \rangle$ is to cancel the Coulomb field in U_{1A}. If we therefore neglect the difference between the dipole field and the acceleration field which is of the same order of magnitude as the acceleration field for neutrons, we may substitute $\langle U_{1A} \rangle$ by $Z_d Z_A e^2 / r$, and the expression (33) therefore becomes formally identical to the expression for neutron transfer if we substitute $\bar{\kappa}_{a_1}^2$ by

$$\left(\bar{\kappa}_{a_1}^{\text{eff}} \right)^2 = \frac{2M}{\hbar^2} \left(-E_{a_1} + \frac{Z_d Z_A e^2}{R(A)} \right),$$ (41)

where $R(A)$ is somewhat larger than the radius of nucleus A.

As a second step in evaluating the form factor from Eq. (33), we substitute the radial wavefunction $R_{a_1}^{(b)}$ by its tail. Because of the slow variation of the Coulomb field, we may again use the expression (35) if we substitute $\bar{\kappa}_{a'_1}$ by

$$\left(\bar{\kappa}_{a'_1}^{\text{eff}} \right)^2 = \frac{2M}{\hbar^2} \left(-E_{a'_1} + \frac{Z_d Z_b e^2}{R(b)} \right).$$ (42)

The expression (40) therefore also applies for proton transfer when the binding energies are substituted by effective binding energies according to (41) and (42) (cf. Trautman and Alder, 1970). For the distance $R(A)$ and $R(b)$, one may use the radius of the Coulomb barrier for a single proton (cf. Eq. (III.2.5))

$$R(i) = 1.07 \left(1 + A_i^{1/3}\right) + 2.72 \text{ fm}. \tag{43}$$

The form factors are equal in post and prior representations if

$$\bar{\kappa}_{a_1}^{\text{eff}} = \bar{\kappa}_{a_1'}^{\text{eff}}, \tag{44}$$

as is also seen directly from the general relation (34). This relation corresponds to a transition with Q–value

$$Q_o = E_b - E_A = \langle U_{1b} \rangle - \langle U_{1A} \rangle. \tag{45}$$

The result of the discussion above is therefore that the tail of the form factor for both neutron and proton transfer and for small recoil effects to a good approximation is given by the expression (40) substituting $\bar{\kappa}$ by $\bar{\kappa}^{\text{eff}}$. It only depends on r through the function $k_\lambda(\bar{\kappa}_1^{\text{eff}} r)$. At smaller distances where r becomes of the order of the sum of the two nuclear radii the form factor (6) goes through a maximum. Although the magnitude and the position of this maximum depends on the single-particle quantum numbers, one may still achieve an essential improvement of (40) by the substitution

$$k_\lambda(\kappa_{a_1'} r) \longrightarrow \frac{k_\lambda(\bar{\kappa}_{a_1'} R)}{1 + \dfrac{k_\lambda(\bar{\kappa}_{a_1'} R)}{k_\lambda(\bar{\kappa}_{a_1'} r)}}, \tag{46}$$

with

$$R = 1.25 \left(A_a^{1/3} + A_A^{1/3}\right) \text{ fm}. \tag{47}$$

The resulting expression for a nucleon stripping (NS) is therefore

$$\left(f_\lambda^{a_1 a_1'}(r)\right)^{NS} = \pi^{3/2} \left((2j_1 + 1)(2\lambda + 1)\right)^{1/2} \delta(\lambda + \pi, \text{even})$$

$$\times \, J \times (-1)^{j_1' + \ell_1' + \frac{1}{2}} \begin{pmatrix} j_1' & j_1 & \lambda \\ \frac{1}{2} & -\frac{1}{2} & 0 \end{pmatrix} \bar{N}_{a_1} \bar{N}_{a_1'}$$

$$\times \, \frac{k_\lambda(\bar{\kappa} R)/\bar{\kappa}}{k_{\ell_1}(\bar{\kappa} R_A) k_{\ell_1'}(\bar{\kappa} R_a)} \times \frac{20}{1 + \dfrac{k_\lambda(\bar{\kappa} R)}{k_\lambda(\bar{\kappa} r)}} \text{MeV} \cdot \text{fm}^2, \tag{48a}$$

where J is the Jacobian (4.28) and where

$$\bar{\kappa} = \begin{cases} \bar{\kappa}^{\text{eff}}_{a'_1} & \text{in prior representation} \\ \bar{\kappa}^{\text{eff}}_{a_1} & \text{in post representation,} \end{cases} \tag{48b}$$

and $\bar{N}_{a_1}$ and $\bar{N}_{a'_1}$ are measured in $\text{fermi}^{-3/2}$. The expression (48) reproduces the form factors down to distances of the order of $r = R$.

The coefficients $\bar{N}$ and the decay constants $\bar{\kappa}^{\text{eff}}$ are plotted for the standard shell model potential (7) in Figs. 10 and 11. Examples of the form factor (48) are displayed in Fig. 12 in comparison with the "exact" form factors $f^{a_1 a_1}_\lambda(r)$ (cf. Eq. (19)).

It is noticed that for $\kappa R_i > \ell_i(\ell_i + 1)$, so that the functions k_ℓ can be approximated by $k_\ell(x) = \pi/2x \; e^{-x}$, the form factor is proportional to

$$f^{a_1 a'_1}_\lambda(r) \sim \frac{R_a R_A}{R_a + R_A} \frac{1}{1 + r/R \; \exp\big(\bar{\kappa}(r - R)\big)}, \tag{49}$$

which has the same form as the ion-ion potential (cf. Eq. III.1.40)).

For the evaluation of the form factors, we give in Table 1 explicit expressions for the 3j–symbols. To determine the possible value of λ, the condition $\delta(\ell_1 + \ell'_1 + \lambda, \text{even})$ has to be used. One may distinguish between three different cases depending on the orientation of the spin with respect to ℓ. These situations are also displayed in Table 1.

$$\begin{pmatrix} j_1 & j_1' & \lambda \\ \tfrac{1}{2} & -\tfrac{1}{2} & 0 \end{pmatrix}$$

$$= \begin{cases} \dfrac{(-1)^{S/2}}{[(2j_1+1)(2j_1'+1)]^{1/2}} \left[\dfrac{S!!}{(S+1)!!} \dfrac{(S-2\lambda-1)!!}{(S-2\lambda)!!} \dfrac{(S-2j_1)!!}{(S-2j_1-1)!!} \dfrac{(S-2j_1')!!}{(S-2j_1'-1)!!} \right]^{1/2}, \\ \qquad \text{for } S = j_1 + j_1' + \lambda = \text{even} \\[1em] \dfrac{(-1)^{(S-1)/2}}{[(2j_1+1)(2j_1'+1)]^{1/2}} \left[\dfrac{(S+1)!!}{S!!} \dfrac{(S-2\lambda)!!}{(S-2\lambda-1)!!} \dfrac{(S-2j_1-1)!!}{(S-2j_1)!!} \dfrac{(S-2j_1'-1)!!}{(S-2j_1')!!} \right]^{1/2} \\ \qquad \text{for } S = j_1 + j_1' + \lambda = \text{odd}. \end{cases}$$

$$\approx \begin{cases} \dfrac{(-1)^{S/2}}{[(2j_1+1)(2j_1'+1)]^{1/2}} \left[\dfrac{2}{\pi} \left(\dfrac{(S-2j_1+1/2)(S-2j_1'+1/2)}{(S+3/2)(S-2\lambda+1/2)} \right)^{1/2} \right]^{1/2}, \\ \qquad \text{for } S \text{ even} \\[1em] \dfrac{(-1)^{(S-1)/2}}{[(2j_1+1)(2j_1'+1)]^{1/2}} \left[\dfrac{2}{\pi} \left(\dfrac{(S+3/2)(S-2\lambda+1/2)}{(S-2j_1+1/2)(S-2j_1'+1/2)} \right)^{1/2} \right]^{1/2}, \\ \qquad \text{for } S \text{ odd} \end{cases}$$

case	j	j'	$\lambda_{\max}$	S	
1	$\ell_1 + 1/2$	$\ell_1' + 1/2$	$j_1 + j_1' - 1 = \ell_1 + \ell_1'$	$\le 2\ell_1 + 2\ell_1' + 1$	odd
2	$\ell_1 + 1/2$ $\ell_1 - 1/2$	$\ell_1' - 1/2$ $\ell_1' + 1/2$ $\Big\}$ $j_1 + j_1' = \ell_1 + \ell_1'$		$\le 2\ell_1 + 2\ell_1'$	even
3	$\ell_1 - 1/2$	$\ell_1' - 1/2$	$j_1 + j_1' - 1 = \ell_1 + \ell_1' - 2$	$\le 2\ell_1 + 2\ell_1' - 3$	odd

Table 1 Expression for the 3j–symbol appearing in the form factor (5.40). In the first two lines the exact expression of the 3j–symbol is given, while in the second two lines approximate relations are shown where terms of the order of $1/S$ have been neglected, S being given by $j_1 + j_1' + \lambda$. The standard definitions $0!! = (-1)!! = 1$ are to be used in evaluating these quantities. In the lower part of the table the possible values of $\lambda_{\max}$ and of S are given. The maximum values of λ indicated correspond to the case of no recoil.

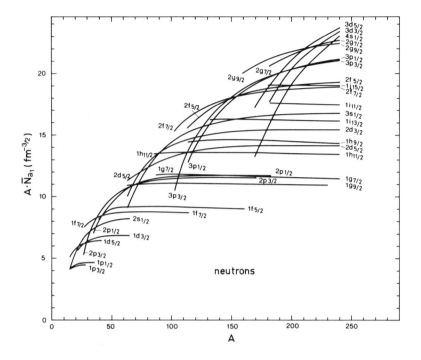

Fig. 10 Normalization constant $\bar{N}_{a_1}$ for proton and neutron single-particle states as defined in Eq. (5.35) times the mass number A, as a function of A. The quantum numbers $a_1 \equiv (n_1 \ell_1 j_1)$ associated with each bound orbital label the different curves. The calculation of the radial single-particle wavefunctions were carried out making use of a Woods-Saxon potential with parameters $r_o = 1.25$ fm and $a = 0.65$ fm. The depth was adjusted for each level individually to reproduce the binding energy obtained by a global fitting and reported in (Myers, 1970 and Døssing et al. 1977).

The dependence of the form factor (48) on the angular momentum transfer λ is contained in the three–j symbol and in the Hankel function. The form factors are largest (cf. Figs. 9 and 12) for the maximum value of

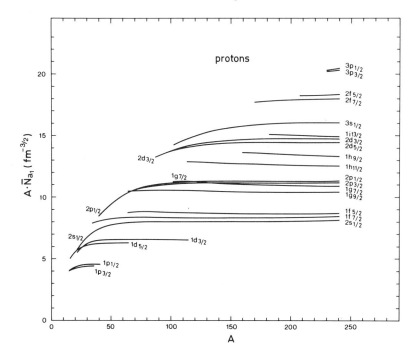

Fig. 10 (cont.) To determine $\bar{N}_{a_1}$, the variable r_{1A} which measures the distance of the particle 1 from the center of mass of the core A was set equal to the Coulomb barrier radius of the $A+1$ system (cf. Eq. (5.43)) in the case of protons and equal to $1.2A_A^{1/3} + 7$ fm in the case of neutrons. In both cases the value $1.25A_A^{1/3}$ fm was chosen for R_A. In the calculations of the modified spherical Hankel function the quantities defined in Eqs. (5.41) were used.

λ, which is either j_1+j_1' or $j_1+j_1'-1$. In fact, one finds from (40) and (48)

$$\frac{f_{\lambda\,\max}}{f_{\lambda\,\max-2}} = \frac{\langle j_1 \tfrac{1}{2} j_1' - \tfrac{1}{2} \mid \lambda_{\max} 0 \rangle\, k_{\lambda\,\max}(\kappa r)}{\langle j_1 \tfrac{1}{2} j_1' - \tfrac{1}{2} \mid \lambda_{\max} - 2\ 0 \rangle\, k_{\lambda\,\max-2}(\kappa r)} \approx \frac{k_{\lambda\,\max}(\kappa r)}{k_{\lambda\,\max-2}(\kappa r)}. \tag{50}$$

Although at large distances this ratio approaches unity, it is at grazing distances often quite large.

The physical picture of Fig. 8 also suggests that since spin is conserved in the transfer, the transition from $j_1' = \ell_1' + 1/2$ to $j_1 = \ell_1 - 1/2$

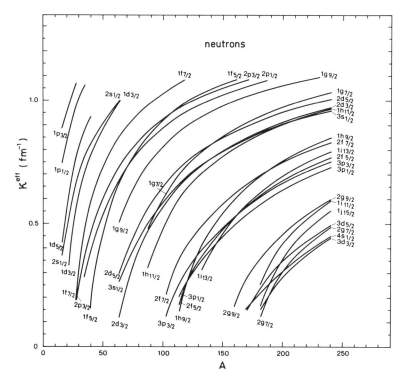

Fig. 11 Effective wavenumber as a function of the mass number A associated with the bound single-particle proton and neutron states. Equations (5.41) and (5.42) have been used with binding energies calculated from a global fit of single-particle levels.

should be favored over the transition to the state $j_1 = \ell_1 + 1/2$. Since, according to Fig. 10, the coefficient $\bar{N}$ is about the same for the states with $j = \ell \pm 1/2$, one finds

$$\frac{f_{\lambda_{\max}}^{j_1=\ell_1-\frac{1}{2},j_1'=\ell_1'+\frac{1}{2}}}{f_{\lambda_{\max}}^{j_1=\ell_1+\frac{1}{2},j_1'=\ell_1'+\frac{1}{2}}} \approx \left(\frac{(2\ell_1+2)(2\ell_1'+1)}{\lambda_{\max}+2} \right)^{1/2} \tag{51}$$

In this, as well as the previous sections, we considered explicitly only stripping reactions. The form factors for pick-up reactions are defined by interchanging the role of target and projectile, i.e. by interchanging

344

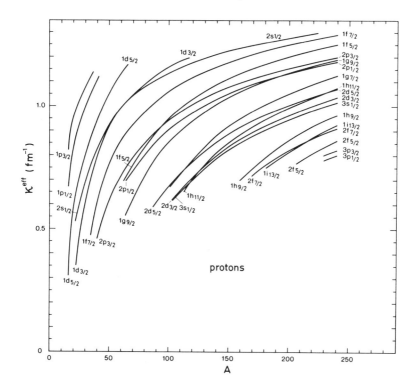

Fig. 11 (cont.) The charge number Z was calculated from the mass number making use of the relation $Z = 0.486A/(1 + A^{2/3}/166)$.

in all expressions a with A and b with B. It is noticed that in the prior representation, which should be used throughout in solving the coupled equations (1.19) the pick-up form factor contains the shell-model

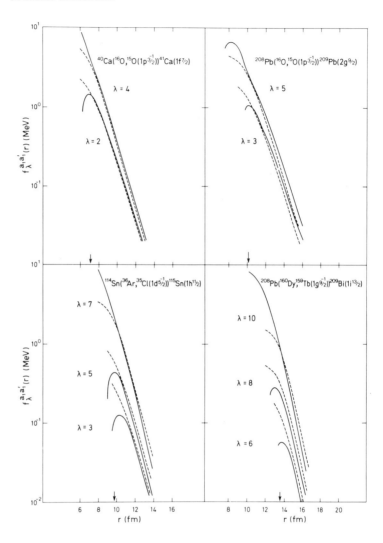

Fig. 12 Single-particle form factors calculated making use of Eq. (5.48) and Figs. 10 and 11, in comparison with the "exact" form factor given by Eq. (5.19) with $\mu = \nu = 0$ (continuous curves). The arrow on the ordinate indicates the value of the sum of the two radii $R_a + R_A$ where $R_i = 1.25\, A_i^{1/3}$ fm $(i = a, A)$.

potential associated with the projectile $U_{1a}(r_{1a})$, i.e.

$$\left(F_{\lambda\mu'}^{(\nu,2n)\alpha}(a_1 a_1' r)\right)^{NP}$$

$$= \frac{4\pi^{3/2}(-1)^n}{2^{\mu'+2n}\,n!(\mu'+n)!\nu!}\sqrt{2j_1'+1}(-1)^{j_1+\lambda+1/2}\begin{Bmatrix} \ell_1 & \ell_1' & \lambda \\ j_1' & j_1 & \frac{1}{2} \end{Bmatrix}$$

$$\times \left(\frac{4m_b m_A}{(m_A+m_B)(m_a+m_b)}\right)^3 \int_{y>o} y\,dy\,dz\,R_{a_1'}^{(a)}(r_{1a})R_{a_1}^{(B)}(r_{1B}) \quad (52)$$

$$\times (U_{1a}-\langle U_{1a}\rangle)\left(\frac{y}{r}\right)^{\mu'+2n}\left(\frac{z-z_\beta}{r}\right)^\nu$$

$$\times \sum_{m_1 m_1'}\langle \ell_1' m_1' \ell_1 m_1 \mid \lambda\mu'\rangle\,Y_{\ell_1' m_1'}(\theta_a,0)Y_{\ell_1 m_1}(\theta_B,0).$$

This form factor is related to the stripping form factor associated with the inverse reaction $b+B \rightarrow a+A$ through the expression

$$\left(F_{\lambda\mu'}^{(\nu,n)\alpha}(a_1' a_1 r)\right)^{NP} = (-1)^{\mu'+j_1-j_1'+\lambda}\left(\frac{2j_1'+1}{2j_1+1}\right)^{1/2}$$

$$\times \left(F_{\lambda\mu'}^{(\nu,n)\beta}(a_1 a_1' r)\right)^{NS}, \quad (53)$$

where the stripping form factor is in the post representation. More generally one may prove the relation

$$\left((f_{\lambda\mu}^{J'J(\alpha)}(\vec{k},\vec{r})\right)^{NP} = (-1)^{J-J'+\lambda}\left(\frac{2J'+1}{2J+1}\right)^{1/2}$$

$$\times \left((f_{\lambda\mu}^{JJ'(\beta)}(\vec{k},\vec{r}))^{NS}\right)^* \quad (53a)$$

The expressions given in the present section refer to spherical nuclei. Single-particle form factors for deformed nuclei are discussed in VI.

The expression (48) is useful for low energies where the effect of the recoil can be incorporated by the "scaling" phase $\bar{\sigma}$. For higher bombarding energies, where it is necessary to use the full form factor given by Eq. (1), one may also obtain a simple parametrization, this time on the basis of Eq. (3). If in this equation we set $k_\parallel = k_\perp = 0$, we can, in fact, determine the proportionality factor comparing the resulting expression with Eq. (32). One obtains the result

$$
\left(f_{m_1 m_1'}^{a_1 a_1'}(k_\parallel k_\perp r) \right)_{(S')}
$$

$$
= e^{i\bar{\sigma}(\alpha)} \times \exp\left\{ -\left(\frac{\Delta_x}{2} k_\parallel\right)^2 - \left(\frac{\Delta_y}{2} k_\perp\right)^2 \right.
$$

$$
\left. + \frac{\Delta_y^2}{2} k_\perp \left(\frac{m_1}{\bar{R}_A} + \frac{m_1'}{r - \bar{R}_A} \right) \right\}
$$

$$
\times \sum_\lambda \langle \lambda\mu j_1 m_1 \mid j_1' m_1' \rangle \, Y_{\lambda\mu}\left(\frac{\pi}{2}, 0\right) f_\lambda^{a_1 a_1'}(r). \tag{54}
$$

One can recover the expression (19a) by expanding (54) for small values of $k_\perp$ and $k_\parallel$. Comparing the results with Eq. (23), one may check the accuracy of the estimates given earlier in § 5.a

$$
\Delta_x = \kappa^{-1}
$$

$$
\Delta_y = [2(r - \bar{R}_A)\bar{R}_A/\kappa r]^{1/2}, \tag{55}
$$

where one may use

$$
\bar{R}_A = 1.25 \, A_A^{1/3} \text{fm}. \tag{56}
$$

These comparisons, as well as the direct numerical evaluation of Eq. (6), show that the parametrization (54) is rather accurate (cf. Sørensen (1988)).

At higher bombarding energies, also the transfer to continuum states can play a role. These processes are described by the same form factors, and one meets no difficulties in their evaluation because the wavefunctions of the unbound states are, in the prior representation, cut off by the potential U_{1A}. One may estimate the form factors to resonant states, i.e. states with energy below the Coulomb and centrifugal barriers, by using the parametrization given in Eq. (48) for $f_\lambda^{a_1 a_1'}(r)$ with normalization factors $\bar{N}_{a_1}$ and $\bar{N}_{a_1'}$ obtained from Fig. 10 extrapolating the curves into the unbound region.

6 TRANSFER PROBABILITY

The rate with which transfer reactions take place in a heavy-ion collision can be estimated in much the same way as the rate of Coulomb excitation or the rate of nuclear inelastic scattering was estimated earlier (cf. §§ II.2–3 and § IV.1, respectively). For optimal kinematic conditions, the reaction rate thus depends on the strength parameter

$$\chi_{\text{trans}}^{(\lambda)} = \frac{1}{\hbar}\tau_{\text{char}}f_\lambda(r_o). \tag{1}$$

The action integral χ (in units of $\hbar$) is estimated as the product of the form factor f_λ at the distance r_o of closest approach and the time (3.2) over which it acts

$$\tau_{\text{char}} = \sqrt{\frac{1}{\kappa \ddot{r}_o}} \approx \frac{\sqrt{2(R_a + R_A)/\kappa}}{v}, \tag{2}$$

where κ is the exponential decay constant of the form factor. Since the characteristic time in units of $\hbar$ is typically of the order of 0.3 MeV^{-1} for low bombarding energies (few MeV/nucleon), and since the transfer form factors at the Coulomb barrier are typically of the order of 1 MeV (cf. Fig. 9), it is seen that the strength parameter is often smaller than unity for grazing collisions. If other reactions were not taking place, this means that lowest order perturbation theory should be applicable for the description of such transfer processes. The probability P_{trans} for the transfer is then given by

$$P_{\text{trans}} = |\chi_{\text{trans}}|^2 g(\tilde{\xi}). \tag{3}$$

Here we have added the factor g to take into account the adiabatic cutoff of the probability. The corresponding adiabaticity parameter $\tilde{\xi}$ is modified from the expression (IV.1.13) for inelastic scattering because of the change in charge and mass which takes place in a transfer reaction.

(a) Optimum Q-value

The adiabaticity of the reaction is determined locally during the transfer process. When a proton is transferred, there is no energy dissipation associated with the formal change in Coulomb energy which occurs because the proton is counted as being a part of B instead of a part of a.

349

V   Transfer Reactions

The energy loss in the interaction region is therefore not $\Delta E = E_\beta - E_\alpha$, but should be corrected by the difference in Coulomb energy

$$\Delta E_{\text{eff}} \approx (E_\beta - E_\alpha) - \left(\frac{Z_a Z_A e^2}{r_o} - \frac{Z_b Z_B e^2}{r_o} \right), \tag{4}$$

where r_o is the distance of closest approach.

The physical significance of this correction to ΔE is illustrated in Fig. 13 (cf. Broglia and Winther 1972). The correction is quite significant for light projectiles on heavy targets where it may amount to ≈ 10 MeV per transferred proton. It means that proton stripping from lighter projectiles is favored to highly excited states in heavy targets, while the corresponding proton pick-up is adiabatically forbidden.

Because of the mass transfer, there are also recoil corrections to the effective energy change. They appear similarly because there is no real energy dissipation associated with the formal change in kinetic energy of relative motion

$$\Delta T = \frac{1}{2}(m_{bB} - m_{aA})v^2 \equiv \frac{1}{2}\delta m_o v^2, \tag{5}$$

which occurs when the transferred particle is counted as being part of B instead of part of a. Also, a formal change in potential energy occurs by an amount

$$\Delta U = (\vec{r}_{bB} - \vec{r}_{aA}) \cdot \vec{\nabla} U \tag{6}$$

because of the redefinition of the center-of-mass coordinate. The latter quantity can be estimated by assuming that the transfer takes place at a distance halfway between the nuclear surfaces, i.e.

$$\vec{r}_{bB} - \vec{r}_{aA} = \delta r \frac{\vec{r}}{r}, \tag{7}$$

with

$$\delta r = \frac{1}{2}\left[\frac{\delta m}{m_A}(\bar{R}_A - \bar{R}_a + r) - \frac{\delta m}{m_b}(\bar{R}_a - \bar{R}_A + r) \right],$$

where $\delta m = m_d$ is the mass of the transferred particle (positive for stripping reactions) and $\bar{R}_A$ is given in (5.56).

The "local" energy change is therefore the sum of (4), (5), and (6), i.e.

$$\Delta E_{\text{loc}} = -\Delta E - \frac{\delta Z(Z_b - Z_A)e^2}{r} + \frac{1}{2}\delta m_o v^2 + \delta r \frac{\partial U}{\partial r}, \tag{8}$$

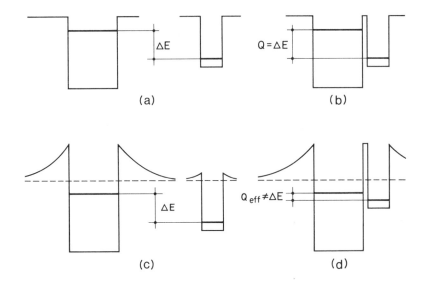

Fig. 13 The figure illustrates effective Q-values to be used in transfer of neutral and charged particles. We represent the two systems by the corresponding single-particle wells in which the particle which is transferred moves. In (a) and (b) we display the situation for a neutral particle for $t = \pm\infty$ and $t = 0$, respectively. The single-particle levels in the initial and final configurations are indicated and the effective Q-value is here equal to the energy difference ΔE between these levels. For the case of charged particles ((c) and (d)) the effective Q-value is the difference between the single-particle levels at $t = 0$, which is equal to the difference in binding energy ΔE, modified by the Coulomb energy (cf. Eq. (6.4)).

where $\delta Z = Z_d$ is the charge of the transferred particle (positive for stripping reactions).

The optimal kinematic condition can be obtained by expressing the energy loss as arising from an effective force $\vec{F}$ (compare §II.2..2) defined by

$$\vec{v} \cdot \vec{F} = \frac{dE_{\text{loc}}}{dt}, \tag{9}$$

or

$$v_r F_r + v_\phi F_\phi = \frac{dE_{\text{loc}}}{dt}, \tag{10}$$

where we decomposed the velocity and the force in a radial and a tangential component. The tangential component F_ϕ of the force is determined from the "local" angular momentum loss ΔL_{loc} perpendicular to the scattering plane, i.e.

$$rF_\phi = \frac{dL_{\mathrm{loc}}}{dt}. \tag{11}$$

Besides the angular momentum transferred to the two nuclei $\hbar\Delta M = (M_b - M_a + M_B - M_A)\hbar$, there is a recoil correction because of the formal change in angular momentum which occurs when the transferred nucleon is counted as being part of B instead of a. The change ΔL_{loc} in angular momentum is therefore

$$\Delta L_{\mathrm{loc}} = -\hbar\Delta M + [(\vec{r}_{bB} - \vec{r}_{aA}) \times m_{\alpha\beta}\vec{v}] \cdot \frac{\vec{L}}{L} = -\hbar\Delta M + \frac{\delta r}{r}L, \tag{12}$$

implying that

$$\frac{dL_{\mathrm{loc}}}{dt} = -\hbar\dot{M} + \frac{1}{r}\frac{\delta r}{\delta t}L. \tag{13}$$

We may now eliminate F_ϕ from (10), (11), and (13) to find

$$v_r F_r = \dot{Q} - \dot{Q}_{\mathrm{opt}} + \hbar\dot{M}\frac{v_\phi}{r}, \tag{14}$$

where $Q = -\Delta E$ is the Q-value of the reaction, while $\dot{Q}_{\mathrm{opt}}$ is defined by

$$\dot{Q}_{\mathrm{opt}} = \frac{(Z_b - Z_A)e^2}{r}\frac{dZ_d}{dt} - \frac{1}{2}\frac{(m_a - m_A)}{m_a + m_A}v^2(0)\frac{dm_d}{dt} + \ddot{r}\,m_{aA}\frac{\delta r}{\delta t}, \tag{15}$$

where dm_d/dt and dZ_d/dt are the rates of mass and charge transfer. In deriving (15), we made use of the relation

$$-\frac{\partial U}{\partial r} + m_{aA}\frac{v_\phi^2}{r} = -\frac{\partial}{\partial r}\left(U + \frac{L^2}{2m_{aA}r^2}\right) = m_{aA}\ddot{r}. \tag{16}$$

In a grazing collision where the reaction only takes place very close to the turning point in the radial motion, the total work done by the radial for $\int v_r F_r dt$ is very small because $v_r(0) = 0$. This implies that the Q-value approximately fulfills the relation

$$Q \approx Q_{\mathrm{opt}} + \mu\hbar\dot{\phi}(0) \tag{17}$$

where

$$Q_{\text{opt}} = \frac{Z_d(Z_b - Z_A)e^2}{r_o} - \frac{1}{2}\frac{m_d(m_b - m_A)}{m_a + m_A}v^2(0) + m_{aA}\ddot{r}_o\delta r \qquad (18)$$

with (cf.(7))

$$\delta r = \frac{m_d}{2}\left(\frac{\bar{R}_A - \bar{R}_a + r}{m_A} - \frac{\bar{R}_a - \bar{R}_A + r}{m_a}\right). \qquad (19)$$

We used here that the total angular momentum transfer $\Delta M\hbar = -\mu\hbar$ and the total mass and charge transfer m_d and Z_d also take place close to $t = 0$.

For more intimate collisions where multistep transfer take place, we approach the situation in which the frictional forces change the trajectory in a continuous way through the "local" average rate of energy, mass, charge, and angular momentum transfer (cf. Chapters 6 and 8).

The argumentation used here is similar to the one used in Section II.2 and the result (17) agrees with the result (II.2.8) since for inelastic scattering $Z_d = m_d = 0$, i.e., $Q_{\text{opt}} = 0$. The parameter $\tilde{\xi}$ introduced in (3) above is correspondingly defined as (cf. II.2.7)

$$\tilde{\xi} = \frac{1}{2}\left(\mu\,\dot{\phi}(0) - \frac{Q - Q_{\text{opt}}}{\hbar}\right)\tau_{\text{char}}. \qquad (20)$$

The adiabatic cut-off function $g(\tilde{\xi})$ will be calculated from the expression for the transfer amplitude in first-order perturbation theory below.

(b) Evaluation of transfer probability

For a stripping reaction the first-order expression (3.12) in the coordinate system A can be written in the form

$$a(t) = -i\sum_{JJ'MM'}\langle I_b M_b J'M' \mid I_a M_a\rangle\,\langle I_A M_A JM \mid I_B M_B\rangle\, I_{MM'}^{JJ'}$$

$$(21a)$$

with

$$I_{MM'}^{JJ'}(t) = \frac{1}{\hbar}\int_{-\infty}^{t} dt'\, e^{i\left(\gamma_{\beta\alpha}(t') + \left(E_\beta - E_\alpha\right)t'\right)/\hbar} f_{MM'}^{JJ'}\big(\vec{k}(t'), \vec{r}(t')\big), \qquad (21b)$$

V Transfer Reactions

where we used (4.2) and (4.31). Utilizing (4.52) and (4.53), we find in the coordinate system A

$$a(t) = -i \sum_{\substack{JJ'MM' \\ \lambda\mu\mu'}} \langle I_A M_A J M \mid I_B M_B \rangle \, \langle I_b M_b J'M' \mid I_a M_a \rangle$$

$$\times \langle \lambda\mu J M \mid J'M' \rangle \, I_{\lambda\mu}(t) \tag{22a}$$

with

$$I_{\lambda\mu}(t) = \sum_{\mu'} D^\lambda_{\mu\mu'}\left(0, \frac{\pi}{2}, \pi\right) \times \frac{1}{\hbar} \int_{-\infty}^t dt' \, \tilde{f}^{JJ'}_{\lambda\mu'}\left(k_\|(t')k_\perp(t')r(t')\right)$$

$$\times \exp\left\{\frac{i}{\hbar}\left[(E_\beta - E_\alpha)t' + \gamma_{\beta\alpha}(t')\right] + i\mu\phi(t')\right\}. \tag{22b}$$

We may estimate the integral over time by utilizing that the form factor in the region of interest is an exponential function. Since the integrand is nonvanishing only for a short period of time around $t = 0$, we may use the approximation

$$r(t) = \frac{1}{2}\ddot{r}_o t^2 + r_o, \tag{23}$$

where r_o is the distance of closest approach for the average trajectory, while $\ddot{r}_o$ is the average acceleration at the turning point. In order to evaluate (22), we use furthermore the approximation (5.20) with

$$k_\| = \frac{m_d \ddot{r}_o}{\hbar} t, \tag{24}$$

and

$$\phi(t) = \dot{\phi}_o t, \tag{25}$$

and the approximation

$$\gamma_{\beta\alpha} = \left(U_\beta(r_o) - U_\alpha(r_o) - \frac{1}{2}(m_\beta - m_\alpha)\left(v(0)\right)^2\right)t \tag{26}$$

to the phase (4.16). The orbital integral in the prior representation is to first order in the recoil

$$I_{\lambda\mu}(t) = \sum_{\mu'=0,\pm1} \frac{1}{\hbar}\sqrt{\frac{2\lambda+1}{4\pi}} \left(\frac{Lm_d}{\hbar m_{\alpha\beta}}\right)^{\mu'} F^{(0,0)\alpha}_{\lambda\mu'}(JJ', r_o) D^\lambda_{\mu\mu'}\left(0, \frac{\pi}{2}, 0\right)$$

$$\tag{27}$$

$$\times \int_{-\infty}^t dt' \exp\left\{-\frac{1}{2}\kappa\ddot{r}_o t'^2 - \frac{i}{\hbar}\left(Q - Q_{\text{opt}} + \Delta^{(\alpha)}\right)t' + i\mu\dot{\phi}(o)t'\right\},$$

where we used (4.11) and (5.21) except that we include the spectroscopic factors according to (4.48). Furthermore, we have defined

$$\Delta^{(\alpha)} = \frac{1}{2} m_d \, \ddot{r}_o (r_o - \bar{R}_a - \bar{R}_A),\tag{28}$$

and have used that the main part of U_α and U_β is the Coulomb interaction.

It is noted that because of the reflection symmetry in the scattering plane, the orbital integral satisfies the simple relation

$$I_{\lambda\mu}^{(A)} = (-1)^{\pi+\mu} I_{\lambda\mu}^{(A)},\tag{29}$$

in the coordinate system A (cf. also Eq. (5.17)). This implies that $\pi+\mu$ must be even, π being the parity change.

The integral can be evaluated in terms of an error function. For $t = \infty$ we find for low recoil the result

$$I_{\lambda\mu} = \left(\frac{2\lambda+1}{2\hbar^2 \kappa \ddot{r}_o}\right)^{1/2} \sum_{\mu'=0,1} \left(\frac{Lm_d}{\hbar m_{\alpha\beta}}\right)^{\mu'} D_{\mu\mu'}^{\lambda}\left(0,\frac{\pi}{2},0\right)\delta_{\mu'}$$

$$\times F_{\lambda\mu'}^{(00)\alpha}(JJ',r_o)\exp\left\{-\frac{\left(Q-Q_{\text{opt}}+\Delta^{(\alpha)}-\mu\hbar\dot{\phi}(0)\right)^2}{2\hbar^2\kappa\ddot{r}_o}\right\}.\tag{30}$$

where

$$\delta_{\mu'} = \delta(\pi+\mu'\text{ even}) \times \begin{cases} 2 & (\mu'>0) \\ \delta(\lambda+\pi,\text{even}) & (\mu'=0) \end{cases}\tag{30a}$$

The terms with $\mu' \geq 2$ are of similar order of magnitude as the ones arising from the expansion in (5.19) for $n \neq 0$.

The result (30) when inserted into (22) offers a simple expression for the one-particle transfer amplitude. It is quite analogous to the expression (IV.9.29) describing inelastic processes. In the application of these formulae to grazing collisions, one has to remember that the turning points, as well as the accelerations, are complex quantities. They are related in a simple way to the quantal amplitude in the distorted-wave Born approximation (cf. Section IV.9 and Section 8 below).

It is seen that the transfer probability

$$p = |a(\infty)|^2\tag{31}$$

V   Transfer Reactions

is given by the expression (3) with

$$g(\tilde{\xi}) = e^{-2\tilde{\xi}^2}, \qquad (32)$$

and with $\tilde{\xi}$ given by (20) except for the term $\Delta^{(\alpha)}$. The occurrence of this term is associated with the post-prior symmetry (3.15) of the transfer amplitude. In changing from prior to post representation, we have shown in Section 5 above (cf. Eqs. (5.52) and (5.53)) that the single-particle form factor in the post representation is approximately given by

$$f_\lambda^{(\beta)}(r) = f_\lambda^{(\alpha)}(r) \exp\left\{(\kappa_{a_1'}^{\text{eff}} - \kappa_{a_1}^{\text{eff}})(r - \bar{R}_a - \bar{R}_A)\right\}, \qquad (33)$$

in terms of the form factor in the prior representation. In the expression for the adiabatic cut-off function g, one should similarly substitute the term $\Delta^{(\alpha)}$ by

$$\Delta^{(\beta)} = -\frac{1}{2} m_d \ddot{r}_o (r_o - \bar{R}_a - \bar{R}_A) = -\Delta^{(\alpha)}. \qquad (34)$$

For optimum Q-value ($\tilde{\xi} = 0$), one finds, according to (5.48), that $\kappa_{a_1}^{\text{eff}} \approx \kappa_{a_1'}^{\text{eff}}$ and the two expressions for the amplitude are the same, except for terms quadratic in the recoil. For non-optimum Q-values, one may show that the difference (33) in the form factor is cancelled by the difference in the adiabatic cut-off function (32) to first order in $Q - Q_{\text{opt}}$.

The stripping probability to a given level for unpolarized target and projectile is according to (22) and (30) given by

$$p_\beta^{NS} = (2I_a + 1)^{-1}(2I_A + 1)^{-1} \sum_{M_a M_b M_A M_B} |a(\infty)|^2$$

$$= \frac{2I_B + 1}{2I_A + 1} \sum_{\substack{\lambda \mu \\ JJ'}} (2\lambda + 1)^{-1}(2J + 1)^{-1} |I_{\lambda\mu}|^2$$

$$\qquad (35)$$

$$= \frac{2I_B + 1}{2I_A + 1} \sum_{\substack{\lambda \mu \\ JJ'}} \frac{1}{2J + 1} \frac{1}{2\hbar^2 \kappa |\ddot{r}_o|} \left| \sum_{\mu'=0,1} \left(\frac{Lm_d}{\hbar m_{\alpha\beta}}\right)^{\mu'} \delta_{\mu'} D_{\mu\mu'}\left(0, \frac{\pi}{2}, 0\right) \right.$$

$$\left. \times F_{\lambda\mu'}^{(00)\alpha}(JJ', r_o) \right|^2 \times \exp\left\{-Re \frac{\left(Q - Q_{\text{opt}} + \Delta^{(\alpha)} - \mu\hbar\dot{\phi}(0)\right)^2}{\hbar^2 \kappa \ddot{r}_o}\right\}.$$

356

In the no-recoil approximation ($\mu' = 0$) we may perform the summation over μ and write the probability in the form

$$P_\beta^{NS} = \frac{2I_B + 1}{2I_A + 1} \sum_{JJ'\lambda} (2J+1)^{-1} \frac{|F_{\lambda 0}^{(00)\alpha}(JJ',r_o)|^2}{2\hbar^2\kappa|\ddot{r}_o|} g(a,b), \qquad (36)$$

where the adiabatic cut-off function is defined by

$$g(a,b) = \sum_{\mu=-\lambda}^{\lambda} \left| D_{\mu_0}^\lambda\left(0,\frac{\pi}{2},0\right)\right|^2 \exp\left(-Re(a - \mu b/\lambda)^2\right) \qquad (37)$$

with

$$a = \left(\hbar^2\kappa\ddot{r}_o\right)^{-1/2}\left(Q - Q_{\text{opt}} + \Delta^{(\alpha)}\right), \qquad (38)$$

and

$$b = \left(\hbar^2\kappa\ddot{r}_o\right)^{-1/2}\lambda\hbar\dot{\phi}(0) = \left(\kappa\ddot{r}_o\right)^{-1/2}\frac{L}{m_{aA}r_o^2}\lambda. \qquad (39)$$

One can estimate the function g by utilizing the approximation

$$\left|D_{\mu_0}^\lambda\left(0,\frac{\pi}{2},0\right)\right|^2 = \frac{2}{\pi}\frac{1}{\sqrt{\lambda^2 - \mu^2}}, \qquad (40)$$

and substituting the sum over μ by an integral. One thus obtains

$$g(a,b) = \frac{1}{\pi}\int_0^\pi \exp[-Re(a - b\cos v)^2]dv. \qquad (41)$$

This function has been evaluated for real values of a and b [Broglia et al., 1981], and the result is given in Fig. 14. It is noted that the imaginary part of a and b only give corrections of second order in the ratio between the imaginary and real parts.

It is seen from this figure how a reaction with a Q-value far from the optimum may be enhanced by a large angular momentum transfer λ. For high excitation energies, there is therefore a strong tendency to populate states with large angular momenta [Bond et al., 1981].

The form factor appearing in (36) can be written in terms of the single-particle form factor evaluated in § 5 through

$$F_{\lambda 0}^{(0,0)}(JJ',r_o)$$

$$= C^*(I_A a_1; I_B)C(I_b a_1'; I_a)f_\lambda^{a_1 a_1'}(r_o) \times \delta(J,j_1)\delta(J',j_1'), \qquad (42)$$

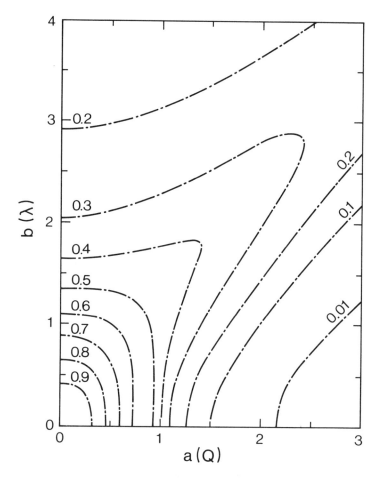

Fig. 14 The adiabatic cut-off function $g(a, b)$ which describes the ratio of the actual transition probability to the same quantity calculated in the sudden approximation. It is defined in (6.37) and is given as a function of the dimensionless parameters a and b. The parameter a which depends on the Q-value is defined in (6.38), while b, which depends on the angular momentum transfer λ, is defined in (6.39).

where we used (4.48) and (4.40). We therefore find

$$p_\beta^{NS} = \frac{2I_B + 1}{2I_A + 1} \sum_\lambda \frac{1}{2j_1 + 1} S(I_A, a_1; I_B) S(I_b a_1'; I_a)$$

$$\left(2\hbar^2 \kappa \mid \ddot{r}_o \mid \right)^{-1} \mid f_\lambda^{a_1 a_1'} (r_o) \mid^2 g(a, b), \tag{43}$$

where

$$S(I_A a_1; I_B) = \mid C(I_A a_1; I_B) \mid^2, \tag{44}$$

is the spectroscopic factor.

The magnitude of S depends on details of the structure of nuclear states involved but is limited by model-independent sum rules. Thus,

$$\sum_{I_B} \frac{2I_B + 1}{(2I_A + 1)(2j_1 + 1)} S(I_A a_1; I_B)$$

$$= \frac{1}{(2I_A + 1)(2j_1 + 1)} \sum_{\substack{I_B M_B \\ M_A m_1}} \mid \langle I_B M_B \mid a_{a_1 m_1}^\dagger \mid I_A M_A \rangle \mid^2$$

$$= \frac{1}{(2I_A + 1)(2j_1 + 1)} \sum_{M_A m_1} \langle I_A M_A \mid a_{a_1 m_1} a_{a_1 m_1}^\dagger \mid I_A M_A \rangle$$

$$= U^2(a_1, I_A) = 1 - \frac{n_{a_1}}{2j_1 + 1}, \tag{45}$$

where $U^2(a_1, I_A)$ is the probability that the j_1-shell is empty in nucleus A. The quantity n_{a_1} is the number of particles in the shell a_1.

Similarly, we find

$$\frac{1}{2j_1' + 1} \sum_{I_b} S(I_b a_1'; I_a)$$

$$= \frac{1}{2j_1' + 1} \sum_{I_b M_b m_1'} \mid \langle I_a M_a \mid a_{a_1' m_1'}^\dagger \mid I_b M_b \rangle \mid^2$$

$$= \frac{1}{2j_1' + 1} \sum_{m_1'} \langle I_a M_a \mid a_{a_1' m_1'}^\dagger a_{a_1' m_1'} \mid I_a M_a \rangle \tag{46}$$

$$= V^2(a_1', I_a) = \frac{n_{a_1'}}{2j_1' + 1}$$

where $V^2(a_1', I_a)$ is the probability that the j_1'-shell is filled in the nucleus a.

At higher bombarding energies where recoil becomes important, one may generalize the expression for the orbital integral (27) by introducing the form factor (5.54) instead of the low-recoil form factor.

If one assumes again the form factor $f_\lambda(r)$ to be exponential and expands $k_\parallel, k_\perp$ and r up to second order in t, one may again evaluate the integral over time to obtain the result (cf. Eq. (21.b))

$$
I^{a_1 a_1'}_{m_1 m_1'} = \left(\frac{2\pi}{\hbar^2 \kappa_{\text{eff}} \ddot{r}_o}\right)^{1/2} \sum_{\lambda,\mu} \langle \lambda \mu j_1 m_1 \mid j_1' m_1' \rangle Y_{\lambda\mu}\left(\frac{\pi}{2},0\right)
$$

$$
\times f^{a_1 a_1'(\alpha)}_\lambda(r_o) \exp\left\{-\frac{(Q - Q_{\text{opt}} + \Delta^{(\alpha)} - \mu\hbar\dot{\phi}_o)^2}{2\hbar^2 \kappa_{\text{eff}} \ddot{r}_o}\right\} \qquad (47a)
$$

$$
\times \exp\left\{\frac{\Delta_y^2 m_d L}{2r_o m_{aA} \hbar}\left[\left(\frac{m_1}{\bar{R}_A} + \frac{m_1'}{r - \bar{R}_A}\right) - \frac{m_d L}{2r_o m_{aA} \hbar}\right]\right\}
$$

with

$$
\kappa_{\text{eff}} = \kappa + \left(\frac{\Delta_x m_d}{\hbar}\right)^2 \frac{\ddot{r}_o}{2} \qquad (47b)
$$

With this expression the post-prior symmetry is satisfied to first order in both $Q - Q_{\text{opt}}$ and $\hbar\dot{\phi}_o \sim k_\perp(t = 0)$.

(c) Spin polarization

The angular momentum dependence of the cross section often gives rise to strong polarization of the two nuclei after the transfer reaction.

The state of polarization after the collision is described by the statistical tensor product for target and projectile defined by

$$
\rho_{k_b \kappa_b, k_B \kappa_B}(I_b, I_B) = (2I_a + 1)^{-1}(2I_A + 1)
$$

$$
\times \sum_{\substack{M_a M_A M_b M_b' \\ M_B M_B'}} \langle I_b M_b k_b \kappa_b \mid I_b M_b' \rangle \langle I_B M_B k_B \kappa_B \mid I_B M_B' \rangle
$$

$$
\times a_{I_a M_a I_A M_A \rightarrow I_b M_b I_B M_B} \times a^*_{I_a M_a I_A M_A \rightarrow I_b M_b' I_B M_B'}. \qquad (48)
$$

It can be written in the form

$$\rho_{k_b \kappa_b, k_B \kappa_B}(I_b, I_B) = \sum_{k\kappa} \langle k_b \kappa_b k_B \kappa_B \mid k\kappa \rangle \, \rho_{k\kappa}(\lambda\lambda')$$

$$F(I_a I_A I_b I_B J J' \lambda\lambda' k_b k_B k), \tag{49}$$

where F is a geometrical coefficient (15-j symbol) and $\rho_{k\kappa}$ is the tensor

$$\rho_{k\kappa}(\lambda\lambda') = \sum_{\mu\mu'} \langle \lambda\mu\lambda' - \mu' \mid k\kappa \rangle \, I_{\lambda\mu}(\infty) I^*_{\lambda'\mu'}(\infty) \tag{50}$$

for the angular momentum transfer. In general, this tensor determines, according to (48), the correlation between the polarization tensors of the two nuclei after the collision. If the state of polarization of b is not measured, it means $k_b = 0$ and the polarization tensor for the target is then proportional to $\rho_{k_B \kappa_B}(\lambda\lambda')$.

The transfer tensor has the general property in coordinate system A that k can take any value up to $2\lambda_{\max}$ while κ according to (29) is even. This means that the target is usually polarized perpendicular to the plane of the motion. This vector polarization is especially pronounced when the reaction has a badly matched Q-value such that large values of μ are dominant in the sum (50).

For well-matched Q-values, it is convenient to consider the orbital integrals in the coordinate system B (cf. Fig. II.2) where the z-axis is pointing along the symmetry axis of the average trajectory. This is approximately the same direction as the direction of motion of the recoiling target. In the no-recoil approximation the orbital integral has in this coordinate system $\mu = 0$, and since the tensor (50) contains the factor $\langle \lambda\, 0\lambda'\, 0 \mid k0 \rangle$, it means that k must be even and κ must be zero. This implies that the state of polarization of the target is an alignment perpendicular to the direction of motion of this nucleus (compare with discussion following Eq. (II.3.6)).

7   ABSORPTION DUE TO TRANSFER

Through particle transfer reactions, the entrance channel is depopulated as a function of time during a heavy-ion collision. We shall in this section consider the case where Coulomb excitation can be neglected. The reactions where surface modes are strongly excited, or reactions with deformed nuclei, are dealt with in Chapter VI. The depopulation is then a function of the impact parameter which rather suddenly sets in when the distance of closest approach becomes comparable to the sum of the nuclear radii. Usually, many transfer channels will open up at about these distances. The situation is schematically depicted in Fig. 15.

As the figure indicates, to evaluate the probability P_o of remaining in the entrance channel after the collision, one must envisage the possibility that there are so many transfer channels n that the sum $p = \Sigma_n p_n$ of the associated probabilities p_n is larger than 1 although every single p_n is very small. The probability P_o can thus not be estimated by $1 - p$, but one should rather use the expression

$$P_o = \prod_n (1 - p_n) \approx \exp\left(-\sum_n p_n\right). \tag{1}$$

In the first relation the assumption that the transfer processes are independent of each other was used, while in the second we assumed $p_n^2 \ll p_n$. The quantities p_n are the first-order transition probabilities for one-particle transfer processes (cf. Eq. (6.43)).

For impact parameters where $p_n \geq 1$ and where perturbation theory breaks down the quantity (1) is so small that the error implied by the use of perturbation theory is of no consequence.

It is noted that in this approximation multi-nucleon transfer is included insofar as it can be considered as the successive transfer of nucleons. The specific effects arising from the two- or four-nucleon correlation energy may, in special cases, be of importance. Generally, it is more important to include the depopulation of the entrance channel due to excitation of surface modes. Assuming that these excitations take place independently of the transfer processes, we may generalize Eq. (1) to include such excitation, i.e.

$$\begin{aligned}
P_o &= P_o^{\text{inel}} \prod_n (1 - p_n) \\
&= \exp\left\{-\sum_n \left(p_n^{NS} + p_n^{NP} + p_n^{TE} + p_n^{PE}\right)\right\},
\end{aligned} \tag{2}$$

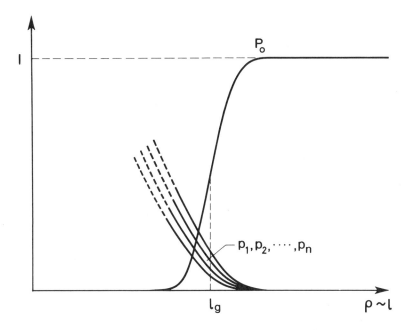

Fig. 15 The probability P_o (transmission function) for the system to remain in the entrance channel as a function of the impact parameter ρ (or the angular momentum ℓ of relative motion). For ℓ-values close to the grazing angular momentum ℓ_g many transfer channels, n, open and the entrance channel is correspondingly depopulated. The quantities p_n indicate the transfer probability as evaluated in perturbation theory.

where NS and NP indicate nucleon stripping and pick-up respectively, while p_n^{TE} and p_n^{PE} are the excitation probabilities in perturbation theory for target and projectile excitation respectively. We used here the result (IV.7.9). The accuracy of Eq. (2) is studied in Chapter VI.

To evaluate the absorptive potential $W(r)$, we use that P_o can also be written as

$$P_o = \exp\left\{ \frac{2}{\hbar} \int_{-\infty}^{\infty} W\big(r(t)\big)\,dt \right\}, \tag{3}$$

V   Transfer Reactions

(see, e.g., III.2.10 or IV.7.17) and obtain

$$\int_{-\infty}^{\infty} W(r(t))\, dt = -\frac{\hbar}{2} \sum_{i,n} p_n^{(i)}. \tag{4}$$

We may therefore write the imaginary part of the potential as a sum of contributions arising from inelastic scattering and from transfer

$$W(r) = W_{\text{in}}(r) + W_{\text{tr}}(r). \tag{5}$$

The quantity $W_{\text{in}}(r)$ was calculated in Section IV.7 from the subsidiary condition that the equation (4) should be satisfied as an identity in the distance of closest approach. This condition essentially amounts to the requirement that $W(r)$ be a local (ℓ-independent) function.

We may use the same approach to calculate W_{tr}. We thus assume W to have a short range so that we may use the approximation

$$r = r_o + \frac{1}{2}\ddot{r}_o t^2, \tag{6}$$

where r_o is the distance of closest approach, $\ddot{r}_o$ being the acceleration at this point. Equation (4) may then be written (Broglia et al., 1981)

$$\sqrt{\frac{2r_o}{\ddot{r}_o}} \int_1^{\infty} \frac{W_{\text{tr}}(r_o x)}{\sqrt{x-1}}\, dx = -\frac{\hbar}{2} \sum_n \left[p_n^{NS}(r_o) + p_n^{NP}(r_o)\right], \tag{7}$$

where p_n^{NS} and p_n^{NP} are the probabilities for one-nucleon stripping and one-nucleon pick-up (cf., e.g., (6.43)). If we approximate furthermore the form factors entering in the expressions locally by an exponential function, we may solve the integral equation (7) for W_{tr} with the result

$$W_{tr}(r) = -\sum_{\beta} \left(\frac{\hbar^2 \ddot{r}_o \kappa_\beta}{4\pi}\right)^{1/2} \left[p_\beta^{NS}(r) + p_\beta^{NP}(r)\right]$$

$$= -\sum_{\beta \lambda a_1 a_1'} \left(16\pi\hbar^2 \kappa_\beta \mid \ddot{r}_0 \mid\right)^{-1/2} \left\{ \frac{2I_B + 1}{(2I_A + 1)(2j_1 + 1)} \right.$$

$$\times \mathcal{S}(I_A a_1; I_B)\mathcal{S}(I_b a_1'; I_a) \left| \left(f_\lambda^{a_1 a_1'(\alpha)}\right)^{NS}(r)\right|^2 g(a_\beta, b_\beta)$$

$$+ \frac{2I_b + 1}{(2I_a + 1)(2j_1' + 1)}$$

$$\left. \times \mathcal{S}(I_a a_1'; I_b)\mathcal{S}(I_B a_1; I_A) \left| \left(f_\lambda^{a_1' a_1(\beta)}\right)^{NP}(r)\right|^2 g(a_\beta, b_\beta) \right\}. \tag{8}$$

For a given single-particle transition specified by the quantum numbers a_1 and a_1', one may, in general, populate different states in the nuclei b and B. If one neglects the dependence of the function g on the spread in energy of the corresponding states, one may carry out the summation over β according to (6.45) and (6.46). One thus finds

$$W_{tr}(r) = - \sum_{a_1 a_1' \lambda} (16\pi \mid \ddot{r}_o \mid \hbar^2 \kappa_\beta)^{-1/2} g(a_\beta b_\beta)$$

$$\times \left\{ (2j_1' + 1) U^2(a_1 I_A) V^2(a_1' I_a) \mid \left(f_\lambda^{a_1 a_1'(\alpha)} \right)^{NS} (r) \mid^2 \right.$$

$$\left. + (2j_1 + 1) V^2(a_1 I_A) U^2(a_1' I_a) \mid \left(f_\lambda^{a_1' a_1(\beta)} \right)^{NP} (r) \mid^2 \right\}$$

$$\approx - \sum_{a_1 a_1' \lambda} (16\pi \hbar^2 \kappa_\beta \mid \ddot{r}_0 \mid)^{-1/2} \mid \left(f_\lambda^{a_1 a_1'(\alpha)} \right)^{NS} (r) \mid^2 g(a_\beta b_\beta)$$

$$\times (2j_1' + 1)[U^2(a_1 I_A) V^2(a_1' I_a) + V^2(a_1 I_A) U^2(a_1' I_a)], \tag{9}$$

where, in the last expression, we used the symmetry relation (5.53) assuming that there is no great difference between single-particle orbitals in b and a or in B and A, and no great difference between post and prior representations for the important transitions close to optimum Q-value.

One may easily include recoil in the absorptive potential by using the expression (6.47) for the orbital integrals. In terms of these quantities, one finds the expression

$$W_{tr}(r) = \sum_{a_1 a_1'} \left(\frac{\kappa \ddot{r}_o \hbar^2}{4\pi} \right)^{1/2} \left\{ U^2(a_1 I_A) V^2(a_1' I_a) \sum_{m m_1'} \left| \left(I_{m_1 m_1'}^{a_1 a_1'} \right)_{r=r_o}^{NS} \right|^2 \right.$$

$$\left. + V^2(a_1 I_A) U^2(a_1' I_a) \sum_{m_1 m_1'} \left| \left(I_{m_1' m_1}^{a_1' a_1} \right)_{r=r_o}^{NP} \right|^2 \right\} \tag{9a}$$

A simple expression is obtained making use of the approximation given by Eq. (5.49) for the form factors. One finds

$$W_{tr}(r) = W_o e^{-2\bar{\kappa}(r-R)}, \tag{10}$$

with

$$W_o = -\left(\frac{\hbar^2}{2M}\right)^2 \left(\frac{R_a R_A}{R_a + R_A}\right)^2 \left(\frac{\pi}{|\ddot{r}_o|\hbar^2 \bar{\kappa}}\right)^{1/2} \sum_{a_1 a_1' \lambda} (2j_1 + 1)(2j_1' + 1)$$

$$\times (2\lambda + 1) \begin{pmatrix} j_1 & j_1' & \lambda \\ -\frac{1}{2} & \frac{1}{2} & 0 \end{pmatrix}^2 \bar{N}_{a_1}^2 \bar{N}_{a_1'}^2 \left(U_{a_1}^2 V_{a_1'}^2 + V_{a_1}^2 U_{a_1'}^2\right) g(a_\beta b_\beta).$$

(11)

This function has a diffuseness parameter $a_W \approx 1/2\bar{\kappa} \approx 0.6$ fm, where κ is the wavenumber associated with a nucleon bound by 8 - 10 MeV in either target or projectile. This diffuseness is approximatel, two times as large as that associated with the absorptive potential due t, inelastic processes (cf. Eq. (IV.7.23)).

The imaginary ion-ion potential thus contains two components. A long-range part due to the depopulation of the entrance channel because of single-particle transfer processes, and a short-range component associated with inelastic scattering (cf. also Baltz et al. (1975)). The potential (10) depends on the surface-surface distance in a similar way as the ion-ion potential. However, in contrast to the ion-ion potential, W_{tr} is proportional to the square of the reduced radius, as was also the case for W_{in} (cf. Eq. (IV.8.1)).

Making use of the form factor (5.19) for $\nu = \mu' = 0$ and a set of single-particle levels obtained from a standard Woods-Saxon potential, the imaginary part of the potential was calculated for a number of cases in Pollarolo et al. (1983). Some of the results are shown in Table 2 and Figs. 16 and 17.

The absorption due to inelastic scattering (cf. Eq. (IV.7.23))

$$W_{in}(r) = -\sigma^2 \sqrt{\frac{\pi a}{|\ddot{r}_o|\hbar^2}} \left(\frac{\partial U_{aA}^N(r)}{\partial r}\right)^2 = K(E)\left(\frac{\partial U_{aA}^N(r)}{\partial r}\right)^2 \quad (12)$$

has also been calculated for the same reactions and bombarding energies. The results are also displayed in Table 2. Situations are encountered where the transfer channels are essentially closed for the given bombarding conditions and the nuclear system becomes rather transparent. A case of this type is found in the reaction $^{16}O + ^{28}Si$ at low bombarding energies. In Fig. 18 the corresponding $W_{tr}(r)$ for a variety of bombarding energies are shown. A very strong variation is observed for the lower

$$^{15}\text{O} + {}^{208}\text{Pb}$$

E_L	$E_{\text{c.m.}}/E_B$	$W_o(E)$ (MeV)	a_W (fm)	$K(E)$ (fm$^2\cdot$ MeV^{-1})
88	1.07	-27.40	0.54	-0.043
96	1.17	-28.16	0.56	-0.045
102	1.25	-28.04	0.57	-0.046
104	1.27	-27.90	0.58	-0.046
129.5	1.58	-24.68	0.60	-0.046
138.5	1.69	-23.60	0.60	-0.045

$$^{16}\text{O} + {}^{40}\text{Ca}$$

E_L	$E_{\text{c.m.}}/E_B$	$W_o(E)$ (MeV)	a_W (fm)	$K(E)$ (fm$^2\cdot$ MeV^{-1})
40	1.21	$-\ 9.54$	0.54	-0.055
50	1.51	-13.67	0.53	-0.059
74	2.24	-17.23	0.52	-0.055
104	3.14	-17.11	0.52	-0.050
140	4.23	-15.92	0.52	-0.045

Table 2 The absorptive potential of $^{16}\text{O} + {}^{208}$ Pb and $^{16}\text{O} + {}^{40}$ Ca at different bombarding energies according to Eqs. (7.10) and (7.12). For the results of (7.10) a Woods-Saxon parametrization $W_{\text{tr}} = W_o(E)\big(1 + \exp(r - R_W)/a_W\big)^{-1}$ was used with $R_W = 1.20(A_a^{1/3} + A_A^{1/3}) - 0.18$.

bombarding energies. This is because under these circumstances all particle transfer channels become closed due to poor Q-value matching. In fact, for $E_{\text{lab}} = 33$ MeV, there are only a couple of transitions which play an important role while four transitions give 90% of the contributions to $W_{\text{tr}}(r)$ at 36 MeV. For higher bombarding energies, many channels open up and the absorption essentially saturates. A detailed discussion of the elastic scattering of $^{16}\text{O} + {}^{28}$ Si also requires a coupled-channel treatment of the Coulomb and inelastic excitation because ^{28}Si is strongly deformed. We thus postpone the discussion of the backward rise phenomenon observed in elastic scattering of $^{16}\text{O} + {}^{28}$ Si until Chapter VI.

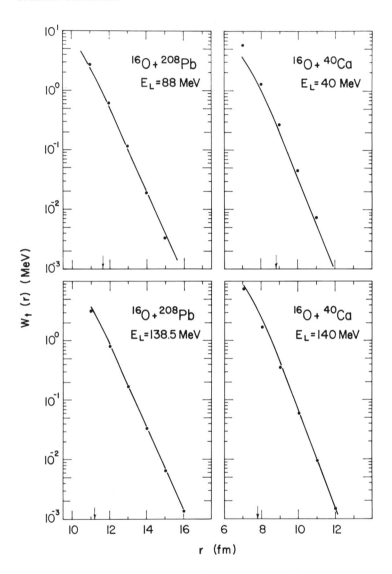

Fig. 16 Absorptive potential associated with the reactions $^{16}\text{O} + ^{40}\text{Ca}$ and $^{16}\text{O} + ^{208}\text{Pb}$ at two bombarding energies. The continuous lines show the quantity (7.9) calculated making use of the parametrized form factors (5.48). The results indicated with dots were obtained making use of the "exact" form factors (5.19). The arrow indicates the grazing distance (III.3.25).

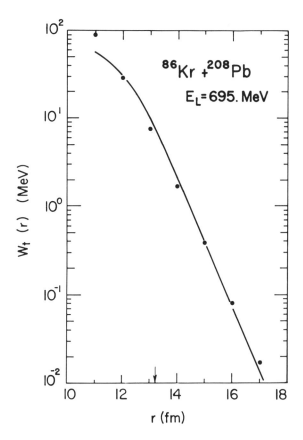

Fig. 17 The absorptive potential W_t associated with the transfer of particles for the reaction ^{86}Kr $+^{208}$ Pb calculated making use of the single-particle experimental spectrum. The full drawn curves indicate the Woods-Saxon fit to the calculations. Approximately 500 transitions contribute to the absorptive potential. The arrow on the abcissa indicates the radius $1.25\left(A_a^{1/3} + A_A^{1/3}\right)$fm. (Quesada *et al.* 1985).

The absorptive potential can vary strongly from one isotope to the other as is shown in Fig. 18 where $W_{tr}(r)$ has been evaluated for the collisions of ^{17}O on ^{28}Si and ^{16}O on ^{29}Si and ^{30}Si.

The adiabatic cutoff function $g(a, b)$ is obtained by summing over the component μ of the angular momentum transfer perpendicular to the scattering plane (cf. (6.37)). The μ-dependence signifies that the depopulation is different for different spin orientations of the colliding nuclei. In fact, one expects that odd-A projectiles with spin opposite to the orbital angular momentum normally are more readily depopulated than projectiles with spin parallel to this direction. Therefore, the elastic scattering in the complex field will lead to scattered particles which are polarized along the angular momentum.

The imaginary potential may thus be considered as the average of separate potentials for each value M_a of the magnetic quantum number of the projectile in coordinate system A. One finds

$$W_{\mathrm{tr}}(M_a, r) = -\sum_{\beta} \left(\frac{\hbar^2 \ddot{r}_o \kappa_\beta}{4\pi} \right)^{1/2} \left(p_\beta^{NS}(M_a; r) + p_\beta^{NP}(M_a; r) \right), \quad (13)$$

where

$$p_\beta^{NS}(M_a; r) = \sum_{M_b m_1 m_1' \lambda \mu} \langle I_b M_b j_1' m_1' \mid I_a M_a \rangle^2 \langle \lambda \mu j_1 m_1 \mid j_1' m_1' \rangle^2$$

$$\times \frac{2\lambda+1}{2\hbar^2 \kappa_\beta |\ddot{r}_o|} U^2(a_1 I_A) \mathcal{S}(I_b a_1'; I_a) \left| D_{\mu o}^\lambda \left(0, \frac{\pi}{2}, 0 \right) \right|^2$$

$$\times \left| f_\lambda^{a_1 a_1'(\alpha)}(r)^{NS} \right|^2 \exp \left\{ -\frac{(Q - Q_{\mathrm{opt}} + \Delta^{(\alpha)} - \mu \hbar \dot{\phi}(o))^2}{\hbar^2 \kappa_\beta \ddot{r}_o} \right\} \quad (14)$$

In this expression we only included the largest value of λ for each transition. A similar expression holds for p_β^{NP}.

The net polarization of the elastically scattered projectile is

$$\frac{\langle M_a \rangle}{I_a} = \frac{1}{I_a} \sum_{M_a} M_a \exp \left\{ -\sum_{\beta} p_\beta^{NS}(M_a, r_o) + p_\beta^{NP}(M_a, r_o) \right\}. \quad (15)$$

These effects would also modify the polarization (cf. §6c) expected in transfer reactions.

Before concluding this section, we note that the absorptive potentials calculated above constitute an upper limit since we have assumed that no repopulation of the entrance channel exists. Effects of coherence in the transfer channels have been studied in coupled-channel calculations by Baldo et al. 1987.

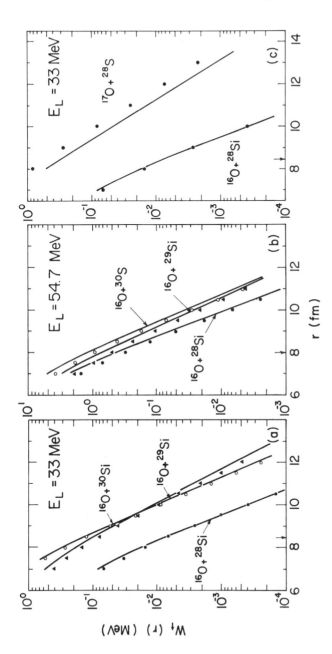

Fig. 18 Absorptive potential (cf. Eq. (7.9)) arising from particle transfer processes associated with different reactions and bombarding conditions. In (a) and (b) the potentials associated with the reaction $^{16}O + ^{28,29,30}Si$ at bombarding energies E_{lab}=33 MeV and 54.7 MeV respectively are shown. The results displayed with continuous curves were obtained making use of the approximated form factors given in Eq. (5.48). In both cases, the results indicated by dots, open circles, and crosses were obtained making use of the "exact" form factors given in Eq. (5.19). In (c) the absorptive potential associated with the reactions $^{17}O + ^{28}Si$ at E_{lab}=33 MeV is shown in comparison with the one associated with the reaction $^{16}O + ^{28}Si$.

8   CROSS SECTIONS FOR TRANSFER

In typical heavy-ion collisions transfer reactions will take place together with strong Coulomb excitation processes. The description of such reactions can only be made by a coupled-channel treatment, which will be the subject of the next chapter.

Even if Coulomb excitation is weak, it is not obvious that first-order perturbation theory is adequate. For bombarding energies above the Coulomb barrier, there will necessarily occur collisions with impact parameters which lead to strong interactions where the transfer process can no longer be treated to lowest order (cf. introduction to § 6). The situation is, however, such that in most cases a large number of transfer channels will open at about the same distance of closest approach. The same argument that was used in connection with the introduction of the absorption in the previous section leads to the following probability for populating a definite transfer channel i

$$
\begin{aligned}
P_i &= p_i \prod_{n \neq i}(1 - p_n), \\
&\approx p_i \exp\left(-\sum_n p_n\right), \\
&\approx p_i P_o,
\end{aligned}
\tag{1}
$$

where again we considered all transfer (and inelastic) channels to be independent. We also used the assumption that in the product (1), there are many channels such that $1 - p_i \approx 1$.

Since the damping factor P_o can be included by using an imaginary part in the potential, the probability p_i can be evaluated from first-order perturbation theory.

(a)   Quantal description

The advantage of the semiclassical treatment discussed in the previous sections is that it allows for simplifications that are useful for higher order processes. In situations where first-order perturbation theory applies, it is quite straightforward to use the corresponding quantal result, the so-called distorted-wave Born approximation (DWBA). The cross section is here given by (cf., e.g., Satchler, 1983)

$$\left(\frac{d\sigma}{d\Omega}\right)_\beta = \frac{k_\beta}{k_\alpha} \frac{m_{aA} m_{bB}}{(2\pi\hbar^2)^2} \mid T_{\beta\alpha} \mid^2, \tag{2}$$

in terms of the T-matrix, which is the matrix element of the interaction

$$T_{\beta\alpha} = \langle \beta \vec{k}_\beta \mid V_{aA} - U_{aA} \mid \alpha \vec{k}_\alpha \rangle \tag{3}$$

between the entrance and exit channels. The total Hamiltonian can be written in two forms, i.e.

$$\begin{aligned} H &= T_{aA} + U_{aA} + H_a + H_A + V_{aA} - U_{aA} \\ &= T_{bB} + U_{bB} + H_b + H_B + V_{bB} - U_{bB}, \end{aligned} \tag{4}$$

where T_{aA} and T_{bB} are the kinetic energies or relative motion in channel α and β, respectively. This implies that the T-matrix can instead be written in the "post" form

$$T_{\beta\alpha} = \langle \beta \vec{k}_\beta \mid V_{bB} - U_{bB} \mid \alpha \vec{k}_\alpha \rangle, \tag{5}$$

the two expressions (3) and (5) being identical because of (4). The relative motion in entrance channel is described by the distorted wave

$$\langle \vec{r}_\alpha \mid \vec{k}_\alpha \rangle = \chi^{(+)}(\vec{k}_\alpha, \vec{r}_\alpha), \tag{6}$$

which is an eigenstate of $T_{aA} + U_{aA}(r_\alpha)$ behaving, at large distances, as a plane wave with wave number $\vec{k}_\alpha$ plus outgoing spherical waves. In the exit channel we have

$$\begin{aligned} \langle \vec{k}_\beta \mid \vec{r}_\beta \rangle &= \chi^{(-)}(\vec{k}_\beta, \vec{r}_\beta)^* \\ &= \chi^{(+)}(-\vec{k}_\beta, \vec{r}_\beta). \end{aligned} \tag{7}$$

It is noted that the identity between (3) and (5) (the post prior symmetry) also holds for complex potentials ($U_{aA} \rightarrow U_{aA} + iW_{aA}$) because of (7). In contrast to the corresponding description of inelastic processes described in § IV.9, the nuclear matrix element $\langle \beta \mid V_{aA} - U_{aA} \mid \alpha \rangle$ cannot be separated from the relative motion. In the integration over the intrinsic degrees of freedom of the nuclear states, $|\alpha\rangle$ and $|\beta\rangle$ one must keep the coordinate $\vec{r}_d$ of the center of mass of the transferred particles fixed since it determines the geometrical relation between the coordinates $\vec{r}_\alpha$ and $\vec{r}_\beta$ appearing in the distorted waves (cf. Fig. 4). The form factor $\langle \beta \mid V_{aA} - U_{aA} \mid \alpha \rangle = f_{\beta\alpha}(\vec{r}, \vec{r}_{d\delta})$ is then a function of two coordinates. Normally, one uses $\vec{r}_{aA}$ and $\vec{r}_{dA}$ or $\vec{r}_{bB}$ and $\vec{r}_{dA}$. For the comparison with the semiclassical description, it is convenient to use instead $\vec{r} = (\vec{r}_\alpha + \vec{r}_\beta)/2$ and $\vec{r}_{d\delta}$ where the point δ is defined in (4.12). Because of the relations (4.9) and (4.10), we may achieve a formal separation by writing the matrix element (3) in the form

$$T_{\beta\alpha} = \int d^3r \; \chi(-\vec{k}_\beta, \vec{r}) \langle \beta \mid e^{i(\vec{K}^\alpha + \vec{K}^\beta)\cdot\vec{r}_{d\delta}} (V_{aA} - U_{aA}) \mid \alpha \rangle \chi(\vec{k}_\alpha, \vec{r}), \quad (8)$$

where the operators

$$\vec{K}^\alpha = \frac{1}{\hbar} \frac{m_d}{2m_o} \vec{p}_\alpha,$$

$$\vec{K}^\beta = \frac{1}{\hbar} \frac{m_d}{2m_o} \vec{p}_\beta,$$

with

$$m_o = \frac{(m_a + m_b)(m_A + m_B)}{4(m_a + m_A)}, \quad (9)$$

are supposed to act on the entrance and exit channel distorted waves respectively. One may now integrate over $r_{d\delta}$ and introduce the form factor

$$f_{\beta\alpha}^{(\alpha)}(\vec{K}, \vec{r}) = \langle \beta \mid e^{i\vec{K}\cdot\vec{r}_{d\delta}} (V_{aA} - U_{aA}) \mid \alpha \rangle_{\vec{r}}, \quad (10)$$

for fixed separation $\vec{r}$. We note that the component of $\vec{K} = \vec{K}^\alpha + \vec{K}^\beta$ perpendicular to $\vec{r}$ is related to the orbital angular momentum, i.e.

$$\vec{K}_\perp = -\hat{r} \times (\hat{r} \times \vec{K}) = -\frac{m_d}{2m_o} \hat{r} \times \frac{(\vec{\ell}_\alpha + \vec{\ell}_\beta)}{r}. \quad (11)$$

It is convenient to evaluate the matrix element (10) in terms of the tensors $f_{\lambda\mu}$ (cf. (4.31) and (4.52)), i.e.

$$f_{\beta\alpha}^{(\alpha)}(\vec{K}, \vec{r}) = \sum_{\substack{JJ' \\ \lambda\mu}} \langle I_A M_A JM \mid I_B M_B \rangle \langle I_b M_b J'M' \mid I_a M_a \rangle$$

$$\times \langle \lambda\mu JM \mid J'M' \rangle f_{\lambda\mu}^{JJ'(\alpha)}(\vec{K}, \vec{r}). \quad (12)$$

Apart from the fact that $\vec{K}$ is an operator while $\vec{k}$ is a c-number, the quantity $f_{\lambda\mu}^{JJ'}$ is identical to the semiclassical form factor. For a single-particle stripping, the form factor is given by

$$f_{m_1 m_1'}^{JJ'}(\vec{K}, \vec{r}) = C^*(I_A a_1; I_B) C(I_b a_1'; I_a)$$
$$\times f_{m_1 m_1'}^{a_1 a_1'}(\vec{K}, \vec{r}) \delta(J, j_1) \delta(J', j_1'), \qquad (13)$$

the C's being the spectroscopic amplitudes. Using the partial wave expansion (cf., e.g., Eq. (IV.9.3))

$$\chi^{(+)}(\vec{k}_\alpha, \vec{r}) = \frac{4\pi}{k_\alpha r} \sum_{\ell_\alpha m_\alpha} i^{\ell_\alpha} \exp(i\beta_{\ell_\alpha}) \chi_{\ell_\alpha}(r)$$
$$\times Y_{\ell_\alpha m_\alpha}^*(\hat{k}) Y_{\ell_\alpha m_\alpha}(\hat{r}), \qquad (14)$$

and correspondingly for the exit channel, one finds

$$T_{\beta\alpha} = \frac{4\pi}{k_\alpha k_\beta} \sum_{\lambda J J'} \langle I_A M_A J M \mid I_B M_B \rangle \langle I_b M_b J' M' \mid I_a M_a \rangle$$
$$\times \langle \lambda m J M \mid J' M' \rangle t_{\beta\alpha}(\lambda J J' m) \qquad (15)$$

where

$$t_{\beta\alpha}(\lambda J J' m) = 4\pi \sum_{\ell_\alpha \ell_\beta} i^{\ell_\alpha - \ell_\beta} \exp(i\beta_{\ell_\alpha} + i\beta_{\ell_\beta})$$
$$\times Y_{\ell_\beta m_\beta}(\hat{k}_\beta) Y_{\ell_\alpha m_\alpha}^*(\hat{k}_\alpha) \langle \ell_\alpha m_\alpha \lambda m \mid \ell_\beta m_\beta \rangle I_{\beta\alpha}, \qquad (16)$$

with

$$I_{\beta\alpha} = (2\ell_\beta + 1)^{-1/2} \int dr \, \chi_{\ell_\beta}(r) \langle \ell_\beta \| f_\lambda \| \ell_\alpha \rangle_r \chi_{\ell_\alpha}(r). \qquad (17)$$

The evaluation of the reduced matrix element was carried out in Broglia et al., 1977. The result may be written

$$\langle \ell_\beta \| f_\lambda \| \ell_\alpha \rangle = \sum_{\mu' > 0} \delta_{\mu'} Z \begin{pmatrix} \ell_\beta & \lambda & \ell_\alpha \\ \tau_\beta & \mu' & \tau_\alpha \end{pmatrix}$$
$$\times C^*(I_A a_1; I_b) C(I_b a_1; I_a) \tilde{f}_{\lambda\mu'}^{a_1 a_1'(\alpha)}(K_\| K_\perp, r). \qquad (18)$$

375

While the Clebsch-Gordan coefficient in (16) describes the angular momentum coupling in the laboratory frame, the coefficient Z describes the corresponding coupling in the intrinsic frame. Since $\tau_\alpha = \tau_\beta = m_d/2m_o$ (cf. ibid., Eqs. (3.16) and (4.13)), one finds

$$
Z \begin{pmatrix} \ell_\beta & \lambda & \ell_\alpha \\ \tau_\beta & \mu' & \tau_\alpha \end{pmatrix}
$$

$$
= \frac{[(2\ell_\alpha + 1)(2\ell_\beta + 1)]^{1/2}}{[2\ell_\alpha(\ell_\alpha + 1) + 2\ell_\beta(\ell_\beta + 1) - \lambda(\lambda + 1)]\mu'/2}
$$

$$
\times \sum_{n=o}^{\mu'} (-1)^{\ell_\beta + n} \begin{pmatrix} \mu' \\ n \end{pmatrix} \left(\frac{(\ell_\beta + n)!(\ell_\alpha + \mu' - n)!}{(\ell_\beta - n)!(\ell_\alpha - \mu' + n)!} \right)^{1/2}
$$

$$
\times \begin{pmatrix} \ell_\beta & \lambda & \ell_\alpha \\ -n & \mu' & -\mu' + n \end{pmatrix}. \tag{19}
$$

The intrinsic form factor $\tilde{f}_{\lambda\mu'}$ is given by (5.6) with $k_\| = K_\|$ and $k_\perp = K_\perp$, while the quantity $\delta_{\mu'}$ is defined by

$$
\delta_{\mu'} = \delta(\ell_\alpha + \ell_\beta + \pi, \text{even}) \times \begin{cases} \delta(\lambda + \pi, \text{even}) & \text{for } \mu' = 0 \\ 2 & \text{for } \mu' > 0 \end{cases}. \tag{20}
$$

The argument of the Bessel function in (5.6) is related to the angular momenta ℓ_α and ℓ_β through (11). Since $\vec{\ell}_\beta = \vec{\ell}_\alpha + \vec{\lambda}$, we find that the magnitude of the transverse momentum is given by

$$
K_\perp r = \frac{m_d}{2m_o}[2\ell_\alpha(\ell_\alpha + 1) + 2\ell_\beta(\ell_\beta + 1) - \lambda(\lambda + 1)]^{1/2}. \tag{21}
$$

The longitudinal recoil effect can be evaluated through a series expansion in powers of $K_\|$. The powers of $K_\| = K_\|^\alpha + K_\|^\beta$ which appear should then be evaluated by letting $K_\|^\alpha$ act on the distorted radial wave $\chi_{\ell_\alpha}(r)$ and respectively $K_\|^\beta$ on $\chi_{\ell_\beta}(r)$, e.g.,

$$
(iK_\|^\alpha)^n \chi_{\ell_\alpha}(r) = \left(\frac{m_d}{2m_o} \right)^n \frac{\partial^n}{\partial r^n} \chi_{\ell_\alpha}(r). \tag{22}
$$

In the prior representation one may include the main part of the longitudinal recoil effect by extracting the average phase

$$
e^{i\bar{\vartheta}(\alpha)} = \exp\left\{ i(K_\|^\alpha + K_\|^\beta)\left(R_A - \frac{m_A + m_B}{2(m_a + m_A)}r \right) \right\} \tag{23}
$$

from the intrinsic form factor. When this shift operator acts on the radial wave functions, it leads to a radial-dependent scaling of the argument of the distorted waves. The radial integral (17) thus takes the form

$$
I_{\beta\alpha} = \int dr\, \chi_{\ell_\beta} \left(\frac{2m_a}{m_a + m_b} r - \frac{m_d}{2m_o} R_A \right) \frac{\langle \ell_\beta \| \bar{f}_\lambda \| \ell_\alpha \rangle}{\sqrt{2\ell_\beta + 1}}
$$

$$
\times \chi_{\ell_\alpha} \left(\frac{2m_b}{m_a + m_b} r + \frac{m_d}{2m_o} R_A \right), \tag{24}
$$

where $\bar{f}_\lambda$ is the form factor (5.6) with the phase factor $\exp(ik_\parallel(z - R_A))$ instead of $\exp(ik_\parallel(z - (m_A + m_B)r/(2(m_a + m_A))))$. The possibility of choosing different origins for the integration over the coordinate of the transferred particle and the scaling of the radial wavefunctions which it implies, has been widely used in DWBA calculations (Buttle and Goldfarb, 1971; Broglia et al., 1977).

The above formulation of the DWBA for transfer reactions was the basis for the computer code DIWRI (Nilsson, 1977). This program calculates the cross sections for unpolarized target and projectile which is given by (cf. (2))

$$
\frac{d\sigma}{d\Omega} = (2I_a + 1)^{-1}(2I_A + 1)^{-1} \frac{k_\beta}{k_\alpha} \frac{m_{aA} m_{bB}}{(2\pi\hbar^2)^2} \sum_{\substack{M_a M_A \\ M_b M_B}} |T_{\beta\alpha}|^2
$$

$$
= \frac{2I_B + 1}{2I_A + 1} \frac{1}{E(\alpha)E(\beta)} \frac{k_\beta}{k_\alpha} \sum_{JJ'\lambda m} \frac{|t_{\beta\alpha}(\lambda J J' m)|^2}{(2J + 1)(2\lambda + 1)}, \tag{25}
$$

where $E(\alpha)$ and $E(\beta)$ are the energies of relative motion in entrance and exit channels, respectively. The results for a number of reactions are displayed in Figs. 19–21. In all calculations, one has used for the potential U_{aA} the expression (III.1.44–45), and for the absorptive potential the one derived in the previous section.

(b) Classical limit of radial matrix elements

The connection between DWBA and the semiclassical description for the transfer cross section can be established in much the same way as was done in Section IV.9 for inelastic scattering.

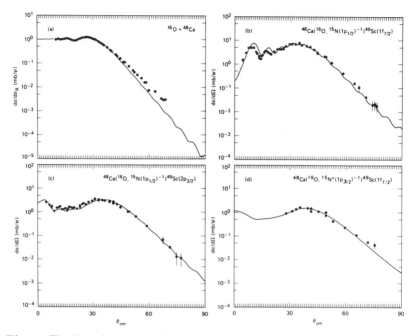

Fig. 19 Elastic and one-particle-transfer cross sections associated with the collision of ^{16}O impinging on ^{48}Ca at $E_{\text{Lab}} = 56$ MeV. In (a) the ratio of the elastic to the Rutherford cross section is shown in comparison with the data (Kovar *et al.* 1978). The standard real potential given in Eqs. (III.1.14–III.1.45) was used in the calculations. The resulting parameters are $V_o =$52.34 MeV, $(r_o)_V =$1.1708 fm, $(a_o)_V =$0.6302 fm. The imaginary optical potential has been calculated making use of the prescription given in (7.8). Approximately 50 single-particle transitions contribute to this imaginary potential. A self-consistent calculation was carried out determining the spectroscopic factors by normalizing the theoretical transfer cross sections to the experimental ones. After three steps, the spectroscopic factors given in Table 3 were obtained.

We first introduce the WKB radial wavefunctions (III.6.12). This implies that the operators $\vec{K}$ can be substituted by c-numbers, i.e.,

$$\vec{K}^{\,\alpha} \rightarrow \frac{m_d}{2m_o}\vec{k}_\alpha(r) \tag{26a}$$

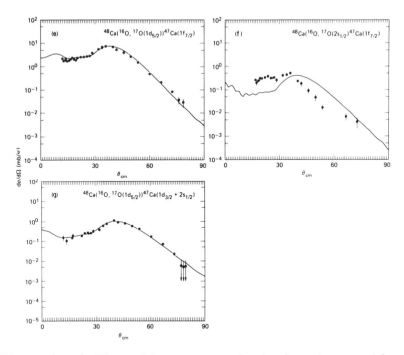

Fig. 19 (cont.) The resulting parameters for the absorptive potential are W_o =31.55 MeV, $(r_o)_W$ =9.289 fm, $(a_o)_W$ =0.572 fm. In (b) and (c) the angular distributions associated with the reactions ^{48}Ca$(^{16}$O,^{15}N$(1p_{1/2}^{-1}))^{49}$Sc(j) are shown, ^{15}N$(1p_{1/2}^{-1})$ indicating the ground state of the outgoing particle. Figure (d) shows the differential cross section associated with the reaction ^{48}Ca$(^{16}$O,^{15}N$(1p_{3/2}^{-1}))^{49}$Sc$(1f_{7/2})$ where the outgoing particle is left in an excited state, while the residual nucleus is in the ground state. In (e), (f), and (g) are displayed angular distributions of one neutron pick-up processes. This figure is due to J. M. Quesada (Quesada 1955).

and

$$\vec{K}^\beta \;\longrightarrow\; \frac{m_d}{2m_o}\,\vec{k}_\beta(r). \tag{26b}$$

These substitutions with the local wavenumbers also hold inside the classical turning point, where they are imaginary.

The radial matrix element appearing in (17) may be evaluated as

Reaction	E_x(MeV)	$\mathcal{S}'$	$\mathcal{S}$	$(\mathcal{S}\mathcal{S}')^{a)}$
$^{15}N(1p_{1/2})^{-1}\ {}^{49}Sc(1f_{7/2})$	0.00	1.0	0.74	0.89
$^{15}N(1p_{1/2})^{-1}\ {}^{49}Sc^*(2f_{3/2})$	3.08	1.0	0.32	0.43
$^{15}N^*(1p_{3/2})^{-1}\ {}^{49}Sc(1f_{7/2})$	6.32	1.08	0.74	1.36
$^{17}O(1d_{5/2})\ {}^{47}Ca(1f_{7/2})$	0.00	1.0	0.50	0.54
$^{17}O^*(2s_{1/2})\ {}^{47}Ca(1f_{7/2})$	0.89	0.56	0.50	0.32
$^{17}O(1d_{5/2})\ {}^{47}Ca^*(2s_{1/2}+1d_{3/2})$	2.60	1.0	0.25	0.36

$^{a)}$Kovar et al. 1978

Table 3 Spectroscopic factors determined self-consistently from the analysis described in caption to Fig. 19 for the proton stripping and neutron pick-up in the collision of ^{16}O on ^{48}Ca. The results listed in column 3 and 4 are spectroscopic factors for projectile and target, respectively. In the last column is given the corresponding product as we have extracted it from the analysis of Kovar et al. 1978, (cf. Tables 5 and 6). This table is due to J. M. Quesada (Quesada 1985).

described in Eq. (IV.9.6) leading to the result

$$I_{\beta\alpha} = (2\ell_\beta + 1)^{-1/2} \int \chi_{\ell_\beta}(r)\,\langle\ell_\beta\|f_\lambda\|\ell_\alpha\rangle\,\chi_{\ell_\alpha}(r)dr$$

$$= -\frac{\sqrt{k_\alpha k_\beta}}{4(2\ell_\beta + 1)^{1/2}} \int_C \langle\ell_\beta\|f_\lambda\|\ell_\alpha\rangle\,(k_\alpha(r)k_\beta(r))^{-1/2}$$

$$\times \exp\left\{i\int_{r_{o\beta}}^r k_\beta(r')dr' - i\int_{r_{o\alpha}}^r k_\alpha(r')\,dr'\right\}dr. \qquad (27)$$

The path C which is indicated in Fig. IV.15 starts at $r = \infty$ and circumvents the branch points at the turning points $r_{o\alpha}$ and $r_{o\beta}$ returning to $r = +\infty$. For inelastic scattering, the outermost turning point is mostly the one, $r_{o\beta}$, corresponding to the exit channel, and the turning points should then, as indicated in Fig. IV.15, be circumvented anti-clockwise. For transfer reactions it may well happen (for $Q - Q_{\mathrm{opt}} > 0$) that $r_{o\beta}$ is the innermost turning point. The path of integration should then be reversed such that wavefunction χ_{ℓ_α} between the turning points is regular.

The great accuracy of the WKB approximation (27) to the radial matrix element has been checked by Landowne et. al., 1976.

^{208}Pb $+^{16}$O $\rightarrow$^{207}Pb(j^{-1}) $+^{17}$O				^{208}Pb $+^{16}$O $\rightarrow$^{209}Bi(j) $+^{15}$N			
$(n\ell j)$	E_x(MeV)	S	$S^{a)}$	$(n\ell j)$	E_x(MeV)	S	$S^{a)}$
$3p_{1/2}$	0.00	0.73	0.95	$1h_{9/2}$	0.00	1.58	0.71
$2f_{5/2}$	0.57	0.61	0.94	$1f_{7/2}$	0.90	1.34	0.85
$3p_{3/2}$	0.88	0.93	0.92	$1i_{13/2}$	1.61	1.12	0.52
$1i_{13/2}$	1.60	2.08	0.87	$2f_{5/2}$	2.84	1.18	0.49
$2f_{7/2}$	2.34	0.58	0.70	$3p_{3/2}$	3.12	0.97	0.54
$1h_{9/2}$	3.35	0.52	0.84	$3p_{1/2}$	3.64	0.68	0.40

$^{a)}$Pieper et al. 1978

Table 4 Spectroscopic factors determined self-consistently from the analysis described in connection with Fig. 19. The results listed in column 3 are based on a spectroscopic factor unity for the transitions ^{16}O $\rightarrow$^{17}O $(d_{5/2})$ and ^{16}O $\rightarrow$^{15}N $(1p_{1/2}^{-1})$. The numbers listed in the fourth column are the results of the analysis by Pieper et al. 1978, where normalization factors of 0.81 and 0.348 were used for the neutron pick-up and proton stripping, respectively. This table is due to J. M. Quesada (Quesada 1985).

If the two turning points $r_{o\alpha}$ and $r_{o\beta}$ are rather close, we introduce the average wavenumber $k(r)$ and velocity $v(r)$ and may write

$$\int_{r_{o\alpha}}^r k_\alpha(r')\,dr' - \int_{r_{o\beta}}^r k_\beta(r')\,dr'$$

$$= \int_{r_o}^r dr' \left\{ (E(\alpha) - E(\beta))\frac{\partial k}{\partial E} + (U_\alpha(r') - U_\beta(r'))\frac{\partial k}{\partial U} \right.$$

$$\left. + (m_{aA} - m_{bB})\frac{\partial k}{\partial m} + (\ell_\alpha - \ell_\beta)\frac{\partial k}{\partial \ell} \right\}. \tag{28}$$

Since the quantities $k_\alpha(r)$ and $k_\beta(r)$ are positive at the starting point of the path C, the time along the trajectory can be defined as

$$t(r) = -\int_{r_o}^r \frac{dr'}{v_r(r')}. \tag{29}$$

where $v_r(r') = \hbar k(r')/m_o$ is the radial velocity on the average trajectory with (complex) turning point r_o.

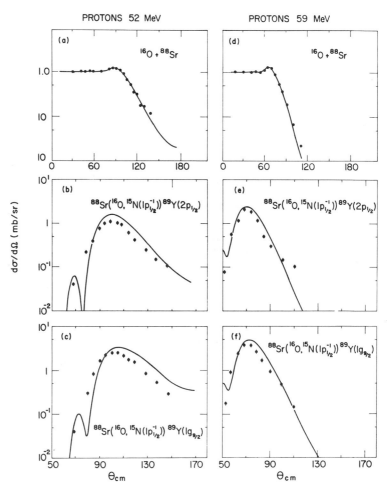

Fig. 20 Angular distribution associated with one-proton stripping reaction ^{88}Sr(^{16}O,^{15}N)^{89}Y at E_{Lab} = MeV and 59 MeV. The standard real potential (III.1.40–45) is used in the calculations leading to parameters V_o =57.79 MeV, $(r_o)_V$ =1.174 fm and $(a_o)_V$ =0.643 fm. Because of the small number of transitions studied, the occupation factors U^2 and V^2 were set to either 0 or 1. The value of the parameters for the imaginary potential (7.9) calculated in this way are: E_{Lab} =52 MeV : W_o =5.821 MeV, $(R_o)_W$ =8.182 fm, and $(a_o)_W$ =0.552 fm; E_{Lab} =59 MeV : W_o =15.201 MeV, $(R_o)_W$ =8.182 fm, and $(a_o)_W$ =0.557 fm. The data are from Anantaraman 1973. The figure is due to J. M. Quesada (Quesada 1985).

PROTONS 138.5 MeV

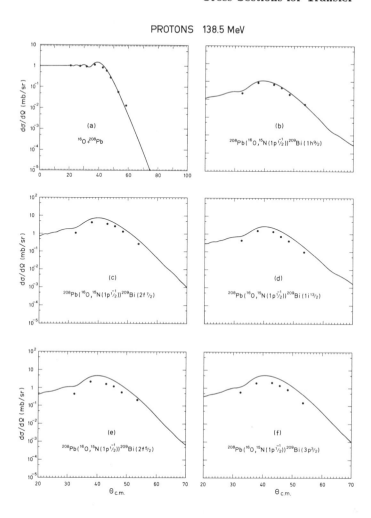

Fig. 21 Elastic and one-particle transfer angular distributions in the collision process of ^{16}O with ^{208}Pb at E_{Lab} =138.5 MeV. The standard real potential (III.1.40–45) is used in the calculations. The value of the parameters are V_o =64.97 MeV, $(r_o)_V$ =1.179 fm, and $(a_o)_V$ =0.658 fm. The imaginary potential due to transfer processes was calculated as explained in the caption to Fig. 19. The resulting spectroscopic factors are given in Table 4. The associated parameters for the absorptive potential are (W_o) =176.65 MeV, $(R_o)_W$ =9.103 fm, and $(a_o)_W$=0.638 fm. In (a) the ratio of the elastic to Rutherford cross section is given.

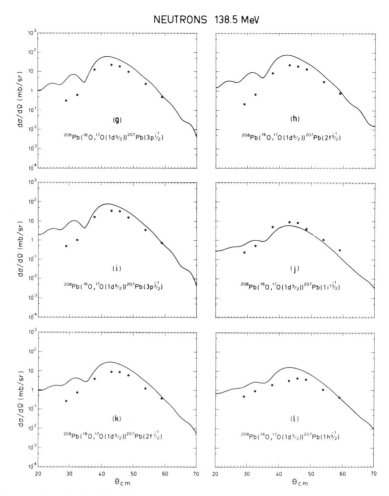

NEUTRONS 138.5 MeV

Fig. 21 (cont.) In (b)–(f) the angular distribution associated with the seven lowest bound proton states closer to the Fermi energy are shown. In (g)–(l) the differential cross section associated with the excitation of the six lowest one-hole neutron states of ^{208}Pb are displayed. The data for the different processes were taken from Pieper *et al.* 1978. One may wonder why the spectroscopic factors cannot be adjusted so as to reproduce the data. This is because decreasing them to obtain a better fit of the transfer data results in a reduction of the absorptive potential which essentially compensates the reduction introduced through the spectroscopic factors. The figure is due to J. M. Quesada (Quesada 1985).

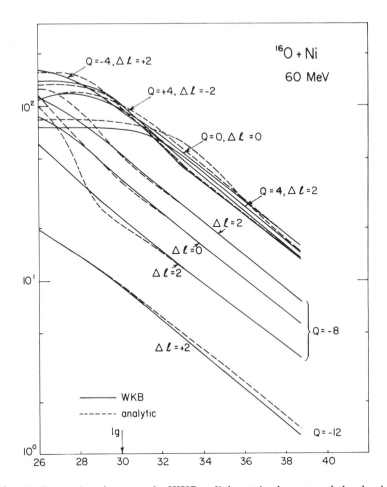

Fig. 22 Comparison between the WKB radial matrix element and the classical analytic expression (6.30) for a proton stripping reaction. For the bombarding condition of 60 MeV ^{16}O on Ni with an optical potential given in Fig. III.9, the radial matrix element (8.27) was evaluated with an exponential form factor $\sim \exp(-\kappa r)$ with $\kappa = 1.17$ fm^{-1}, including the recoil only through $\Delta^{(\alpha)}$. The results for the modulus of the matrix elements are given as functions of $\bar{\ell} = (\ell_\alpha + \ell_\beta)/2$ with full drawn curves for different Q-values and different values of $\Delta \ell = \ell_\beta - \ell_\alpha$. The corresponding classical results in the parabolic approximation are given by dashed curves. The grazing angular momentum is indicated by an arrow. The optimum Q-value for the proton stripping reaction is about -4 MeV, and the results cover a Q-value mismatch of less

When we insert (28) in the expression (27) for the radial integral, we find

$$I_{\beta\alpha} = -\frac{v}{4} \int_C \frac{dr}{v_r(r)} \frac{\langle \ell_\beta \| f_\lambda \| \ell_\alpha \rangle_r}{\sqrt{2\ell_\beta + 1}} e^{-i(\ell_\alpha - \ell_\beta)\phi(t)} e^{i\left((E_\beta - E_\alpha)t + \gamma_{\beta\alpha}(t)\right)/\hbar}$$

where the quantity $\gamma_{\beta\alpha}(t)$ is defined in (4.16), while $\phi(t)$ is the azimuthal angle in the scattering plane defined so that $\phi(0) = 0$. The difference $\ell_\beta - \ell_\alpha$ is equal to the component of $\vec{\ell}_\beta - \vec{\ell}_\alpha$ along the average orbital angular momentum $(\vec{\ell}_\alpha + \vec{\ell}_\beta)/2$, i.e.

$$(\vec{\ell}_\beta - \vec{\ell}_\alpha) \cdot \frac{(\vec{\ell}_\alpha + \vec{\ell}_\beta)}{(\ell_\alpha + 1/2 + \ell_\beta + 1/2)} = \frac{\ell_\beta(\ell_\beta + 1) - \ell_\alpha(\ell_\alpha + 1)}{\ell_\alpha + \ell_\beta + 1}$$

$$= \ell_\beta - \ell_\alpha = \mu, \tag{30}$$

where μ is the angular momentum transferred along the z-axis in the coordinate system A. The radial matrix element is thus proportional to the orbital integral (6.22) in this coordinate system. The coefficient Z defined in (19) can in the limit where ℓ_α and ℓ_β are much larger than λ be approximated by

$$Z = (-1)^{\ell_\alpha + \ell_\beta + \lambda} \sqrt{2\ell_\beta + 1} \, D^\lambda_{\ell_\alpha - \ell_\beta, \mu'} \left(0, \frac{\pi}{2}, 0\right), \tag{31}$$

as can be seen from the asymptomic expression (IV.9.15) for the 3–j symbol. We may therefore write the radial matrix element

$$I_{\beta\alpha} = \frac{v}{4} \int_{-\infty}^{\infty} dt \sum_{\mu'} D^\lambda_{\mu\mu'} \left(0, \frac{\pi}{2}, 0\right) (-1)^{\lambda + \pi}$$

$$\times \tilde{f}^{JJ'}_{\lambda\mu'}(k_\parallel k_\perp r) e^{i\mu\phi(t)} e^{i\left((E_\beta - E_\alpha)t + \gamma_{\beta\alpha}(t)\right)/\hbar}, \tag{32}$$

where $f_{\lambda\mu'}$ is the semiclassical form factor entering in (4.52). For single-particle stripping it takes the explicit form (5.6). Comparing (32) with (6.22) and using (5.17), we see that the radial integral is given by

$$I_{\beta\alpha} = (-1)^\mu \frac{\hbar v}{4} I^{(A)}_{\lambda\mu}(\infty), \tag{33}$$

---

than 8 MeV. For larger mismatches, and especially for $\ell < \ell_g$, the discrepancies become significant. The agreement between the phases of the radial matrix elements is better than 5 degrees for the cases indicated. The results can be compared to the corresponding results for inelastic scattering shown in Fig. IV.18. The calculations are due to M. Guidry and C. Dasso.

in terms of the classical orbital integral in coordinate system A. The definition of the orbital integral has here been generalized in that the classical motion takes place in the complex potential $U + iW$ and that the "time" t also follows a complex path starting at $(-\infty + i0)$ and ending at $(+\infty - i0)$ in such a way that r through the definition (29) circumvents the turning point r_o.

For the evaluation of the orbital integral, we may use the approximation discussed in § 6. The result (6.30) thus applies if one inserts for the quantities entering, the corresponding complex numbers. For $\ddot{r}_o$ one should use

$$\ddot{r}_o = -\frac{1}{m_o}\left(U'(r_o) + iW'(r_o)\right) + \frac{L^2}{m_o^2 r_o^3}, \tag{34}$$

where the primes indicate derivatives, and for $\dot{\phi}(0)$ the complex quantity

$$\dot{\phi}(0) = \frac{1}{m_o}\frac{L}{r_o^2}. \tag{35}$$

A numerical comparison of the radial matrix elements and the analytic expression for the orbital integral is given in Fig. 22. This comparison is analogous to the comparison made in Fig. IV.18 for the inelastic radial matrix elements.

It is noted that the symmetry relation (6.29) is equivalent to the selection rule

$$\pi + \ell_\alpha + \ell_\beta = \text{even} \tag{36}$$

i.e., parity conservation.

It is seen from the present derivation that the matching condition $\tilde{\xi} = 0$ is equivalent to the condition that the turning points in entrance and exit channel coincide. Thus the integrand in (28) is proportional to the shift in r_o because of the change in energy, potential, mass, and angular momentum. An additional change is due to the change in the definition of the center of mass. This change will appear explicitly when in (32) one extracts from the form factor the average phase (23) with $K_\parallel^{(\alpha)} + K_\parallel^{(\beta)} = k_\parallel(t)$ or if one used the scaled radial wavefunctions entering in (24). The total shift is therefore

$$\Delta r_o = \frac{1}{m_o \ddot{r}_o}\left(Q - Q_{\text{opt}} + \Delta^{(\alpha)} + \left(\ell_\alpha - \ell_\beta\right)\hbar\dot{\phi}(0)\right). \tag{37}$$

While in most cases the condition $\Delta r_o = 0$ will lead to the largest matrix element and thus to the largest cross section, this rule may be violated for bombarding energies well below the Coulomb barrier. This happens when the integrand in (27) receives the largest contribution well inside the turning point because the exponential decay of the radial wavefunction is slower than the exponential increase in the form factor. This situation of "transfer below the barrier" is discussed in Appendix E and will be neglected henceforth.

In the remainder of this section we shall complete the semiclassical limit for the cross section by the method of steepest descent as used in § III.4 and § IV.9.

(c) Classical limit of cross sections

Using the laboratory coordinate system (cf. Fig. III.21) with z-axis along the beam direction, we find for large values of ℓ_α and ℓ_β that (16) can be written ($\mu = \ell_\beta - \ell_\alpha$)

$$t_{\beta\alpha}(\lambda J J' m) = \sqrt{4\pi} \sum_{\ell_\alpha \ell_\beta} e^{i\frac{\pi}{2}\mu + 2i\bar{\beta}_\ell}(2\ell + 1)^{1/2}$$

$$\times \langle \ell_\alpha 0 \lambda m | \ell_\beta m \rangle \frac{\hbar v}{4} I_{\lambda\mu}^{(A)}(\infty) Y_{\ell_\beta m}(\theta, \phi), \tag{38}$$

where we introduced the average $\ell = (\ell_\alpha + \ell_\beta)/2$ and the average phase shift $\bar{\beta}_\ell = (\beta_{\ell_\alpha} + \beta_{\ell_\beta})/2$. Neglecting the μ dependence of this quantity and using the asymptotic expression (IV.9.22) and (IV.9.15) for the spherical harmonics and for the Clebsch-Gordan coefficient, we may write (38) in the form

$$t_{\beta\alpha}(\lambda J J' m) = \frac{\hbar v}{\sqrt{8\pi}\sqrt{\sin\theta}}\sum_{\ell}(\ell+1/2)^{1/2}$$

$$\times\left\{\sum_{\mu}\left(D_{\mu m}^{\lambda}\left(\frac{\pi+\theta}{2},\frac{\pi}{2},-\frac{\pi}{2}-\phi\right)\right)^{*}I_{\lambda\mu}^{(A)}(\theta>0)\right.$$

$$\times\exp\left(i\left[2\bar{\beta}_{\ell}-(\ell+1/2)\theta+\frac{\pi}{4}\right]\right)$$

$$+\sum_{\mu}\left(D_{\mu m}^{\lambda}\left(\frac{\pi-\theta}{2},\frac{\pi}{2},\frac{\pi}{2}-\phi\right)\right)^{*}I_{\lambda\mu}^{(A)}(\theta<0)$$

$$\left.\times\exp\left(i\left[2\bar{\beta}_{\ell}+(\ell+1/2)\theta-\frac{\pi}{4}\right]\right)\right\}. \tag{39}$$

For the evaluation of the sum over ℓ we use the Poisson formula (cf. Eq. (III.4.5)). Since β_{ℓ} is a complex function, the "stationary phase" points determined by

$$\Theta(\bar{\ell}) = 2\left(\frac{d\bar{\beta}_{\ell}}{d\ell}\right)_{\ell=\bar{\ell}} = \pm\theta - 2\pi p, \tag{40}$$

lead to complex values of $\bar{\ell}$ (cf. Knoll and Schaeffer 1976 and Schaeffer 1978). The path of integration over ℓ should therefore be deformed to go through these saddle points. The result is

$$t_{\beta\alpha}(\lambda J J' m) = \frac{\hbar v}{2\sqrt{\sin\theta}}\sum_{\bar{\ell}}\frac{(\bar{\ell}+1/2)^{1/2}}{(\Theta'(\bar{\ell}))^{1/2}}$$

$$\times I_{\lambda m}^{(\text{lab})}(\infty)e^{i\left(2\bar{\beta}_{\bar{\ell}}-(\bar{\ell}+1/2)\Theta(\bar{\ell})+\frac{\pi}{2}S(\bar{\ell})\right)}, \tag{41}$$

where $\Theta'(\bar{\ell}) = (d\Theta(\ell)/d\ell)_{\ell=\bar{\ell}}$ and where $S(\bar{\ell}) = +0$ for $0 < \theta(\bar{\ell}) < \pi$ and $S(\bar{\ell}) = -1$ for $-\pi < \theta(\ell) < 0$, corresponding to the use of the upper and the lower sign in (40) respectively. The D-functions in (39) transform the orbital integrals to the frame used in the evaluation of the scattering amplitude (cf. Fig. III.21), the transformation being different for positive and negative scattering angles.

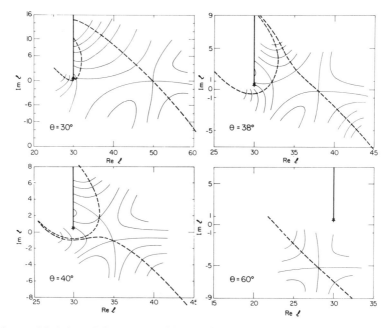

Fig. 23 Modulus of the integrand in the elastic scattering amplitude for a few positive scattering angles (first term in Eq. (III.4.5)) in the complex ℓ-plane for the collision of ^{16}O on ^{48}Ca at 56 MeV bombarding energy. For the nuclear potential we used the Woods-Saxon parametrization (III.1.40) and (III.2.13) with $V_o = 52.34$ MeV, $R_V = 7.693$ fm, $a_V = 0.50$ fm, $W_o = 17.6$ MeV, $R_W = 7.205$ fm, and $a_W = 0.543$ fm. The integrand has a branch point at the orbiting (grazing) angular momentum, and the associated branch cut has been indicated with a vertical line. It is the solution of the equations (III.3.22–23) and is, for the complex potential above, equal to $\ell_g = 30.0 + i\, 0.52$. While the integration in (III.4.5) follows the real axis, one may deform the path of integration below the cut to follow the path of steepest descent, which has been indicated by a dotted curve. For each angle, there are two saddle points indicated by crosses. They determine the main contributions to the integral while the path which was chosen in the valleys is unimportant. For $\theta < 40°$ one has to surpass two saddles in order to go from $\ell = 0$ to $+\infty$. The innermost is associated with nuclear scattering, while the other corresponds to Coulomb scattering. For angles $\theta > 40°$ the topology which is indicated by contour lines changes such that in order to reach $\ell = +\infty$ from $\ell = 0$ one should only surpass one saddle. It is topologically associated with the Coulomb scattering and leads to diffractive scattering (cf. also discussion in § III.7). In the saddle

Inserting (41) in (15), we find the T-matrix

$$T_{\beta\alpha} = \frac{2\pi\hbar v i}{k^2\sqrt{\sin\theta}} \sum_{\bar{\ell}} \frac{(\bar{\ell}+1/2)^{1/2}}{(\Theta'(\bar{\ell}))^{1/2}} a_{\alpha\to\beta}^{(\text{lab})}(\infty)$$

$$\times \exp\left\{i\left(2\bar{\beta}_{\bar{\ell}} - (\bar{\ell}+1/2)\Theta(\bar{\ell}) + \frac{\pi}{2}S(\bar{\ell})\right)\right\}, \tag{42}$$

where $a_{\alpha\to\beta}$ is the excitation amplitude (6.21) evaluated in the same coordinate system in which $T_{\beta\alpha}$ is evaluated. This expression is equivalent, though more general, than the expression (IV.9.20) for inelastic scattering. In (42) we have included the fact that a strong absorptive potential gives rise to an ℓ-dependence of the imaginary part of β_ℓ which cannot always be neglected. This leads to the complication that $\bar{\ell}$ is a complex number (cf. Eq. (III.7.4)). It also means that the orbital integrals which appear in (41) are different from the ones appearing in (38) since they are evaluated with a trajectory with complex ℓ.

For the evaluation of the cross section from (42), it may be convenient to use a coordinate system (D) with z-axis along $\vec{k}_f \times \vec{k}_i$ and with y-axis in the $\vec{k}_i, \vec{k}_f$ plane bisecting the angle between these two vectors. This system coincides with the system A for positive scattering angles while for negative θ it is obtained from the system A by a rotation of $180°$ around the y-axis. The scattering amplitude is in this coordinate system explicitly given by

$$f_{\alpha\to\beta}(\theta, \phi = 0) = -\frac{m_\beta}{2\pi\hbar^2}T_{\beta\alpha}$$

$$= -\sum_{JJ'\lambda} \langle I_A M_A J M \mid I_B M_B\rangle \langle I_b M_b J'M' \mid I_a M_a\rangle \langle \lambda\mu J M \mid J'M'\rangle$$

$$\times \frac{\sqrt{m_{bB}/m_{aA}}}{k\sqrt{\sin\theta}} \sum_{\bar{\ell}} \frac{(\bar{\ell}+1/2)^{1/2}}{(\Theta'(\bar{\ell}))^{1/2}} \left\{ I_{\lambda\mu}^{(A)}(\theta > 0)e^{i[2\bar{\beta}_{\bar{\ell}} - (\bar{\ell}+1/2)\theta]} \right.$$

$$\left. +(-1)^{\lambda+\mu}I_{\lambda-\mu}^{(A)}(\theta < 0)e^{i[2\bar{\beta}_{\bar{\ell}}+(\bar{\ell}+1/2)\theta - \frac{\pi}{2}]}\right\} \tag{43}$$

point approximation one substitutes the contributions from the saddles by a Gaussian integral along a straight line. This is obviously a bad approximation for the innermost saddle at small angles. For negative angles the second term in (III.4.5) gives the main contribution. There is here only one saddle which corresponds to nuclear scattering close to orbiting.

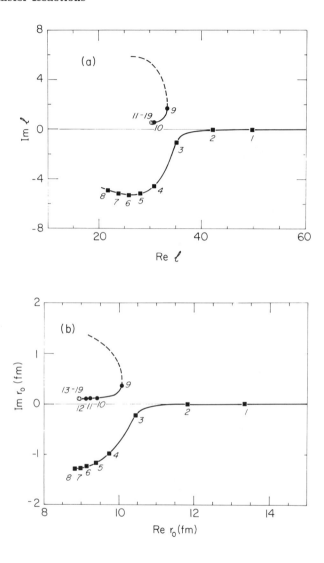

Fig. 24 Values of the angular momenta associated with the saddle points shown in the previous figure are displayed in (a), while in (b), (c), and (d) we show the associated distance of closest approach, the accelerations and the angular velocities at the turning point. For angles less than or equal to 40°, there are two solutions indicated with the numbers 1–3 and 9–22. The associated quantities are almost real.

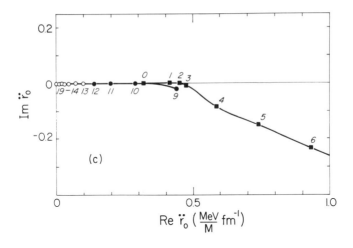

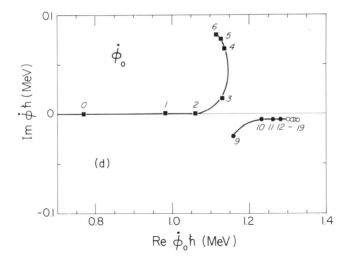

Fig. 24 (cont.) For angles larger than 40°, indicated by the points 4–8, only one saddle point contributes, and the associated diffractive scattering leads to quantities with large imaginary parts. The inner saddle points at 50° and 60°, indicated on the dashed curve, will contribute when one makes an improved calculation through the use of the uniform approximation (cf. § III.E). The figure has been prepared by Enrico Vigezzi.

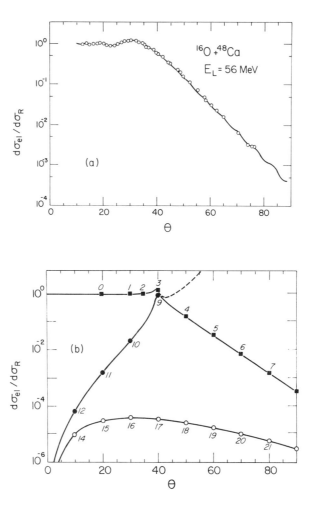

Fig. 25 Angular distributions for elastic scattering and proton transfer reaction in the collision of ^{16}O on ^{48}Ca at a bombarding energy of 56 MeV analyzed in terms of the saddle point approximation (cf. Figs 23 and 24). The experimental data are from Kovar *et al.* 1978. In (b) the ratio of elastic scattering to Rutherford scattering, displayed in (a), is decomposed in contributions arising from the different saddle points. The points which are labelled with the numbers 1–9 are associated with Coulomb and diffractive scattering. They arise from the ℓ-values indicated on Fig. 24(a) with the same labels. The branch labelled with the numbers 10–12 is associated with nuclear scattering

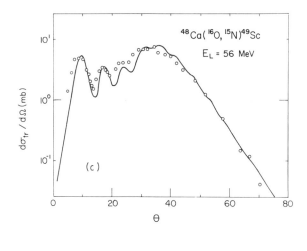

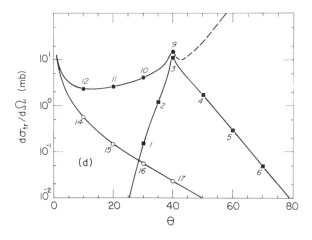

Fig. 25 (cont.) leading to positive angles, while the points 14–22 arising from the second term in (III.4.5) are associated with negative angles. The oscillations in the elastic cross section shown in (a) are for small angles mainly due to the interference of the Coulomb and the nuclear branch, while the oscillations for large angles arise from the interference between the diffractive and the orbiting branch. The results shown include the uniform approximation for angles close to $\theta = 40°$. The angular distribution for the proton transfer reaction $^{48}\text{Ca}(^{16}\text{O},^{15}\text{N}(p_{1/2}^{-1}))^{49}\text{Sc}(f_{7/2})$, shown in (c), and its decomposition (d),

395

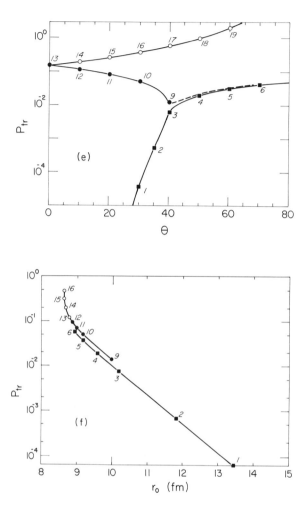

Fig. 25 (cont.) is based on the expression (8.44) and the analytic expression (6.30) for $I_{\lambda\mu}$. Only the no-recoil form factor with $\lambda = 4$, $(10.15\exp(-1.1(r - 7.7))$ was included. The values of $\ell, r_0, \ddot{r}_0$, and $\dot{\phi}_0$, also used in the calculation, are given in Fig. 24. The labelling of the different contributions 1–23 is the same in both figures. The oscillations in forward angles are here mainly associated with the interference between the nuclear branch 11–13 and the orbiting branch 14–16 while the Coulomb branch only contributes rather close $(\theta > 30°)$ to the rainbow angle. The agreement with the experiments (and the DWBA calculation of Fig. 19) close to zero degrees is obtained by including

The cross section for unpolarized target and projectile is therefore

$$\left(\frac{d\sigma}{d\Omega}\right)_{\alpha\to\beta} = \frac{v_\beta}{v_\alpha}\frac{1}{(2I_a+1)(2I_A+1)}\sum_{\substack{M_aM_A\\M_bM_B}}|f_{\alpha\to\beta}(\theta\phi)|^2$$

$$= \frac{2I_B+1}{2I_A+1}\frac{1}{k_\alpha^2\sin\theta}\sum_{\lambda JJ'\mu}(2\lambda+1)^{-1}(2J+1)^{-1}$$

$$\times\left|\sum_{\bar{\ell}}\frac{(\bar{\ell}+1/2)^{1/2}}{(\Theta'(\bar{\ell}))^{1/2}}e^{i[2\bar{\beta}_\ell-(\bar{\ell}+1/2)\Theta(\bar{\ell})]}I_{\lambda\mu}(\bar{\ell})\right|^2 \tag{44}$$

where

$$I_{\lambda\mu}(\ell) = \begin{cases} I_{\lambda\mu}^{(A)} & \text{for } \Theta(\bar{\ell}) = \theta > 0, \\ -i(-1)^{\lambda+\mu}I_{\lambda-\mu}^{(A)} & \text{for } \Theta(\bar{\ell}) = -\theta < 0. \end{cases} \tag{45}$$

The results obtained by means of the formalism discussed above are illustrated for the ^{48}Ca $(^{16}$O, ^{15}N $(p_{1/2}^{-1}))^{49}$Sc$(f_{7/2})$ reaction in Figs 23–25 (Vigezzi and Winther, 1989).

The total cross section is most easily obtained from expressions (16) and (25) by integration over the angles. One obtains

$$\sigma_\beta = \frac{2I_B+1}{2I_A+1}\frac{4\pi}{E(\alpha)E(\beta)}\frac{k_\beta}{k_\alpha}\sum_{\substack{\lambda JJ'\\\ell_\alpha\ell_\beta}}\frac{(2\ell_\beta+1)}{(2J+1)(2\lambda+1)}$$

$$\times e^{-2Im(\beta_{\ell_\alpha}+\beta_{\ell_\beta})}|I_{\beta\alpha}|^2. \tag{46}$$

---

the glory effect. In (e) we display the ratio of the transfer and elastic cross sections for the various branches. This transfer probability shows a monotonic increase from the distant Coulomb scattering 1–3 via the nuclear scattering 9–14 to the orbiting 15–20. This is a direct consequence of the exponential dependence of the transfer probability on the distance of closest approach shown in Fig. 24(b). The dependence is displayed in (f). In the orbiting branch there is a significant increase in collision time. For the present case, where one of the five possible magnetic substates fulfills the condition of optimum Q-value, this leads to an increase in transfer probability. The figure is due to Enrico Vigezzi.

The classical limit is obtained by inserting the expression (33) for $I_{\beta\alpha}$. We thus obtain

$$\sigma_\beta = \frac{2\pi}{k_\alpha^2} \sum_\ell (\ell + 1/2) p_\beta(\ell) e^{-4Im\bar\beta_\ell}, \qquad (47)$$

where we used the definition (6.35) for the transfer probability, $p_\beta(\ell)$, on the average trajectory with impact parameter

$$\rho = \frac{\ell + 1/2}{k}. \qquad (48)$$

Conferring with Eq. (1), we note that

$$p_\beta e^{-4Im\bar\beta_\ell} = P_\beta, \qquad (49)$$

since the absorption is included in the phase shift.

For reactions below the Coulomb barrier where the absorption can be neglected, one may evaluate the expression (47) by utilizing the connection

$$\rho^2 = (r_o - a_o)^2 - a_o^2 \qquad (50)$$

between the impact parameter ρ and the distance r_o of closest approach, where

$$2a_o = \frac{Z_a Z_A e^2}{2E(\alpha)} + \frac{Z_b Z_B e^2}{2E(\beta)}, \qquad (51)$$

is the average distance of closest approach in a head-on collision. Substituting the summation over ℓ by an integral over ρ, one finds

$$\sigma_\beta = 2\pi \int_{2a_o}^\infty (r_o - a_o) P_\beta(r_o) dr_o. \qquad (52)$$

Assuming that the main dependence of the transfer probability (6.36) on r_o is associated with the exponential decay of the form factor $f(r) \sim \exp(-r/a_{tr})$, we may evaluate (52) to find

$$\sigma_\beta = \pi a_{tr}(a_o + \frac{1}{2}a_{tr}) P_\beta(2a_o). \qquad (53)$$

For nearly identical target and projectile, special complications arise because it is not possible to distinguish the scattered particle and the recoil in the exit channel. This indistinguishability gives rise to special interference phenomena between different reaction channels which are discussed in Appendix D.

9 TWO-NUCLEON TRANSFER REACTIONS

In the present and following sections we study the transfer of two nucleons and the special correlation aspects introduced by the pairing interaction.

As discussed in the introduction to Section 8, grazing reactions can be discussed in terms of perturbation theory which, in the case of two-nucleon transfer processes, should be carried out to second order. According to (3.21)–(3.24), the transfer amplitudes are

$$a(\infty) = (a_\beta)_{(1)} + (a_\beta)_{\mathrm{orth}} + (a_\beta)_{\mathrm{succ}}, \tag{1}$$

where

$$(a_\beta)_{(1)} = \frac{1}{i\hbar} \int_{-\infty}^{\infty} dt \, \langle \psi_\beta \mid V_\alpha - \langle V_\alpha \rangle \mid \psi_\alpha \rangle_{\vec{R}_{\beta\alpha}} e^{i(E_\beta - E_\alpha)t/\hbar}, \tag{2}$$

$$(a_\beta)_{\mathrm{succ}} = \left(\frac{1}{i\hbar}\right)^2 \sum_{\gamma \neq \beta} \int_{-\infty}^{\infty} dt \, \langle \psi_\beta \mid V_\gamma - \langle V_\gamma \rangle \mid \psi_\gamma \rangle_{\vec{R}_{\beta\gamma}(t)} e^{i(E_\beta - E_\gamma)t/\hbar}$$

$$\times \int_{-\infty}^{t} dt' \, \langle \psi_\gamma \mid V_\alpha - \langle V_\alpha \rangle \mid \psi_\alpha \rangle_{\vec{R}_{\gamma\alpha}(t')} e^{i(E_\gamma - E_\alpha)t'/\hbar}, \tag{3}$$

and

$$(a_\beta)_{\mathrm{orth}} = -\frac{1}{i\hbar} \sum_{\gamma \neq (\beta,\gamma)} \int_{-\infty}^{\infty} dt \, \langle \psi_\beta \mid \psi_\gamma \rangle_{\vec{R}_{\beta\gamma}(t)}$$

$$\times \langle \psi_\gamma \mid V_\alpha - \langle V_\alpha \rangle \mid \psi_\alpha \rangle_{\vec{R}_{\gamma\alpha}(t)} e^{i(E_\beta - E_\alpha)t/\hbar}. \tag{4}$$

The intermediate states γ are the channels in which one nucleon has been transferred between a and A. The summation should be extended over the same states in (3) and (4). If the summation were extended over all states of the two nuclei f and F (including continuum states), then we may perform the closure in (4) and the non-orthogonality term $(a)_{\mathrm{orth}}$ would then exactly cancel out the first-order term $(a)_{(1)}$. In this limit the two-nucleon transfer is solely given by the second-order term describing the time-ordered successive transfer of the two nucleons.

The time-ordered integral can be written in terms of the first-order, single-particle amplitude by introducing the step function

$$\Theta(t - t') = \frac{1}{2}\left(1 + \varepsilon(t - t')\right) = \begin{cases} 1 & t > t' \\ 0 & t < t' \end{cases}. \tag{5}$$

We thus find

$$(a)_{\text{succ}} = \sum_\gamma \frac{1}{2} \left(\frac{1}{i\hbar}\right)^2 \int_{-\infty}^{\infty} dt \, \langle \psi_\beta \mid V_\gamma - \langle V_\gamma \rangle \mid \psi_\gamma \rangle e^{i(E_\beta - E_\gamma)t/\hbar}$$

$$\times \int_{-\infty}^{\infty} dt \, \langle \psi_\gamma \mid V_\alpha - \langle V_\alpha \rangle \mid \psi_\alpha \rangle e^{i(E_\gamma - E_\alpha)t/\hbar} dt$$

$$+ \sum_\gamma \frac{1}{2} \left(\frac{1}{i\hbar}\right)^2 \int_{-\infty}^{\infty} dt \int_{-\infty}^{\infty} dt' \, \varepsilon(t - t') \langle \psi_\beta \mid V_\gamma - \langle V_\gamma \rangle \mid \psi_\gamma \rangle$$

$$\times \langle \psi_\gamma \mid V_\alpha - \langle V_\alpha \rangle \mid \psi_\alpha \rangle \exp\left\{ \frac{i}{\hbar}(E_\beta - E_\gamma)t + \frac{i}{\hbar}(E_\gamma - E_\alpha)t' \right\}. \quad (6)$$

Furthermore, if we introduce the following integral representation

$$\varepsilon(t - t') = \frac{i}{\pi} \mathcal{P} \int_{-\infty}^{\infty} \frac{dq}{q} \exp\{-i(t - t')q\}, \quad (7)$$

where $\mathcal{P}$ indicates the principal part of the integral, we may write both terms in (6) in terms of the first-order amplitudes $a_{\alpha \to \gamma}(\omega_{\gamma\alpha}) = (a_\gamma)_{(1)}$ of single-particle transfer. We thus find

$$(a)_{\text{succ}} = \sum_\gamma \frac{1}{2} a_{\gamma \to \beta}(\omega_{\beta\gamma}) a_{\alpha \to \gamma}(\omega_{\gamma\alpha})$$

$$+ \frac{i}{2\pi} \mathcal{P} \int_{-\infty}^{\infty} \frac{dq}{q} a_{\gamma \to \beta}(\omega_{\beta\gamma} - q) a_{\alpha \to \gamma}(\omega_{\gamma\alpha} + q). \quad (8)$$

In the first term the energies entering in the amplitudes are equal to those of the physical transition, i.e.

$$\omega_{\beta\gamma} = \frac{1}{\hbar}(E_\beta - E_\gamma), \quad (9)$$

and

$$\omega_{\gamma\alpha} = \frac{1}{\hbar}(E_\gamma - E_\alpha). \quad (10)$$

This part of the second-order process thus corresponds to transitions on the energy shell.

The other part, where the energies are shifted by the amount q corresponds to off-the-energy shell transitions. For real trajectories where the first-order amplitude are purely imaginary, the two terms also offer a separation between the real and the imaginary part of the reaction amplitude. The last term interferes with the terms $(a)_{(1)}$ and $(a)_{\text{orth}}$ which in this case are also purely imaginary.

The first term in (8) indicates the independent transfer of the two nucleons and should be the only one contributing in the independent particle limit. This can be seen from the fact that in this limit the two transitions $\alpha \to \gamma$ and $\gamma \to \beta$ can be interchanged, which means that for each intermediate channel γ_1 there exists another channel γ_2 such that $\omega_{\gamma_1 \alpha} = \omega_{\beta \gamma_2}$ and

$$a_{\alpha \to \gamma_1}(\omega_{\gamma_1 \alpha}) = a_{\gamma_2 \to \beta}(\omega_{\beta \gamma_2}). \tag{11}$$

This implies that

$$\sum_{\gamma_1} a_{\alpha \to \gamma_1}(\omega_{\gamma_1 \alpha} + q) a_{\gamma_1 \to \beta}(\omega_{\beta \gamma_1} - q)$$

$$= \sum_{\gamma_2} a_{\alpha \to \gamma_2}(\omega_{\gamma_2 \alpha} - q) a_{\gamma_2 \to \beta}(\omega_{\beta \gamma_2} + q). \tag{12}$$

The summation over the integrands in the second term in (8) therefore leads to an odd function of q and the principal part integral (summed over γ) vanishes.

It is in this limit that the absorption evaluated in § 7 includes the two-nucleon transfer.

If, on the other hand, the two nucleons are paired, and have an associated correlation energy, it means that significant contributions to the amplitude arise from intermediate states of high excitation energy. For such intermediate states the first term in (8) can be neglected because the amplitudes $a(\omega)$ vanish exponentially when the energy difference ΔE becomes large. The second term in (8), however, may give an important contribution. The integrand in the principal part integral has a pronounced maximum for $q = q_m \equiv (\omega_{\beta \gamma} - \omega_{\gamma \alpha})/2$. Thus for large values of $\omega_{\gamma \alpha}$ we may approximately write

$$\frac{i}{2\pi} P \int_{-\infty}^{\infty} \frac{dq}{q} \sum_{\gamma} a_{\gamma \to \beta}(\omega_{\beta \gamma} - q) a_{\alpha \to \gamma}(\omega_{\gamma \alpha} + q)$$

$$= \frac{i}{2\pi} \sum_{\gamma} \frac{1}{q_m} \int_{-\infty}^{\infty} dq \, a_{\gamma \to \beta}(\omega_{\beta \gamma} - q) a_{\alpha \to \gamma}(\omega_{\gamma \alpha} + q)$$

$$= \frac{1}{i\hbar} \int_{-\infty}^{\infty} dt \, e^{i\omega_{\beta \alpha} t} \sum_{\gamma} \frac{\langle \beta \mid V_\gamma - \langle V_\gamma \rangle \mid \gamma \rangle_t \, \langle \gamma \mid V_\alpha - \langle V_\alpha \rangle \mid \alpha \rangle_t}{\frac{1}{2}(E_\beta + E_\alpha) - E_\gamma}, \tag{13}$$

where we used the explicit form of the single-particle transfer amplitudes in terms of orbital integrals. The contribution from high-lying intermediate states thus only vanish as $(\Delta E_{\gamma\alpha})^{-1}$ in the principal part integral.

Because these transitions of high intermediate energy are virtual transitions, they do not influence the trajectory or relative motion. In fact, one may prove by a quantal calculation (cf., e.g., Alder and Winther, 1975, Eq. (IX.2.26)) that the energy of relative motion during the collision is $\big(E(\alpha)+E(\beta)\big)/2$ because the effective transition energies associated with the two steps are $\Delta E_{\gamma\alpha}+q_m\hbar$ and $\Delta E_{\beta\gamma}-q_m\hbar$.

In the limit (13) we may consider the virtual transitions as giving rise to an induced interaction

$$V^{\text{pol}} = \sum_\gamma \frac{(V_\gamma - \langle V_\gamma \rangle) \mid \gamma\rangle \langle \gamma \mid (V_\alpha - \langle V_\alpha \rangle)}{\frac{1}{2}(E_\beta + E_\alpha) - E_\gamma}, \qquad (14)$$

that causes a first-order transition amplitude

$$a_{\alpha\to\gamma} = \frac{1}{i\hbar} \int_{-\infty}^{\infty} dt\, e^{i\omega_{\beta\alpha}t} \langle \beta \mid V^{\text{pol}} \mid \alpha \rangle. \qquad (15)$$

If we consider the unrealistic limiting situation where the nucleon pair forms a strongly bound cluster such that (13) applies for all intermediate states, i.e.

$$(a_{\alpha\to\beta})_{\text{succ}}$$

$$= \frac{1}{i\hbar} \int_{-\infty}^{\infty} dt\, e^{i\omega_{\beta\alpha}t} \sum_\gamma \frac{\langle \beta \mid V_\gamma - \langle V_\gamma \rangle \mid \gamma\rangle \langle \gamma \mid V_\alpha - \langle V_\alpha \rangle \mid \alpha\rangle}{\frac{1}{2}(E_\alpha + E_\beta) - E_\gamma}, \quad (16)$$

we may in this situation estimate

$$(V_\gamma - \langle V_\gamma \rangle) \mid \gamma\rangle = (H - T_\gamma - H_f - H_F - U_\gamma) \mid \gamma\rangle$$

$$= \left(\frac{1}{2}(E_\alpha + E_\beta) - E_\gamma \right) \mid \gamma\rangle, \qquad (17)$$

by assuming that the energy of relative motion $T_\gamma + U_\gamma$ in the intermediate state according to the discussion above is $\frac{1}{2}\big(E(\alpha)+E(\beta)\big)$. Inserting (17) into (16), we find that the second-order contribution in the limit of strong pairing cancels the non-orthogonality term (4) and the pair transfer is then caused by the first-order amplitude (2) only.

10 POLARIZATION EFFECTS

In the successive transfer of two nucleons, the initial and final states may belong to the same nuclei. Such processes contribute coherently to elastic and inelastic scattering. In this case the excitation goes via the transfer of a nucleon followed by the transfer back to the ground state or to an excited state.

The inelastic excitation amplitude is given, to second order, by Eq. (9.1). The first term (Eq. (9.2)) describes the first-order process ($\beta = \alpha'$) dealt with in Chapter IV. The last terms describe the successive transfer back and forth as the label γ indicates channels where one nucleon has been transferred between a and A.

The non-orthogonality term (9.4) gives a contribution which has to be included even in the independent particle limit. In fact, one has to treat the contributions (9.3) and (9.4) together, which is more economically done using the post-prior representation of the second-order term (cf. Eq. (3.27)). The physical reason for the appearance of the non-orthogonality term is associated with the definition (3.17) of transfer probabilities.

The second-order amplitude leading to the excitation of the target is therefore given by

$$
a_{(2)} = -\frac{1}{\hbar^2} \sum_{\substack{a_1 a_2 a_1' a_2' f F \\ m_1 m_2 m_1' m_2'}} \left\{ \langle \psi^{A'} \mid a(a_2 m_2) \mid \psi^F \rangle \langle \psi^F \mid a^\dagger(a_1 m_1) \mid \psi^A \rangle \right.
$$

$$
\times \langle \psi^a \mid a^\dagger(a_2' m_2') \mid \psi^f \rangle \langle \psi^f \mid a(a_1' m_1') \mid \psi^a \rangle
$$

$$
\times \int_{-\infty}^{\infty} \left(f_{m_2' m_2}^{a_2' a_2(\beta)} \right)^{NP} e^{i(E_A' + E_a - E_F - E_f)t/\hbar} dt
$$

$$
\times \int_{-\infty}^{t} \left(f_{m_1 m_1'}^{a_1 a_1'(\alpha)} \right)^{NS} e^{i(E_F + E_f - E_A - E_a)t'/\hbar} dt'
$$

$$
+ \langle \psi^{A'} \mid a^\dagger(a_2 m_2) \mid \psi^F \rangle \langle \psi^F \mid a(a_1 m_1) \mid \psi^A \rangle
$$

$$
\times \langle \psi^a \mid a(a_2' m_2') \mid \psi^f \rangle \langle \psi^f \mid a^\dagger(a_1' m_1') \mid \psi^a \rangle
$$

$$
\times \int_{-\infty}^{\infty} \left(f_{m_2 m_2'}^{a_2 a_2'(\beta)} \right)^{NS} e^{i(E_A' + E_a - E_F - E_f)t/\hbar} dt
$$

$$
\left. \times \int_{-\infty}^{t} \left(f_{m_1' m_1}^{a_1' a_1(\alpha)} \right)^{NP} e^{i(E_F + E_f - E_A - E_a)t'/\hbar} dt' \right\}. \tag{1}
$$

In what follows, we will be interested in the excitation of collective vibrations, which we describe in the harmonic approximation discussed in § IV.2.

We assume that the single-particle strength associated with, e.g., the product $\langle \psi^{A'} \mid a^\dagger(a_2 m_2) \mid \psi^{F'} \rangle \langle \psi^{F'} \mid a(a_1 m_1) \mid \psi^A \rangle$ is concentrated at one energy $E_{F'}$. We may thus perform the summation over F' and obtain

$$\sum_{F'} \langle \psi^{A'} \mid a^\dagger(a_2 m_2) \mid \psi^{F'} \rangle \langle \psi^{F'} \mid a(a_1 m_1) \mid \psi^A \rangle$$

$$= \langle \psi^{A'} \mid a^\dagger(a_2 m_2) a(a_1 m_1) \mid \psi^A \rangle$$

$$= (-1)^{j_i - m_i + \pi_i} \langle j_k m_k j_i - m_i \mid \lambda \mu \rangle$$

$$\times \begin{cases} X_\lambda(a_k a_i) & (\text{for } a_2 = a_k, \ a_1 = a_i) \\ (-1)^\mu Y_\lambda(a_k a_i) & (\text{for } a_2 = a_i, \ a_1 = a_k) \end{cases}. \tag{2}$$

We have, furthermore, assumed that A labels an even nucleus, $(I_A = 0)$, while $I_{A'}, M_{A'} = \lambda, \mu$.

For the projectile we similarly assume that $I_a = 0$ and that the single-particle strength is concentrated in a single level. We thus obtain (cf. Eq. (6.45)):

$$\langle \psi_a \mid a(a_2' m_2') \mid \psi_f \rangle \langle \psi_f \mid a^\dagger(a_1' m_1') \mid \psi_a \rangle$$

$$= U^2(a_1', I_a) \delta(a_1' a_2') \delta(m_1' m_2'). \tag{3}$$

Performing the angular momentum algebra and using the symmetry relation (5.53), we obtain

$$a_{(2)} = \sum_{\substack{a_1' a_k a_i \\ \lambda_1 \lambda_2 \lambda}} (-1)^{j_1' + j_k + \pi_i + \mu} (2j_1' + 1) \begin{Bmatrix} \lambda_1 & \lambda_2 & \lambda \\ j_i & j_k & j_1' \end{Bmatrix}$$

$$\times \left\{ V^2(a_1' I_a) \left[X_\lambda(a_k a_i) I^{(1)}_{\lambda - \mu} - Y_\lambda(a_k a_i) I^{(2)}_{\lambda \mu} \right] \right.$$

$$\left. - U^2(a_1' I_a) \left[X_\lambda(a_k a_i) I^{(3)}_{\lambda \mu} - Y_\lambda(a_k a_i) I^{(4)}_{\lambda - \mu} \right] \right\} \tag{4}$$

where the quantities $I_{\lambda\mu}^{(i)}$ are second-order orbital integrals of the type

$$I_{\lambda\mu}^{(i)} = \sum_{\mu_1\mu_2} (-1)^{\mu_2} \langle \lambda_1\mu_1\lambda_2\mu_2 \mid \lambda\mu \rangle$$

$$\times \frac{1}{\hbar^2} \int_{-\infty}^{\infty} dt \left(\bar{f}_{\lambda_2-\mu_2}^{a_i a_1'(\alpha)} \right)^{NS^*} e^{i(\hbar\omega - E_{a_k} + E_{a_i})t/\hbar}$$

$$\times \int_{-\infty}^{t} dt' \left(\bar{f}_{\lambda_1\mu_1}^{a_k a_1'(\alpha)} \right)^{NS} e^{i(E_{a_k} - E_{a_1'})t'/\hbar} \tag{5}$$

We utilized here the relation (5.53) between pick-up and the inverse stripping reaction. We may evaluate the double integral analytically within the approximation introduced in § 6.b. In fact, one finds in coordinate system A

$$\int_{-\infty}^{\infty} \left(\frac{dt}{\hbar} \right) f_{\lambda_2\mu_2}^{(2)}(\vec{k}(t), \vec{r}(t)) e^{-iQ_2 t/\hbar} \int_{-\infty}^{t} \frac{dt'}{\hbar} f_{\lambda_1\mu_1}^{(1)}(\vec{k}(t'), \vec{r}(t')) e^{-iQ_1 t'/\hbar} \tag{6}$$

$$\approx \frac{\pi}{\sqrt{\kappa_1\kappa_2}\ddot{r}_o\hbar^2} Y_{\lambda_2\mu_2}\left(\frac{\pi}{2}, 0 \right) Y_{\lambda_1\mu_1}\left(\frac{\pi}{2}, 0 \right) f_{\lambda_2}^{(2)}(r_o) f_{\lambda_1}^{(1)}(r_o) e^{-q^2} w(s),$$

where κ_2 and κ_1 measure the exponential slopes of the two form factors $f^{(2)}$ and $f^{(1)}$, respectively. The function $w(s)$ is defined by

$$w(s) = \frac{i}{\pi} \int_{-\infty}^{\infty} \frac{dt}{s - t + i\varepsilon} e^{-t^2} = g(s) + i\, h(s). \tag{7}$$

The real part of this function is

$$g(s) = e^{-s^2}, \tag{8}$$

while the imaginary part is related to the Dawson integral

$$h(s) = \frac{2}{\pi} \int_{0}^{\infty} \frac{dt}{t} e^{-(t^2+s^2)} \sinh(2ts) = \frac{2}{\sqrt{\pi}} \mathrm{Daw}(s). \tag{9}$$

The argument is

$$s = \sqrt{\frac{\kappa_1\kappa_2}{\kappa_1 + \kappa_2}} \left(\frac{Q_1 - Q_{\mathrm{opt}}^{(1)} + \Delta^{(1)} + \mu_1\hbar\dot{\phi}_0}{\kappa_1\sqrt{2\ddot{r}_o\hbar^2}} \right.$$

$$\left. \frac{Q_2 - Q_{\mathrm{opt}}^{(2)} + \Delta^{(2)} + \mu_2\hbar\dot{\phi}_0}{\kappa_2\sqrt{2\ddot{r}_o\hbar^2}} \right). \tag{10}$$

The functions $g(s)$ and $h(s)$ are displayed in Fig. 26. It is noted that while $g(s)$ is peaked at $s = 0$, $h(s)$ has a maximum at $s = 0.92414$. For large values of s, one may use the expression

$$h(s) \rightarrow \frac{1}{\sqrt{\pi}}\frac{1}{s}, \tag{11}$$

while for small s

$$h(s) \approx \frac{2}{\sqrt{\pi}}s. \tag{12}$$

The factor $\exp(-q^2)$ appearing in (6) is the same adiabatic cutoff function that would appear in a first-order transition with the form factor $f^{(1)} \times f^{(2)}$ and with the Q-value $Q_1 + Q_2$, i.e.

$$q = \frac{Q_1 + Q_2 - Q_{\text{opt}}^{(1)} - Q_{\text{opt}}^{(2)} + \Delta^{(1)} + \Delta^{(2)} - (\mu_1 + \mu_2)\hbar\dot{\phi}_0}{\sqrt{2(\kappa_1 + \kappa_2)\ddot{r}_0\hbar^2}} \tag{13}$$

From this discussion, it follows that the second-order correction (4) to the inelastic scattering can be dealt with by introducing an effective complex form factor. To discuss the physical contents of this form factor, we first investigate the result for the simple case of elastic scattering. The second-order correction can be obtained from (1) by setting $A' = A$ and since

$$\begin{aligned} \langle \psi_A \mid a(a_2 m_2)a^\dagger(a_1 m_1) \mid \psi_A \rangle &= U^2(a_1, I_A)\delta(a_1, a_2)\delta(m_1, m_2) \\ \langle \psi_A \mid a^\dagger(a_2 m_2)a(a_1 m_1) \mid \psi_A \rangle &= V^2(a_1, I_A)\delta(a_1, a_2)\delta(m_1, m_2), \end{aligned} \tag{14}$$

one obtains

$$a_{(2)} = \frac{1}{\hbar^2} \sum_{a_1 a_1' \lambda \mu} \left\{ \frac{2j_1'+1}{2\lambda+1} V^2(a_1', I_a) U^2(a_1, I_A) \right.$$

$$\times \int_{-\infty}^{\infty} dt \left(f_{\lambda\mu}^{a_1 a_1'(\alpha)}(r) \right)^{NS*} e^{i(E_{a_1}-E_{a_1'})t/\hbar}$$

$$\times \int_{-\infty}^{t} dt' \left(f_{\lambda\mu}^{a_1 a_1'(\alpha)}(r) \right)^{NS} e^{-i(E_{a_1}-E_{a_1'})t'/\hbar}$$

$$+ \frac{2j_1+1}{2\lambda+1} U^2(a_1', I_a) V^2(a_1, I_A)$$

$$\times \int_{-\infty}^{\infty} dt \left(f_{\lambda\mu}^{a_1' a_1(\alpha)}(r) \right)^{NP*} e^{i(E_{a_1'}-E_{a_1})t/\hbar}$$

$$\times \int_{-\infty}^{t} dt' \left(f_{\lambda,\mu}^{a_1' a_1(\alpha)}(r) \right)^{NP} e^{-i(E_{a_1'}-E_{a_1})t'/\hbar} \right\}. \tag{15}$$

This expression can be interpreted as a correction to the optical potential by writing the total elastic amplitude in second order as

$$a_0 = 1 + a_{(2)} = 1 + \frac{1}{i\hbar} \int_{-\infty}^{\infty} (\Delta U + iW) dt. \tag{16}$$

The result for W coincides with the expression (7.9) for the imaginary potential. The correction to the real part of the potential ΔU is the part of the polarization potential arising from transfer reactions. It can be obtained from the same expression as (7.9) when one substitutes the function $g(a_\beta b_\beta)$ by the function

$$h(a,b) = \sum_{\mu=-\lambda}^{\lambda} |D_{\mu 0}^\lambda \left(0, \frac{\pi}{2}, 0\right)|^2 \frac{2}{\sqrt{\pi}} \text{Daw}(a - \mu b/\lambda). \tag{17}$$

The function $h(a,b)$ is displayed in Fig. 27 together with the function $g(a,b)$. It is seen that while $g(a,b)$ gives an effective cutoff in $W(r)$ for high intermediate energies, this cutoff is more delicate for the real part ΔU. A discussion of the microscopic calculation of ΔU on this basis is given in Dasso et al., 1986. The effect of the polarization potential has been observed experimentally (Baeza et al., 1984 and Lilley et al., 1985)

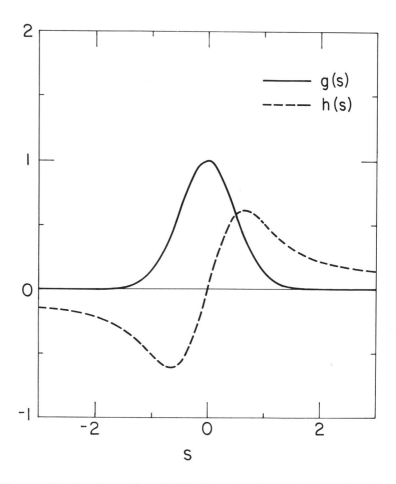

Fig. 26 The function $g(s)$ and $h(s)$ defined in Eqs. (10.7) and (10.8) as a function of s.

and has been discussed on the basis of dispersion relations by Nagarajan *et al.*, 1985 and Mahaux *et al.*, 1986.

The second-order effects (4) in inelastic scattering are quite similar though more elaborate to evaluate. The real part of the second-order amplitude can be described by giving the form factors for inelastic scat-

408

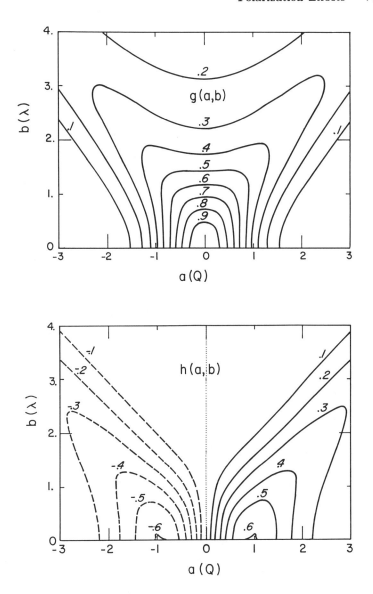

Fig. 27 Contour plots of the adiabatic cutoff function g (top) and the polarization function h (bottom). The axes relate to the energy and angular momentum of the intermediate states. The parameters a and b are defined in (6.38) and (6.39).

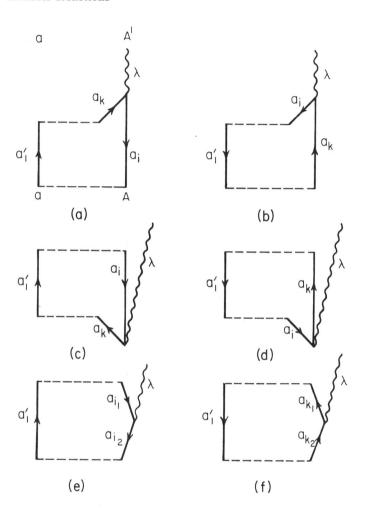

Fig. 28 Graphical representation of second-order corrections to the excitation of collective modes (cf. Fig. IV.4(b)) in the target nucleus. A wavy line describes a phonon, while an arrowed line indicates a particle or a hole of the quantum numbers indicated. States of the projectile are shown in the left part of each graph while those of the target are given to the right. The transfer vertex has thus been represented by horizontal, dashed lines. The first four graphs (a)–(d) describe the RPA results given in Eq. (10.4). Graphs (e) and (f) describe scattering events.

tering an imaginary component. For collective surface vibrations, the origin of this component is associated with the fact that the rate of transfer reactions depends on the instantaneous position of the surface of the target nucleus. If we parametrize the absorptive potential, as in Eq. (7.10), we must expect that R is a function of the deformation parameters $\alpha_{\lambda\mu}$ (cf. Eq. (IV.3.1)) and that one obtains a form factor for inelastic scattering (cf. Eq. (IV.3.15))

$$\langle 1_{\lambda\mu} \mid V_{aA} \mid 0 \rangle = \sqrt{\frac{\hbar\omega_\lambda}{2C_\lambda}} \, R_A^{(o)} \frac{d}{dR_A} (U_{aA} + i\,W_{aA}) Y_{\lambda\mu}^*(\hat{r}), \qquad (18)$$

which contains an imaginary part related to the absorptive potential.

The result (18) for the transfer part of W_{aA} can be proved directly from (1) by assuming that the excitation energy ω of the collective mode can be neglected, i.e. in the sudden approximation. It is noted that a similar correction to the form factor also arises from higher-order effects in inelastic scattering and that this effect for the inelastic part of W_{aA} can also be written in the form (18) in the sudden approximation (cf. Sect. IV.7b).

If the period of oscillation (ω^{-1}) is comparable to or shorter than the collision time, one expects deviations from the simple expression (18).

For a systematic discussion of higher-order effects in heavy-ion collisions, it is useful to take advantage of the graphical representation of the elementary processes as was also suggested in Chapter IV, Fig. 4. The graphs for the four processes in (1) contributing to the simple inelastic scattering are displayed in Fig. 28. Also shown are two scattering graphs which go beyond RPA, but which are expected to contribute significantly.

Transfer reactions are also expected to be influenced by second-order effects from inelastic processes. The strong effects coming from low-lying states mostly have to be included by a coupled channel treatment, as will be discussed in the following chapter. Effects arising from the coupling to giant resonances can, however, be treated in perturbation theory as a polarization effect.

In Fig. 29 we show with graph (1) the simple transfer of a nucleon, while (2a) and (2b) indicate the second-order corrections. Because of the strong coupling, one expects that the four third-order processes are very important. The graphs $(3a_1)$ and $(3a_2)$ can be incorporated in graph (2a) by introducing an effective charge. Similarly, the graphs $(3b_1)$ and $(3b_2)$ can be incorporated in (2b). The two graphs (3c) lie outside this picture and have to be treated explicitly.

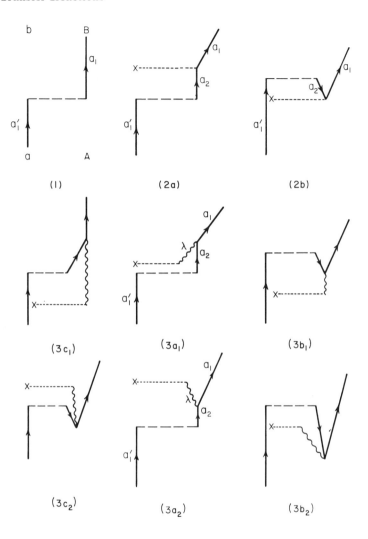

Fig. 29 Graphical representation of a first order stripping reaction and higher-order renormalization processes arising from inelastic scattering. Graphs (2a) and (2b) indicate the corrections from single-particle inelastic events, while (3a) and (3b) can be incorporated in these graphs through an effective charge. The two graphs (3c) have to be calculated explicitly because they are separated by a transfer event.

11 MICROSCOPIC DESCRIPTION OF PAIR TRANSFER

Correlation between pairs of identical particles play an important role in a systematic description of the nuclear structure. These correlations are responsible for the existence of striking regularities in the spectrum of both nuclei far away from and around closed shells. These phenomena, denoted as pairing rotations and vibrations, can be specifically studied making use of two-nucleon transfer reactions (cf., e.g., Broglia *et al.*, 1973 and reference therein).

There is essentially a single number which contains all the nuclear structure information associated with a given pairing rotation or vibrational band. It is the enhancement of the two-nucleon transfer cross section as compared to the cross section associated with a pure two-particle configuration. The enhacement is proportional to the square of the zero-point amplitude in the case of pairing vibrations observed near closed shells. For pair rotations observed in superfluid nuclei far away from closed shells, the enhancement is proportional to the square of the deformation parameter (pairing gap).

The pairing modes have their counterpart in the collective surface vibrations and rotations discussed in Chapters II and IV. In particular, the ratio between the inelastic cross sections calculated making use of the collective form factors to the one resulting from an average particle-hole excitation measures the collectivity of the vibrational excitation in single-particle units. This quantity is proportional to the square of the static quadrupole moment for the excitation of rotational states and to the square of the zero-point amplitudes for the excitation of multipole surface vibrations.

The single-particle form factors and overlaps discussed in Sections 4 and 5 (cf. especially Eq. (4.48)) can be viewed as the building blocks needed to describe multi-nucleon transfer processes. In particular, for the simultaneous (first-order) transfer of the two nucleons

$$a(= b + 2 \text{ nucleons}) + A \rightarrow b + B(= A + 2 \text{ nucleons})$$

we may use the fractional expansion (4.40) with $C^{(A)}$ given by (4.43a), i.e.

$$\tilde{\phi}_{JM}^{B(A)}(\vec{r}_{dA}, \zeta_d) = \sum_{a_1 a_2} \frac{\langle I_B \| [a_{j_1}^\dagger(a_1) a_{j_2}^\dagger(a_2)]_J \| I_A \rangle^*}{(2I_B + 1)^{1/2}}$$

$$\times \frac{1}{\sqrt{2}} \sum_{m_1 m_2} \langle j_1 m_1 j_2 m_2 \mid JM \rangle \, \phi_{j_1 m_1}^{(A)}(a_1; \vec{r}_{1A}\zeta_1) \, \phi_{j_2 m_2}^{(A)}(a_2; \vec{r}_{2A}\zeta_2)$$

$$= \sum_{a_1 > a_2} B^{(A)}(a_1 a_2; I_A J I_B) \, \phi_{JM}^{(A)}(a_1 a_2; \vec{r}_{1A}\zeta_1, \vec{r}_{2A}\zeta_2). \tag{1}$$

We have here defined the two-nucleon spectroscopic amplitude

$$B^{(A)}(a_1 a_2; I_A J I_B) = \frac{\langle I_B \| [a_{j_1}^\dagger(a_1) a_{j_2}^\dagger(a_2)]_J \| I_A \rangle^*}{(2I_B + 1)^{1/2}(1 + \delta(a_1, a_2))^{1/2}}, \tag{2}$$

such that the two-particle wavefunction

$$\phi_{JM}^{(A)}(a_1 a_2; \vec{r}_{1A}\zeta_1, \vec{r}_{2A}\zeta_2) = [2(1 + \delta(a_1, a_2))]^{-1/2}$$

$$\times \{ [\phi_{j_1}^{(A)}(a_1; \vec{r}_{1A}\zeta_1) \, \phi_{j_2}^{(A)}(a_2; \vec{r}_{2A}\zeta_2)]_{JM}$$

$$- [\phi_{j_1}^{(A)}(a_2; \vec{r}_{2A}\zeta_2) \, \phi_{j_2}^{(A)}(a_2; \vec{r}_{1A}\zeta_1)]_{JM} \}, \tag{2a}$$

is normalized (and antisymmetric).

The quantity $\tilde{\phi}$ is related to the transition pair density $\rho(\vec{r}\sigma, \vec{r}'\sigma')$ through

$$\langle I_B M_B \mid \rho_2(\vec{r}\sigma, r'\sigma') \mid I_A M_A \rangle \equiv \langle I_B M_B \mid a^\dagger(\vec{r}'\sigma') a^\dagger(\vec{r}\sigma) \mid I_A M_A \rangle$$

$$= \sqrt{2} \sum_{JM} \langle I_A M_A J M \mid I_B M_B \rangle \, \tilde{\phi}_{JM}^{B(A)}(\vec{r}_{dA}, \zeta_d) \tag{3}$$

and is thus not necessarily normalized to unity.

(a) Pairing vibration

In the case of the excitation of monopole pairing vibrations in even nuclei, one introduces the boson operators for the pair addition and the pair removal modes $\Gamma^\dagger(\alpha, n)$, where $\alpha = \pm 2$ indicates the particle transfer quantum number. The index n labels the different states of the pairing modes. The boson operators $\Gamma^\dagger(\alpha, n)$ are bilinear combinations of the fermion operators

$$\Gamma^\dagger(a_k) = \frac{1}{\sqrt{2}}[a_{j_k}^\dagger(a_k)a_{j_k}^\dagger(a_k)]_0,$$

and

$$\Gamma^\dagger(a_i) = \frac{1}{\sqrt{2}}[b_{j_i}^\dagger(a_i)b_{j_i}^\dagger(a_i)]_0, \tag{4}$$

which create pairs of particles in the orbitals a_k and pairs of holes in the orbitals a_i, both coupled to spin 0 (and even parity) (cf. notation in § IV.2.c).

The behavior of the different operators with respect to gauge transformations

$$\mathcal{G}(\phi_o) = \exp\{-iN\phi_o/2\}, \tag{5}$$

where

$$N = \sum_{a_1 m_1} a_{j_1 m_1}^\dagger(a_1)a_{j_1 m_1}(a_1), \tag{6}$$

is the number operator, is given by

$$a_{j_1 m_1}'^\dagger(a_1) = \mathcal{G}(\phi_o)a_{j_1 m_1}^\dagger(a_1)\mathcal{G}^{-1}(\phi_o) = e^{-i\phi_o/2}a_{j_1 m_1}^\dagger(a_1), \quad (\alpha = +1)$$

$$b_{j_1 m_1}'^\dagger(a_1) = \mathcal{G}(\phi_o)b_{j_1 m_1}^\dagger(a_1)\mathcal{G}^{-1}(\phi_o) = e^{i\phi_o/2}b_{j_1 m_1}^\dagger(a_1), \quad (\alpha = -1)$$
$$\tag{7}$$

and

$$\Gamma'^\dagger(a_k) = \mathcal{G}(\phi_o)\Gamma^\dagger(a_k)\mathcal{G}^{-1}(\phi_o) = e^{-i\phi_o}\Gamma^\dagger(a_k), \quad (\alpha = +2)$$
$$\Gamma'^\dagger(a_i) = \mathcal{G}(\phi_o)\Gamma^\dagger(a_i)\mathcal{G}^{-1}(\phi_o) = e^{i\phi_o}\Gamma^\dagger(a_i), \quad (\alpha = -2) \tag{8}$$

The primes on the operators indicate that they are referred to a system in gauge space related by a rotation of angle ϕ_o to the laboratory system.

The creation operators associated with the pair addition and pair subtraction modes are defined according to

$$\Gamma^\dagger(+2, n) = \sum_{a_k} X_n(a_k)\Gamma^\dagger(a_k) + \sum_{a_i} Y_n(a_i)\Gamma(a_i),$$

$$\Gamma^\dagger(-2, n) = \sum_{a_i} X_n(a_i)\Gamma^\dagger(a_i) + \sum_{a_k} Y_n(a_k)\Gamma(a_k). \tag{9}$$

The real coefficients X and Y are the forward-going and backward-going amplitudes of the corresponding phonons and can, e.g., be calculated by diagonalizing the monopole pairing force in the RPA.

V Transfer Reactions

From the boson commutation relations

$$[\Gamma(\alpha n), \Gamma^\dagger(\alpha' n')] = \delta(n, n')\delta(\alpha, \alpha'),$$

and

$$[\Gamma(\alpha n), \Gamma(\alpha' n')] = [\Gamma^\dagger(\alpha n), \Gamma^\dagger(\alpha' n')] = 0 \qquad (10)$$

one obtains orthonormalization conditions of the type

$$\sum_k X_n^2(a_k) - \sum_i Y_n^2(a_i) = 1, \qquad (11)$$

and

$$\sum_k X_n(a_k)Y_{n'}(a_k) - \sum_i X_n(a_i)Y_{n'}(a_i) = 0. \qquad (12)$$

when one uses the random phase approximation, which amounts to the ansatz $[\Gamma(a_j), \Gamma^\dagger(a_{j'})] = \delta(j, j')$. Making use of these equations, we can invert the relations (9) to obtain

$$\Gamma^\dagger(a_k) = \sum_n (X_n(a_k)\Gamma^\dagger(2, n) - Y_n(a_k)\Gamma(-2, n)) \qquad (13)$$

and

$$\Gamma^\dagger(a_i) = \sum_n (X_n(a_i)\Gamma^\dagger(-2, n) - Y_n(a_i)\Gamma(2, n)). \qquad (14)$$

A pairing vibrational band is characterized by the quantum number N_{-2} and N_{+2} indicating the number of pair removal and the number of pair addition quanta present in each state (cf. Fig. 30). The normalized states describing the members of a monopole pairing band are

$$|N_{-2}N_{+2}\rangle = \frac{1}{\sqrt{N_{-2}!N_{+2}!}}\left(\Gamma^\dagger(-2)\right)^{N_{-2}}\left(\Gamma^\dagger(+2)\right)^{N_{+2}}|\tilde{0}\rangle \qquad (15)$$

where $|\tilde{0}\rangle$ indicates the vacuum state

$$\Gamma(\pm 2n)|\tilde{0}\rangle = 0, \qquad (16)$$

that is, the correlated ground state of the closed shell nucleus.

For transitions between the members (15) of the pairing vibrational band, one can evaluate the spectroscopic amplitudes B in terms of the coefficients X and Y using (13) and (14)

$$\langle 2n|| \frac{[a_{j_1}^\dagger(a_1)a_{j_1}^\dagger(a_1)]_o}{\sqrt{2}}||\tilde{0}\rangle = \begin{cases} X_n(a_k) & a_1 = a_k \\ -Y_n(a_i) & a_1 = a_i, \end{cases} \qquad (17)$$

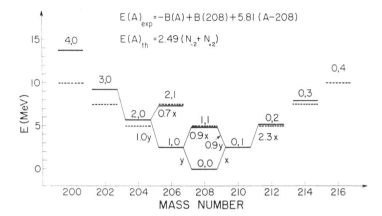

Fig. 30 Pairing vibrational band based on the ^{208}Pb ground state. The long, horizontal level bars indicate the experimental energies of the Pb 0^+ states. Except for a term linear in excess of neutrons relative to ^{208}Pb added for convenience, the energies are the difference between the ^{208}Pb and the ^{A}Pb binding energies. The coefficient in the linear term was chosen to make $E(206) = E(210)$. The broken lines indicate the energies predicted from the harmonic pairing vibrational model, using a phonon energy of 2.49 MeV, i.e. equal to $E(206)$. The pair of numbers above the levels count the number of removal phonons (hole-phonons) N_{-2} (left) and the number of addition phonons (particle-phonons) N_{+2} (right). The quantity x symbolizes the basic ^{208}Pb(t,p) ^{210}Pb$(g.s.)$ cross section. Other (t,p) intensities, where an addition phonon is added, are given relative to the basic intensity. The quantity y has a similar meaning for pickup. In the simple harmonic picture these transitions should be proportional to $N_{\pm 2} + 1$.

and

$$\langle \tilde{0} || \frac{[a_{j_1}^\dagger (a_1) a_{j_1}^\dagger (a_1)]_0}{\sqrt{2}} || -2n \rangle = \begin{cases} X_n(a_i); & a_1 = a_i \\ -Y_n(a_k); & a_1 = a_k. \end{cases} \quad (18)$$

For a stripping reaction connecting the states $|I_a\rangle = |N_{-2}^a, N_{+2}^a\rangle$, and $|I_A\rangle = |N_{-2}^A, N_2^A\rangle$, with $|I_b\rangle = |N_{-2}^a+1, N_{+2}^a\rangle$, and $|I_B\rangle = |N_{-2}^A, N_2^A+1\rangle$, one obtains

$$B^{(A)}(a_1 a_1; I_A 0 \ I_A) = (N_{+2}^A + 1)^{1/2} \times \begin{cases} X_n^{(A)}(a_k); & a_1 = a_k \\ -Y_n^{(A)}(a_i); & a_1 = a_i \end{cases} \quad (19)$$

417

and

$$B^{(b)}(a_1'a_1'; I_b 0 \ I_b) = (N_{-2}^a + 1)^{1/2} \times \begin{cases} X_n^{(b)}(a_i'); & a_1' = a_i' \\ -Y_n^{(b)}(a_k'); & a_1' = a_k'. \end{cases} \tag{20}$$

A quantity which plays an important role in characterizing the collectivity of the pairing vibrational states is the pair moment

$$\begin{aligned} M_2 &= \sum_{jm} a_{jm}^\dagger(a_j) \, a_{j\tilde{m}}^\dagger(\tilde{a}_j) \\ &= \sum_n (\alpha_{+2,n} \Gamma^\dagger(2, n) + \alpha_{-2,n} \Gamma(-2, n)), \end{aligned} \tag{21}$$

where

$$\alpha_{+2,n} = \sum_k (-1)^{\pi_k} \sqrt{j_k + 1/2} \, X_n(a_k) - \sum_i (-1)^{\pi_i} \sqrt{j_i + 1/2} \, Y_n(a_i) \tag{22}$$

and

$$\alpha_{-2,n} = \sum_i (-1)^{\pi_i} \sqrt{j_i + 1/2} \, X_n(a_i) - \sum_k (-1)^{\pi_k} \sqrt{j_k + 1/2} \, Y_n(a_k) \tag{23}$$

are the zero-point amplitudes associated with the pair addition and pair removal modes of the closed-shell nuclei A and b.

For the example given above, we find

$$\begin{aligned} \langle I_B \mid M_2 \mid I_A \rangle &= (N_{+2}^A + 1)^{1/2} \alpha_{+2}(A) \\ \langle I_a \mid M_2 \mid I_b \rangle &= (N_{-2}^a + 1)^{1/2} \alpha_{-2}(a) \end{aligned} \tag{24}$$

The RPA equations which determine the coefficients X and Y are especially simple if one assumes a residual two-body interaction with constant matrix elements. In this case, the Hamiltonian can be written

$$\begin{aligned} H &= \sum_{a_1 m_1} \varepsilon_{a_1} a_{j_1 m_1}^\dagger(a_1) a_{j_1 m_1}(a_1) + \Delta_2 \sum_{a_1 m_1 > 0} a_{j_1 \tilde{m}_1}(\tilde{a}_1) a_{j_1 m_1}(a_1) \\ &\quad + \Delta_{-2} \sum_{a_1 m_1 > 0} a_{j_1 m_1}^\dagger(a_1) a_{j_1 \tilde{m}_1}^\dagger(\tilde{a}_1) \end{aligned} \tag{25}$$

where the pair field is

$$\Delta_{\pm 2} = G \alpha_{\pm 2}. \tag{26}$$

(b) Pairing rotations

In many nuclei the solution of the RPA equations leads to a situation where there is a non-vanishing pair field. This condensation of pairing vibrations (superfluidity) is described by a state that does not have a fixed number of particles and is deformed in gauge space. As for nuclei where the mean field is deformed in ordinary space, we introduce explicitly the degree of freedom which describes the orientation of the intrinsic frame with respect to the laboratory frame (cf. Eqs. (7) and (8)). For the simple Hamiltonian (25) the expectation value of Δ_{+2} and Δ_{-2} are then non-vanishing in the intrinsic frame, i.e.

$$\Delta = \langle \Delta_{+2} \rangle = \langle \Delta_{-2} \rangle = G \alpha_o, \tag{27}$$

where α_o denotes the deformation in the intrinsic frame.

The Hamiltonian (25) becomes diagonal in the quasiparticle basis

$$\begin{aligned}
\alpha^\dagger_{j_1 m_1}(a_1) &= U(a_1) a'^\dagger_{j_1 m_1}(a_1) + V(a_1) b'^\dagger_{j_1 m_1}(a_1), \\
\alpha_{j_1 \widetilde{m}_1}(a_1) &= U(a_1) b'^\dagger_{j_1 m_1}(a_1) - V(a_1) a'^\dagger_{j_1 m_1}(a_1)
\end{aligned} \tag{28}$$

The probability that the state $a_1 m_1$ is filled with a particle is $V^2(a_1)$ and that it is empty is $U^2(a_1)$, the total probability being equal to 1, i.e.

$$U^2(a_1) + V^2(a_1) = 1. \tag{29}$$

The vacuum of the quasiparticles

$$\begin{aligned}
|BCS\rangle &= \prod_{j_1 m_1 > 0} \left(U(a_1) + V(a_1) a'^\dagger_{j_1 m_1}(a_1) a'^\dagger_{j_1 \widetilde{m}_1}(a_1) \right) |0\rangle \\
&= \prod_{j_1 m_1 > 0} \alpha_{j_1 m_1} |0\rangle
\end{aligned} \tag{30}$$

is the BCS-wavefunction. It is the intrinsic state of a pairing rotational band (cf. Fig. 31)

$$|N\rangle = (2\pi)^{-1/2} e^{iN\phi'/2} |BCS\rangle \tag{31}$$

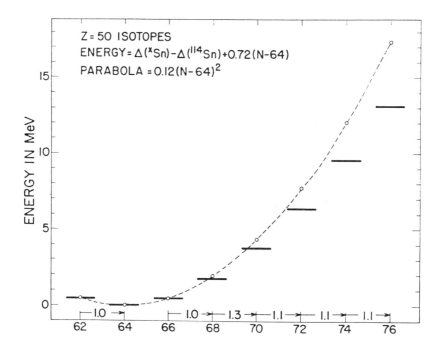

Fig. 31 The Sn pairing rotational band. The rotational energies plotted versus neutron number are the mass excess of the isotope in question minus the ^{114}Sn mass excess plus a term linear in the neutron number minus the ^{114}Sn neutron number. The coefficient of the linear term was chosen so as to put $E(^{112}\text{Sn}) = E(^{116}\text{Sn})$. The choice of $N = 64$ as the zero point for the parabola (marked by open circles) is somewhat arbitrary, since the number N_o in the pairing rotational energy need not represent an actual nucleus. The equation for the parabola defined by forcing it through the three lowest Sn energies, is given in the figure. The numbers given between the arrows at the bottom of the figure represent the observed $\text{Sn}(t, p)g.s.$ maximum cross sections, which in the pairing-rotational scheme should all be identical. The (t, p) intensities were obtained by combining the data of Bjerregaard *et al.* (1968) and (1969).

whose members are the ground states of a series of isotopes or isotones.

Because of the unitary character of the transformation (28), one can invert it obtaining

$$a'^\dagger_{j_1 m_1}(a_1) = U(a_1)a^\dagger_{j_1 m_1}(a_1) - V(a_1)\alpha_{j_1 \tilde{m}_1}(a_1),$$
$$b'^\dagger_{j_1 m_1}(a_1) = V(a_1)a^\dagger_{j_1 m_1}(a_1) + U(a_1)\alpha_{j_1 \tilde{m}_1}(a_1).$$

(32)

The operator (21) can now be expressed in the intrinsic system

$$M'_2 = \sum_{a_1} M'_2(a_1)$$

(33)

where

$$M'_2(a_1) = \sum_{m_1 > 0} a'^\dagger_{j_1 m_1}(a_1)a'^\dagger_{j_1 \tilde{m}_1}(\tilde{a}_1)$$

$$= (j_1 + 1/2)U(a_1)V(a_1) + (-1)^{\pi_{a_1}}\sqrt{j_1 + 1/2}\left\{U^2_{a_1}\frac{[a^\dagger_{j_1}(a_1)a^\dagger_{j_1}(a_1)]_o}{\sqrt{2}}\right.$$

$$\left. +V^2_{a_1}\frac{[\alpha_{j_1}(a_1)\alpha_{j_1}(a_1)]_o}{\sqrt{2}}\right\}.$$

(34)

The B-coefficients connecting the members of the rotational band (31) are

$$B(a_1 a_1; I_A 0 I_B) = \int d\phi' \langle I_B \| \frac{[a^\dagger_{j_1}(a_1)a^\dagger_{j_1}(a_1)]_o}{\sqrt{2}} \| I_A \rangle$$

$$= \frac{1}{2\pi}\int d\phi' e^{i\phi'(N_B - (N_A + 2))/2}\langle BCS \| \frac{[a'^\dagger_{j_1}(a_1)a'^\dagger_{j_1}(a_1)]_o}{\sqrt{2}} \| BCS \rangle$$

$$= \delta(N_B, N_A + 2)\sqrt{j_1 + 1/2}(-1)^{\pi_{a_1}}U(a_1)V(a_1).$$

(35)

The pure two-particle configurations are replaced in the case of superfluid nuclei by two quasiparticle excitations

$$|2qp\rangle = \frac{[\alpha^\dagger_{j_1}(a_1)\alpha^\dagger_{j_1}(a_1)]_o}{\sqrt{2}}|BCS\rangle,$$

(36)

and the associated B-coefficients are given by

$$B(a_1 a_1; I_A 0\, I_B) = (-1)^{\pi_{a_1}}U^2(a_1).$$

(37)

V Transfer Reactions

The intrinsic pair deformation (27)

$$\alpha_o = \langle BCS \mid M_2' \mid BCS \rangle = \sum_{j_1} (j_1 + 1/2) U(a_1) V(a_1) = \frac{\Delta}{G}, \qquad (38)$$

is the ratio between the pairing gap and the strength of the monopole pairing interaction.

(c) Evaluation of cross sections

We can now calculate the amplitudes (9.1)–(9.3) in the shell-model basis. For $I_A = I_B = I_a = I_b = 0$, the first-order amplitude is given by

$$(a)_{(1)} = -i \sum_{a_1 a_1'} B^{(A)}(a_1 a_1; 0) B^{(b)}(a_1' a_1'; 0) \left(\frac{2j_1' + 1}{2j_1 + 1} \right)^{1/2}$$

$$\times \sum_{\lambda \mu'} \frac{(-1)^{\lambda + \mu'}}{2\lambda + 1} 2 \int_{-\infty}^{\infty} \frac{dt}{\hbar} \, \tilde{f}_{\lambda \mu'}^{a_1 a_1'}(k_\parallel k_\perp r) \tilde{g}_{\lambda - \mu'}^{a_1 a_1'}(k_\parallel k_\perp r)$$

$$\times \exp \frac{i}{\hbar} \big((E_\beta - E_\alpha)t + \gamma_{\beta\alpha}(t) \big). \qquad (39)$$

We have here taken full account of the recoil effect in the form factors. For two-nucleon transfer, an estimate for the recoil effect is (compare (5.24))

$$k_\perp^{(2)} \Delta_y = \sqrt{2} \left(\frac{\mathcal{E}_{\text{MeV}}}{5} \right)^{1/2}, \qquad (40)$$

where we used that $k_\perp^{(2)}$, being proportional to the mass of the transferred particles, is twice as large as for single-particle transfer, while Δ_y, according to the estimate preceding Eq. (5.3), is $1/\sqrt{2}$ times the estimate for single-particle transfer. The recoil effects for two-nucleon transfer thus become more important at lower bombarding energies than for one-nucleon transfer.

For the contributions (9.3) and (9.4) to the two-particle transfer amplitude, we need the fractional expansions (cf. (4.43))

$$\tilde{\phi}_{j_1 m_1}^{F(A)}(\vec{r}_{1A} \zeta_1) = C^{(A)}(0 \; a_1; I_F) \phi_{j_1 m_1}^{(A)}(a_1 \vec{r}_{1A} \zeta_1) \qquad (41)$$

and

$$\tilde{\phi}_{j_2m_2}^{B(F)}(r_{2A}\zeta_2) = (-1)^{j_2-m_2}\sqrt{2j_2+1} \times \sqrt{2}$$

$$\int \psi_{j_2-m_2}^{F*}(\zeta_A\vec{r}_{1A}\zeta_1)\psi_0^B(\zeta_A\vec{r}_{1A}\zeta_1\vec{r}_{2A}\zeta_2)d\zeta_A\, d^3r_{1A}d\zeta_1 \qquad (42)$$

$$= \sqrt{2}\, B^{(A)}(a_2a_2;0)C^{(A)}(0\ a_2;I_F)\phi_{j_2m_2}^{(A)}(a_2;\vec{r}_{2A}\zeta_2)$$

and similar for $\tilde{\phi}^{a(f)}$ and $\tilde{\phi}^{f(b)}$.

Inserting these expressions in (9.3) and (9.4) and utilizing the results of § 5, we find

$$(a)_{\text{orth}} = i \sum_{a_1a_1'\gamma} B^{(A)}(a_1a_1;0)B^{(b)}(a_1'a_1';0)$$

$$\left(\frac{2j_1'+1}{2j_1+1}\right)^{1/2} |C^{(A)}(0\ a_1;I_F)|^2\, |C^{(b)}(0\ a_1';I_f)|^2$$

$$\times \sum_{\lambda\mu'} \frac{(-1)^{\lambda+\mu'}}{2\lambda+1}\, 2\int_{-\infty}^{\infty} \frac{dt}{\hbar} \tilde{f}_{\lambda\mu'}^{a_1a_1'}(k_\| k_\perp r)\tilde{g}_{\lambda-\mu'}^{a_1a_1'}(k_\| k_\perp r) \qquad (43)$$

$$\times \exp\frac{i}{\hbar}i((E_\beta - E_\alpha)t + \gamma_{\beta\alpha}(t)),$$

and

$$(a)_{\text{succ}} = -\sum_{a_1a_1'} B^{(A)}(a_1a_1;0)B^{(b)}(a_1'a_1';0)\left(\frac{2j_1'+1}{2j_1+1}\right)^{1/2}$$

$$\times 2\sum_{\gamma\mu\mu'\mu''} \frac{(-1)^{\lambda+\mu}}{2\lambda+1} D_{-\mu\mu''}^\lambda \left(0\frac{\pi}{2}\pi\right) D_{\mu\mu'}^\lambda \left(0\frac{\pi}{2}\pi\right)$$

$$\times |C^{(A)}(0\ a_1;I_F)|^2\, |C^{(b)}(0\ a_1';I_f)|^2 \qquad (44)$$

$$\times \int_{-\infty}^{\infty} \frac{dt}{\hbar} \tilde{f}_{\lambda\mu'}^{a_1a_1'}(k_\| k_\perp r)e^{i\left((E_\beta-E_\gamma)t+\gamma_{\beta\gamma}(t)\right)/\hbar+i\mu\phi(t)}$$

$$\times \int_{-\infty}^{t} \frac{dt'}{\hbar} \tilde{f}_{\lambda\mu''}^{a_1a_1'}(k_\| k_\perp r)e^{i\left((E_\beta-E_\alpha)t'+\gamma_{\gamma\alpha}(t')\right)/\hbar-i\mu\phi(t')}$$

The summation over I_f and I_F runs over all states of the channel γ with spins $I_F = j_1$ and $I_f = j_1'$ and parity $\pi_F = (-1)^{\ell_1}$ and $\pi_f = (-1)^{\ell_1'}$.

In the shell-model basis all wavefunctions $\phi^{(A)}$ appearing in (42) and (41) are eigenstates of a fixed shell-model potential. In this description the form factors and overlaps appearing in (43) and (39) are identical. We may then perform the summation over the quantum numbers of the intermediate states in (43) utilizing the sum rule (6.45). The non-orthogonality contribution (43) to the two-particle transfer then exactly cancels the first-order contribution (39). The two-particle wavefunctions (1) which have been used in actual calculations is constructed on the basis of harmonic oscillator single-particle wavefunctions matched with an exponential tail fixed by half the separation energy of the two nucleons. In this way, one attempts to include effects of high-lying states belonging to the continuum in the single-particle spectrum. A more consistent way to include this effect is to use the Sturm-Liouville method where the depth of the single-particle potential is adjusted for each state to reproduce this binding energy (cf. Bang et al., 1985). If such prescriptions are used, the form factors and overlaps appearing in (39) and (43) are not the same and the cancellation between the two terms is only approximate (cf. Fig. 32 and Maglione et al., 1985).

(d) Estimate of successive versus simultaneous

An estimate of the amplitudes (39) and (43) can be obtained neglecting recoil and using the fact that the form factors and overlaps are exponential functions of r. We may thus write (cf. Eq. (5.19))

$$
\tilde{f}_{\lambda\mu'}^{a_1 a_1'(\alpha)}(k_\parallel k_\perp, r)\tilde{g}_{\lambda-\mu'}^{a_1 a_1'}(k_\parallel k_\perp, r)
$$
$$
= e^{i(\bar{\sigma}^{(\alpha)}+\bar{\sigma}^{(o)})}\frac{2\lambda+1}{4\pi}f_\lambda^{a_1 a_1' \alpha}(r_o)\, g_\lambda^{a_1 a_1'}(r_o)e^{-2\kappa(r-r_o)}
$$

(45)

where $\bar{\sigma}^{(\alpha)}$ is given by (5.4), while $\bar{\sigma}^{(o)}$ is the average phase corresponding to the overlap. If we assume the maximum contribution to the overlap comes from halfway between the nuclear surfaces as defined by $R_i = 1.25\, A_i^{1/3}$, one finds (cf. Broglia et al. (1977), Fig. 2)

$$
\sigma^{(o)} = \frac{M\dot{r}}{\hbar}\left(0.625\,(A_A^{1/3}-A_a^{1/3}) + \frac{1}{2}r - \frac{m_A+m_B}{2(m_a+m_B)}r\right)
$$

(46)

The decay constant κ is determined from the slope of $f(r)\cdot g(r)$ which is approximately given by $2\kappa = \left(4M(E_b-E_a)/\hbar^2\right)^{1/2}$. The orbital integral

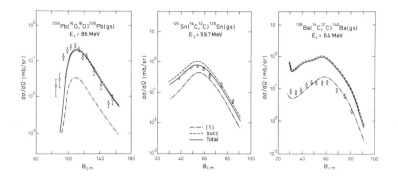

Fig. 32 Calculated angular distributions for two-particle transfer reactions compared with experimental data (von Oertzen et al., 1983 and Lilley, 1984). Also shown are the cross sections obtained by only including the one-step simultaneous transfer and the two-step successive transfer, while the solid lines give the results obtained when both contributions are taken into account, together with the non-orthogonality term. The amplitudes (39), (43), and (44) were calculated (Maglione et al., 1985) for the different partial waves contributing to the reaction and inserted in Eq. (8.39) to obtain the transition amplitude and associated differential cross section.

can then be estimated by the parabolic approximation (6.23). This leads to (cf. (6.30))

$$
(a)_{(1)} = -i \sum_{a_1 a_1'} B^{(A)}(a_1 a_1; 0) \, B^{(b)}(a_1' a_1'; 0)
$$
$$
\left(\frac{2j_1' + 1}{2j_1 + 1}\right)^{1/2} \sum_{\lambda} (-1)^{\lambda} \, f_{\lambda}^{a_1 a_1'(\alpha)}(r_o) g_{\lambda}^{a_1 a_1'}(r_o) \tag{47}
$$
$$
\times \left(4\pi\hbar^2 \kappa \, \ddot{r}_o\right)^{-1/2} \exp\left\{ -\frac{(Q - Q_{\text{opt}} + \Delta)^2}{4\hbar^2 \kappa \ddot{r}_o} \right\},
$$

where

$$
\Delta = \frac{1}{2} M \ddot{r}_o (r_o - R_a - R_A). \tag{48}
$$

The selection rule $\lambda + \pi = $ even for the form factors imply that $(-1)^{\lambda} = (-1)^{\ell_1}(-1)^{\ell_1'}$.

A simple estimate of the coherence associated with the two-particle transfer can be obtained by introducing the average

$$\langle fg \rangle = \sum_\lambda \langle \frac{1}{j+1/2} f_\lambda^{a_1 a_1'}(r_o) g_\lambda^{a_1 a_1'}(r_o) \rangle \qquad (49)$$

of the product $f \cdot g$ over the single-particle configurations. One then obtains

$$(a)_{(1)} = -i(4\pi\hbar^2 \kappa \ddot{r}_o)^{-1/2} \alpha(b) \alpha(A)$$

$$\times \langle fg \rangle_{r_o} \exp\left\{ -\frac{(Q - Q_{\text{opt}} + \Delta)^2}{4\hbar^2 \kappa \ddot{r}_o} \right\}, \qquad (50)$$

where the quantities $\alpha(b)$ and $\alpha(A)$ were defined in Eqs. (23) and (22) for normal systems, and (38) for superfluid nuclei. They measure the coherence of the pair transfer. For the excitation of a pure two-particle configuration

$$\alpha(A) = (j_1 + 1/2)^{1/2}. \qquad (51)$$

For the non-orthogonality term, we obtain, with the same approximations, the same expression as (50) but with opposite sign (Broglia et al., 1973).

The successive transfer can be estimated by the parabolic approximation. Utilizing the result (10.6), we may write the successive amplitude (44) as

$$(a)_{\text{succ}} = -\sum_{a_1 a_1'} B^{(A)}(a_1 a_1; 0) B^{(b)}(a_1' a_1'; 0) \left(\frac{2j_1' + 1}{2j_1 + 1} \right)^{1/2}$$

$$\times |C^{(A)}(0a_1; I_F)|^2 |C^{(b)}(0a_1'; I_f)|^2$$

$$\times \sum_\lambda (-1)^\lambda \frac{1}{4\sqrt{\kappa_1 \kappa_2} \ddot{r}_o \hbar^2} |f_\lambda^{a_1 a_1'(\alpha)}(r_o)|^2 e^{-q^2} \qquad (52)$$

$$\times \sum_\mu \frac{4\pi}{2\lambda + 1} |Y_{\lambda\mu}(\pi/2, 0)|^2 (g(s) + i h(s)).$$

In this expression κ_1 and κ_2 indicate the exponential slopes of the form factors for the first and second step, respectively. The quantity q is defined in Eq. (10.13) and is seen to be essentially independent of the intermediate state. In fact, the factor $\exp(-q^2)$ is similar to the exponential factor appearing in (47) with $\kappa = \sqrt{\kappa_1 \cdot \kappa_2}$. The quantity s which is defined in Eq. (10.10) depends strongly on the intermediate state and is approximately given by

$$s = \frac{\frac{1}{2}(E_\alpha + E_\beta) - E_\gamma + \mu\hbar\dot{\phi}_o}{\sqrt{\kappa}\,\ddot{r}_o\hbar^2}. \tag{53}$$

We note that the quantity $(E_\alpha + E_\beta)/2 - E_\gamma$ is equal to the energy denominator in (9.14). In fact, the expression (9.14) may be derived in the limit of large values of $q_1 - q_2$ utilizing (10.11).

It is interesting to note again that for intermediate states of high energy the real part of (52) vanishes, and the imaginary part tends to cancel the non-orthogonality term according to the discussion in connection with Eq. (9.17). Utilizing the definitions (6.37) and (10.17), we may write the real part of $(a)_{\text{succ}}$ as

$$(a)_{\text{succ}}^R = -\frac{1}{4\kappa\ddot{r}_o\hbar^2}\,\alpha(a)\alpha(A)e^{-\frac{(Q-Q_{\text{opt}}+2\Delta^{(a)})^2}{4\kappa\ddot{r}_o\hbar^2}}$$

$$\times \langle \sum_\lambda \frac{1}{j_1 + 1/2}\left(f_\lambda^{a_1 a_1'}(r_o)\right)^2 (g(a,b) + i\,h(a,b))\rangle, \tag{54}$$

where the average is taken over all the intermediate configurations. The arguments in the adiabatic cut-off functions $g(a,b)$ and $h(a,b)$ are

$$a(Q) = \frac{1}{\sqrt{\kappa}\,\ddot{r}_o\hbar^2}\left\{\frac{1}{2}(E_\alpha + E_\beta) - E_\gamma\right\}, \quad b(\lambda) = \frac{\dot{\phi}_o}{\sqrt{\kappa}\,\ddot{r}_o}\lambda. \tag{55}$$

The functions g and h are illustrated in Fig. 27.

The averaging that is implied by the expression (54) is not as well defined as the averaging in (49). This is because there is in the successive transfer a systematic suppression of the intermediate states with high energy. However, the main components are known to be the ones close to the Fermi surface and the more important states of high angular momentum are not suppressed very much because of the dependence of $g(a,b)$ on $b(\lambda)$. Aside from this, the successive amplitude shows the same coherence property as the one corresponding to simultaneous transfer, and the Q value dependence is almost the same since $2\Delta^{(\alpha)} \approx \Delta$ (cf. Eq. (48)). We may thus estimate the ratio between the successive and the first-order amplitude by

$$
\frac{(a)^{R}_{\mathrm{succ}}}{(a)_{(1)}} = -i\,\frac{\sqrt{\pi}}{\sqrt{4\kappa\ddot{r}_{o}\hbar^{2}}}\,\frac{\langle\sum_{\lambda}\frac{1}{j_{1}+1/2}\left(f^{a_{1}a'_{1}}_{\lambda}(r_{o})\right)^{2}(g+ih)\rangle}{\langle\sum_{\lambda}\frac{1}{j_{1}+1/2}f^{a_{1}a'_{1}}_{\lambda}(r_{o})g^{a_{1}a'_{1}}_{\lambda}(r_{o})\rangle}
$$

$$
\approx -i\,\frac{\tau_{\mathrm{char}}}{\hbar}\,\frac{f^{a_{1}a'_{1}}_{\lambda}(r_{o})}{g^{a_{1}a'_{1}}_{\lambda}(r_{o})}
\tag{56}
$$

where τ_{char} is the collision time defined in (6.2).

From the examples given in Fig. 9, we may estimate the ratio of the form factor to the overlap to be of the order of magnitude 10–20 MeV. For low bombarding energies where $\tau_{\mathrm{char}}/\hbar$ is of the order of magnitude 0.2 MeV^{-1}, we estimate that the ratio $(a)_{\mathrm{succ}}$ to $(a)_{(1)}$ is about 2–4. This means that the probability (and cross section) for simultaneous (first order) pair transfer is only about 10–20 percent of the cross section (Götz et al., 1975). Detailed calculations using the expressions (39)–(44) shown in Fig. 32 support this estimate.

There is an intimate connection between two-particle transfer among superfluid nuclei and the Josephson effect in solid state physics. Within the present formulation, the connection is established by going to the limit of long collision times, where the successive transfer acquires the asymptotic expression (9.16).

We thus approach the situation where the second-order process can be described accurately in first order by a local (energy-independent) form factor. Since the two nuclei are both superfluid, the intermediate states are separated from the initial and final states by twice the energy gap Δ, and the real part of the double integral vanishes if only the collision time τ is much larger than $\hbar/(2\Delta)$.

The effective form factor for the transfer of a Cooper pair is, according to (9.16), given by

$$
F = \sum_{\lambda}\frac{\langle\beta\mid V_{\gamma}-\langle V_{\gamma}\rangle\mid\gamma\rangle\,\langle\gamma\mid V_{\alpha}-\langle V_{\alpha}\rangle\mid\alpha\rangle}{\frac{1}{2}(E_{\alpha}+E_{\beta})-E_{\gamma}}
\tag{57}
$$

In the intermediate state $|\gamma\rangle \approx \alpha^{\dagger}(a'_{1})\alpha^{\dagger}(a_{1})|\alpha\rangle$ a quasiparticle is created in A, as well as in a, and the energy in the intermediate state is therefore

$$
E_{\gamma} = E_{\alpha} + \Delta\lambda + E_{1} + E'_{1},
\tag{58}
$$

where $\Delta\lambda$ is the difference in the Fermi level

$$E_\beta - E_\alpha = 2\Delta\lambda. \tag{59}$$

The quantities E_1 and E_1' are the quasiparticle energies in the nuclei A and a, respectively. The transfer matrix elements for a stripping reactions are, according to (35), given by

$$\langle \gamma \mid V_\alpha - \langle V_\alpha\rangle \mid \alpha \rangle = -U(a_1')V(a_1)\langle a_1' \mid T \mid a_1\rangle, \tag{60}$$

where, for the single-particle form factor, we used a notation which is common in solid state physics. The second matrix element is

$$\langle \beta \mid V_\gamma - \langle V_\gamma\rangle \mid \gamma \rangle = -V(a_1')U(a_1)\langle \tilde{a}_1' \mid T \mid \tilde{a}_1\rangle. \tag{61}$$

Due to time-reversal invariance, the interaction matrix element is here the complex conjugate of $\langle a_1' \mid T \mid a_1\rangle$. The effective form factor F is therefore

$$F = -\sum_{a_1 a_1'} \frac{U(a_1)V(a_1)U(a_1')V(a_1') \mid \langle a_1' \mid T \mid a_1\rangle \mid^2}{E_1 + E_1'} \tag{62}$$

This is identical to the expression for the interaction energy used in the description of the Josephson junction (Josephson (1962), Anderson (1963)).

In solid state physics, however, the initial and final states do not correspond to a well-defined number of particles, but rather to a well-defined value of the gauge angle difference. The electric supercurrent therefore contains term linear in the pair transfer amplitude (cf. Section 13 below). In nuclear physics linear terms only occur in the collision of two nuclei that differ by a neutron or a proton pair (cf. Appendix D).

12 MACROSCOPIC DESCRIPTION OF PAIRING

There is a direct analogy between the RPA treatment of pairing modes as given in the previous section and the RPA treatment of particle-hole excitation given in Section IV.2. Correspondingly, we shall in this section present a macroscopic description of pairing vibrations which is similar to the macroscopic description of surface vibrations given in Section IV.3.

(a) Dynamical deformations in gauge space

We consider a spherical closed shell nucleus with the density ρ_o which corresponds to the self-consistent (Hartree-Fock) solution for the ground state.

In the present context we think of this density in phase space. The information which, in the classical description, is contained through the dependence of the density on coordinate and momentum, is in the quantal description contained in the non-local single-particle density matrix. A connection between the two descriptions is obtained through the Wigner transformation (cf. App. G).

The phase space density is mainly characterized by two parameters, namely the nuclear radius in configuration space and the radius of the Fermi sphere. Distorting these variables, we bring the system away from equilibrium and can set up, in this way, normal modes provided that the induced changes in the potential follow in a self-consistent way the changes that this potential creates in the density.

We may write the distorted density as

$$\rho = \rho_o + \sum_{\lambda\mu} \frac{\delta\rho}{\delta\alpha_{\lambda\mu}} \alpha_{\lambda\mu} + \sum_{\alpha} \frac{\delta\rho}{\delta\alpha_\alpha} \alpha_\alpha, \tag{1}$$

where the second term describes deformations of the radius, and the third term, deformations in the Fermi sphere. The amplitudes $\alpha_{\lambda\mu}$ measure the multipole distortions of order $\lambda\mu$ in space, and one can write

$$\delta\rho_{\lambda\mu} = \frac{\delta\rho}{\delta\alpha_{\lambda\mu}} \alpha_{\lambda\mu} = \frac{\partial\rho}{\partial R} \frac{\partial R}{\partial\alpha_{\lambda\mu}} \alpha_{\lambda\mu} = R^{(o)} \frac{\partial\rho}{\partial R} Y_{\lambda\mu}^*(\theta,\phi)\alpha_{\lambda\mu} \tag{2}$$

where the parametrization (IV.3.1) for the radius was used.

The amplitudes α_α measure the changes in the radius of the Fermi sphere, that is, changes in the Fermi energy λ. This can only be achieved by adding or subtracting particles.

In a similar way as the nuclear radius can be changed differently in different directions, the change in the particle number can be different in different directions in gauge space according to

$$A = A^{(o)} + \sum_\alpha e^{-i\alpha\phi/2}\,\alpha_\alpha. \tag{3}$$

In this expression ϕ is the gauge angle, while α_α measures the magnitude of the distortion of transfer quantum number $\alpha = \pm 2$.

It is natural to use the number of particles A as the macroscopic variable. In fact, following Appendix G [Eq. (G.52)], we find for the transition pair density $\langle 1\,|\,\rho_2\,|\,0\rangle$ between the ground state $|0\rangle$ and the one-phonon state $|\,1_{\alpha=2}\rangle$ of the pair addition mode

$$\langle 1_{\alpha=2}\,|\,\rho_2(\vec{r},\vec{r}')|\,0\rangle = \sum_n \Big\{ \langle 1_{\alpha=2}|\rho_0(\vec{r},\vec{r}')\,|\,n\rangle\,\langle n\,|\,M_2|\,0\rangle$$

$$-\langle 1_{\alpha=2}\,|\,M_2\,|\,n\rangle\,\langle n\,|\,\rho_0(\vec{r},\vec{r}')\,|\,0\rangle\Big\} \tag{4}$$

$$\approx \langle 1_{\alpha=2}\,|\,M_2\,|\,0\rangle\Big\{ \langle 1_{\alpha=2}\,|\,\rho_0(\vec{r},\vec{r}')\,|\,1_{\alpha=2}\rangle - \langle 0\,|\,\rho_0(\vec{r},\vec{r}')\,|\,0\rangle\Big\}.$$

In keeping with Eq. (2) and making use of relation (3), one can write

$$\delta\rho_\alpha = \frac{\delta\rho}{\delta\alpha_\alpha}\alpha_\alpha = \frac{\partial\rho}{\partial A}\frac{\partial A}{\partial\alpha_\alpha}\,\alpha_\alpha = \frac{\partial\rho}{\partial A}e^{-i\alpha\phi/2}\,\alpha_\alpha. \tag{5}$$

We use A as a general particle number variable. In fact, one should, in heavy nuclei, interpret A as either N or Z, while in light nuclei, one should use an isospin invariant formalism.

Under rotations in gauge space where the angle ϕ in (5) is changed to $\phi - \phi_0$, the amplitude α_α transforms as

$$\mathcal{G}(\phi_0)\alpha_\alpha\mathcal{G}^{-1}(\phi_0) = e^{-i\alpha\phi_0/2}\,\alpha_\alpha. \tag{6}$$

This implies that the quantities (1)–(3) are gauge invariant since

$$\mathcal{G}(\phi_0)\alpha_{\lambda\mu}\mathcal{G}^{-1}(\phi_0) = \alpha_{\lambda\mu} \tag{7}$$

implying that $\alpha_{\lambda\mu}$ has transfer quantum number $\alpha = 0$.

Under rotations in ordinary space, $\alpha_{\lambda\mu}$ transforms like a tensor of order $\lambda\mu$, while α_α is a scalar. One might also include the possibility of simultaneous deformations in ordinary and gauge space, where the density is also anisotropic in ordinary space. This would lead to amplitudes $\alpha_{\alpha\lambda\mu}$ of multipole pairing. In what follows, we only consider monopole pairing distortions.

(b) Equations of motion

The normal modes corresponding to monopole pairing should be described by a Hamiltonian which is quadratic in the quantities α_α and $\dot\alpha_\alpha$ and invariant under gauge transformations. The Lagrangian may thus be written

$$L = D\dot\alpha_2\dot\alpha_{-2} - (C - \omega^2 D)\,\alpha_2\alpha_{-2} + i\omega D(\alpha_2\dot\alpha_{-2} - \alpha_{-2}\dot\alpha_2). \quad (8)$$

This Lagrangian is also invariant under time reversal because

$$T\alpha_\alpha T^{-1} = \alpha_\alpha$$

and

$$T\dot\alpha_\alpha T^{-1} = -\dot\alpha_\alpha$$

$$(9)$$

and Hermitean since

$$\alpha_\alpha^\dagger = \alpha_{-\alpha}$$

and

$$(10)$$

$$\dot\alpha_\alpha^\dagger = \dot\alpha_{-\alpha}.$$

For surface deformations, one might write down a term corresponding to the last term in (8). It would be Hermitean, time-reversal invariant, and invariant for rotations around the z-axis. It would, however, not be invariant for rotations of 180 degrees around the y-axis. Therefore, it should disappear for a nucleus which is not rotating. Similarly, it disappears for pairing if the system is invariant for rotations of 180 degrees around the y-axis in gauge space (cf. Appendix G). Since this rotation is equivalent to a substitution of particles with holes, this means that it disappears if the nucleus shows symmetry between adding and removing particles, and we would have

$$L = D\,\dot\alpha_2'\,\dot\alpha_{-2}' - C\,\alpha_2'\,\alpha_{-2}'. \quad (11)$$

The terms proportional to ω in (8) may thus be interpreted as due to the rotation of the nucleus in gauge (quasispin) space. Transforming α_α to a coordinate system which rotates with frequency $\dot\phi_0$ in this space, i.e.

$$\alpha'_\alpha = \mathcal{G}^{-1}(\dot\phi_0 t)\alpha_\alpha \mathcal{G}(\dot\phi_0 t) = \alpha_\alpha\, e^{i\alpha\dot\phi_0 t/2} \tag{12}$$

we find

$$\dot\alpha'_\alpha = (\dot\alpha_\alpha + i\frac{\alpha}{2}\dot\phi_0\,\alpha_\alpha)e^{i\alpha\dot\phi_0 t/2} \tag{13}$$

and the Lagrangian (11) would, in the new variables α_α, take the form (8) with $\omega = \dot\phi$. Since the rotation frequency in gauge space is related to the Fermi energy λ (cf. Eq. (G.62)) through

$$\dot\phi_0 = \frac{2\lambda}{\hbar}, \tag{14}$$

we may always remove the terms proportional to ω for a specific pairing mode by appropriately adjusting the definition of the Fermi surface. Equation (14) also shows that ϕ is Hermitean but odd under time reversal, that is,

$$T\phi T^{-1} = -\phi \tag{15}$$

From the Lagrangian, we find the conjugate momenta

$$\begin{aligned}\pi_{+2} &= D(\dot\alpha_{-2} - i\omega\alpha_{-2})\\ \pi_{-2} &= D(\dot\alpha_2 + i\omega\alpha_2)\end{aligned} \tag{16}$$

and the Hamiltonian

$$H = \frac{1}{D}\pi_{-2}\pi_2 + C\alpha_2\alpha_{-2} - i\omega(\alpha_2\pi_2 - \alpha_{-2}\pi_{-2}). \tag{17}$$

We note that

$$T\pi_\alpha T^{-1} = -\pi_\alpha$$

and

$$\pi^\dagger_\alpha = \pi_{-\alpha}. \tag{18}$$

The equations of motion are

$$D\ddot\alpha_{\mp 2} \mp 2i\omega D\dot\alpha_{\mp 2} + (C - \omega^2 D)\alpha_{\mp 2} = 0 \tag{19}$$

leading to

$$\alpha_{-2} = a_- e^{i\omega_- t} + b_+ e^{-i\omega_+ t}$$
$$\alpha_2 = a_+ e^{i\omega_+ t} + b_- e^{-i\omega_- t}, \qquad (20)$$

where $a_\pm$ and $b_\pm$ are the integration constants, and

$$\omega_\pm = \sqrt{\frac{C}{D}} \mp \omega. \qquad (21)$$

The pair addition ($\alpha = +2$) and the pair removal ($\alpha = -2$) modes thus have different frequencies, although, as remarked earlier, we can always, in dealing with a single pairing mode ($\alpha = \pm 2$), adjust the Fermi level such that $\omega = 0$.

It is noted that the quantity

$$N = N(\alpha; \pi) = \sum_\alpha i \frac{\alpha}{2} \pi_\alpha \alpha_\alpha \qquad (22)$$

appearing in the last term in (17) is related to the angular momentum in gauge space of the pairing mode. If we interpret $\omega = \dot{\phi}$ as a dynamical variable, N is the conjugate variable to ϕ, i.e.

$$N = \frac{dL}{d\dot{\phi}} = \frac{dL}{d\omega} \qquad (23)$$

and should then be interpreted as the number of particles associated with the pairing oscillations. The Hamiltonian can then be written

$$H = \frac{1}{D}\pi_{-2}\pi_2 + C\,\alpha_{-2}\alpha_2 - \frac{2\lambda}{\hbar}N(\alpha_{\pm 2}; \pi_{\pm 2}). \qquad (24)$$

We can now use $a_\pm$ as the new dynamical variables, i.e.

$$A_\pm = e^{i\omega_\pm t}a_\pm = \frac{1}{2}\left(\alpha_{\pm 2} - \frac{i}{\sqrt{CD}}\pi_{\mp 2}\right). \qquad (25)$$

A canonical transformation from the old variables to the new coordinates A_+ and A_- is generated by the function (cf. Goldstein, 1951)

$$F_3 = -2(A_+\pi_{+2} + A_-\pi_{-2}) - \frac{i}{\sqrt{CD}}\pi_{+2}\pi_{-2} + 2i\sqrt{CD}\,A_+A_-. \qquad (26)$$

This leads to the new momenta

$$\Pi_\pm = \pi_{\pm 2} - i\sqrt{CD}\,\alpha_{\mp 2} = -2i\sqrt{CD}\,b_\pm e^{-i\omega_\pm t}. \tag{27}$$

The new Hamiltonian then reads

$$H = i\omega_+ A_+ \Pi_+ + i\omega_- A_- \Pi_-. \tag{28}$$

In a quantal description the quantities A_- and Π_- being conjugate variables satisfy the following relations

$$[A_-, \Pi_-] = i\hbar. \tag{29}$$

In view of the fact that $\Pi_\pm = -2i\sqrt{CD}\,A_\pm^\dagger$, we can therefore identify the operators

$$A_\pm = \beta\,\Gamma^\dagger(\pm 2), \tag{30}$$

and

$$\Pi_\pm = -\frac{i\hbar}{\beta}\Gamma(\pm 2), \tag{31}$$

which satisfy the commutation relations

$$[\Gamma(\pm 2),\, \Gamma^\dagger(\pm 2)] = 1, \tag{32}$$

as boson creation and annihilation operators. We have here introduced the parameter

$$\beta^2 = \frac{\hbar}{2\sqrt{CD}}. \tag{33}$$

In terms of these operators, the Hamiltonian takes the form

$$H = \hbar\omega_+ \Gamma^\dagger(+2)\Gamma(+2) + \hbar\omega_- \Gamma^\dagger(-2)\Gamma(-2). \tag{34}$$

The pair vibrational amplitudes are given by

$$\alpha_{+2} = \beta\big(\Gamma^\dagger(+2) + \Gamma(-2)\big)$$

and

$$\tag{35}$$

$$\alpha_{-2} = \beta\big(\Gamma^\dagger(-2) + \Gamma(+2)\big)$$

while

$$\dot\alpha_{\pm 2} = i\beta\big(\omega_\pm \Gamma^\dagger(\pm 2) - \omega_\mp \Gamma(\mp 2)\big) \tag{36}$$

The quantity β is thus the zero point amplitude of the pairing mode.

(c) Pair density and pair field

A simple estimate of the distorted densities $\delta\rho_\alpha$ is obtained using the Thomas-Fermi expression for the equilibrium density ρ_0 in (1), i.e. (cf. App. G)

$$\bar{\rho}_0(\vec{p},\vec{q}) = \frac{1}{(2\pi\hbar)^3} \, \Theta\left(\lambda - \frac{p^2}{2M} - U(q)\right), \tag{37}$$

where Θ is a step function, while λ is the Fermi energy and U is the single-particle potential. The quantity (37) is zero unless the momentum of the particle is smaller than the local Fermi momentum

$$p_F(q) = \hbar\, k_F(q) = \left(2M\left(\lambda - U(q)\right)\right)^{1/2}. \tag{38}$$

The corresponding non-local equilibrium density is

$$\begin{aligned}
\rho_0(\vec{r},\vec{r}\,') &= \int d^3p\, e^{i\vec{p}\cdot\vec{\xi}/\hbar}\, \bar{\rho}_0(\vec{p},\vec{q}) \\
&= \rho_0(q)\frac{3j_1\left(k_F(q)\xi\right)}{k_F(q)\xi} \\
&= \frac{1}{2\pi^2}\frac{1}{\xi^3}\left(\sin\left(k_F(q)\xi\right) - k_F(q)\xi\,\cos\left(k_F(q)\xi\right)\right)
\end{aligned} \tag{39}$$

with $\vec{\xi} = \vec{r} - \vec{r}\,'$ and $\vec{q} = \frac{1}{2}(\vec{r}+\vec{r}\,')$. We have also introduced the local limit of the density matrix $\rho_0(\vec{q})$ given by

$$\rho_0(q) = \rho_0(q,q) = \frac{k_F^3(q)}{6\pi^2} = \int \bar{\rho}_0(\vec{p},\vec{q})d^3p. \tag{40}$$

The derivatives of ρ_0 with respect to R or A appearing in (2) and (5) may, according to (39) and (40), be written

$$\frac{\partial\rho}{\partial R} = \frac{\partial\rho_0(q)}{\partial R} \cdot \frac{\sin\left(k_F(q)\xi\right)}{k_F(q)\xi} \tag{41}$$

and

$$\frac{\partial\rho}{\partial A} = \frac{\partial\rho_0(q)}{\partial A} \cdot \frac{\sin\left(k_F(q)\xi\right)}{k_F(q)\xi}. \tag{42}$$

A first estimate of the derivative of the local density with respect to the number of particles can be obtained using the empirical expression (III.A.15–16). The quantity $\partial\rho_o/\partial A$ should be substituted, in the case of pairing among neutrons by $\partial\rho_{o\nu}/\partial N$, and by $\partial\rho_{o\pi}/\partial Z$ in the case of pairing among protons. A comparison of these expressions with microscopic calculations based on standard parameters for the single-particle potentials have been carried out by Lotti et al. (1990). The most conspicuous discrepancy between the macroscopic and microscopic results, is that the latter are larger, and extend further away in the surface region. The reason for this discrepancy is still an open question. It may be related to the use of a too large radius for the single-particle potentials. A detailed comparison of Eqs. (42)–(43) with microscopic calculation of the non-local pair density has been made for the pair addition mode of ^{208}Pb (Ferreira et al., 1983).

The microscopic expression for $\delta\rho(\vec{r}\sigma, \vec{r}'\sigma')$ is

$$
\begin{aligned}
\delta\rho(\vec{r}\sigma, \vec{r}'\sigma') = &\sum_{a_k} X(a_k)\,[\phi(a_k, \vec{r}\sigma)\,\phi(a_k, \vec{r}'\sigma')]_o \\
&-\sum_{a_i} Y(a_i)\,[\phi(a_i, \vec{r}\sigma)\,\phi(a_i, \vec{r}'\sigma')]_o,
\end{aligned}
\tag{43}
$$

with $a \equiv (n\ell j)$ labeling the single-particle orbitals.

(d) Static deformation in gauge space. Superfluidity.

In the present paragraph we study the equations of motion when the system is excited with many pair quanta. Such a system achieves a permanent deformation in the intrinsic frame and is therefore superfluid. The transformation to the intrinsic frame is obtained by using, instead of the variables $\alpha_{\pm 2}$ the orientation angle ϕ' and the deformation α in the intrinsic frame in gauge space defined by

$$
\alpha_{\pm 2} = \alpha\, e^{\pm i\phi'}.
\tag{44}
$$

The time derivative of the amplitudes is therefore

$$
\dot{\alpha}_{\pm 2} = (\dot{\alpha} \pm i\,\dot{\phi}'\alpha)e^{\pm i\phi'},
\tag{45}
$$

and the Lagrangian (8) takes the form

$$
L = D\dot{\alpha}^2 - \big(C - D(\omega + \dot{\phi}')^2\big)\alpha^2.
\tag{46}
$$

437

The momentum conjugated to the variable α is

$$\pi = \frac{\partial L}{\partial \dot{\alpha}} = 2D\dot{\alpha}, \tag{47}$$

while the variable conjugate to the gauge angle is the number of pairs times $\hbar$:

$$N = \frac{\partial L}{\partial \dot{\phi}'} = 2D\alpha^2(\dot{\phi}' + \omega). \tag{48}$$

Using these variables, we may write the Hamiltonian in the form

$$H = \frac{1}{4D}\pi^2 + (C - \omega^2 D)\alpha^2 + \frac{(N - 2\omega D\alpha^2)^2}{4D\alpha^2}. \tag{49}$$

For large values of $N = N_0$, the Hamiltonian exhibits a minimum for non-vanishing values of α. The equilibrium value of the pair deformation is

$$\alpha_0 = (4DC)^{-1/4}\sqrt{N_0}. \tag{50}$$

We may now expand the Hamiltonian (49) for small values of

$$\alpha' = \alpha - \alpha_0 \tag{51}$$

keeping $N = N_o$. We then find

$$H = \frac{\pi^2}{4D} + 4C\alpha'^2 + \text{const.} \tag{52}$$

The vibrational spectrum corresponds to a pairing vibration in the intrinsic frame. As was to be expected, it has the frequency $\omega_p = 2(C/D)^{1/2}$ equal to the sum of the frequencies associated with the pair addition and pair subtraction modes of the normal system (cf. Eq. (21)).

If, on the other hand, we keep $\alpha = \alpha_o$ fixed and expand in

$$N' = N - N_o, \tag{53}$$

we find

$$H = \frac{N'^2}{4D\alpha_0^2} - \frac{2\lambda}{\hbar}N', \tag{54}$$

where we introduced the Fermi-level λ at $N = N_0$ and left out a constant.

The Hamiltonian shows a rotational behavior as a function of N', with a "moment of inertia" $\mathcal{I} = 2D\,\alpha_0^2$ superimposed on a linear background which depends on the definition of the Fermi energy.

The density associated with a pair rotation is, according to (5)

$$\delta\rho = \sum_\alpha \frac{\partial\rho}{\partial A}\cdot e^{-i\alpha\phi/2}\,\alpha_0\,e^{i\alpha\phi'/2} = 2\alpha_0\,\frac{\partial\rho}{\partial A}\,\cos(\phi-\phi'). \tag{55}$$

This real "dipole deformation" in gauge space is rotating with frequency $\dot\phi'$. The velocity dependence of $\delta\rho$ which is introduced in the next subsection implies that, in general, there will be also a centrifugal term, i.e.

$$\delta\rho = 2\alpha_0\left(\frac{\partial\rho}{\partial A} + \frac{\partial\rho}{\partial\chi}\,\dot\phi'\right)\cos(\phi-\phi'). \tag{56}$$

In passing, we note that in a quantal treatment the operator N is determined from the commutation relation

$$[N,\phi'] = -i\hbar, \tag{57}$$

i.e.

$$N = -i\hbar\frac{\partial}{\partial\phi}. \tag{58}$$

The rotational eigenstates of the Hamiltonian (54) therefore satisfy the equation

$$\left(-\frac{\hbar^2}{2\mathcal{I}}\frac{\partial^2}{\partial\phi'^2} + 2i\lambda\frac{\partial}{\partial\phi'}\right)\psi_n = E_n\psi_n, \tag{59}$$

leading to

$$\psi_n = \frac{1}{\sqrt{2\pi}}\,e^{in\phi}. \tag{60}$$

Since the density is invariant for rotations of $\phi' \to \phi' + 2\pi$, the pair quantum number n is

$$n = 0, \pm1, \pm2, \ldots \tag{61}$$

and the energy eigenvalues are

$$E_n = \frac{\hbar^2}{2\mathcal{I}}\,n^2 - 2\lambda n. \tag{62}$$

439

(e) Velocity-dependent deformations

When the deformation parameters α change with time, they introduce modifications in the pair density which to lowest order are proportional to $\dot{\alpha}$.

A more general expression for $\delta\rho$ is therefore (Ferreira et al., 1988)

$$\delta\rho_{\lambda\mu} = \frac{\partial\rho}{\partial\alpha_{\lambda\mu}}\,\alpha_{\lambda\mu} + \frac{\partial\rho}{\partial\dot{\alpha}_{\lambda\mu}}\,\dot{\alpha}_{\lambda\mu}, \tag{63}$$

and time reversal have similar effect on ρ except that Hermitean conjugation implies an interchange of $\vec{r}$ and $\vec{r}'(\vec{\xi} \to -\vec{\xi})$. Therefore $\partial\rho/\partial\alpha_{\lambda\mu}$ must be even in $\vec{\xi}$ (cf. also Eq (41)), while $\partial\rho/\partial\dot{\alpha}_{\lambda\mu}$ must be odd. The last term in (63) therefore vanishes for $\vec{r} = \vec{r}'$, i.e., it does not contribute to the local density. It contributes, however, to the current density. The density associated with a pair vibration of amplitude α_α may, in general, also depend on the velocity of the amplitude, i.e.

$$\delta\rho = \sum_\alpha \left(\frac{\partial\rho}{\partial\alpha_\alpha}\,\alpha_\alpha + \frac{\partial\rho}{\partial\dot{\alpha}_\alpha}\,\dot{\alpha}_\alpha \right) = \sum_\alpha \left(\frac{\partial\rho}{\partial A}\,\alpha_\alpha - i\frac{\alpha}{2}\frac{\partial\rho}{\partial\chi}\,\dot{\alpha}_\alpha \right) e^{-i\alpha\phi/2} \tag{64}$$

We have here written $\partial\rho/\partial\dot{\alpha}_\alpha$ in terms of a Hermitean, time-reversal invariant quantity $\partial\rho/\partial\chi$. As we shall see later, it is related to the second derivative $\partial^2\rho/\partial A^2$.

Making use of the relation (35) and (36), we find

$$\delta\rho = \sum_\alpha \left(\frac{\partial\rho}{\partial A} + \frac{\alpha}{2}\omega_{-\alpha}\frac{\partial\rho}{\partial\chi} \right) \left(e^{-i\alpha\phi/2}A_\alpha + e^{i\alpha\phi/2}\frac{i}{2\sqrt{CD}}\,\Pi_\alpha \right)$$

$$= \sum_\alpha \beta \left(\frac{\partial\rho}{\partial A} + \frac{\alpha}{2}\omega_{-\alpha}\frac{\partial\rho}{\partial\chi} \right) \left(e^{-i\alpha\phi/2}\Gamma^\dagger(\alpha) + e^{i\alpha\phi/2}\Gamma(\alpha) \right) \tag{65}$$

where $\omega_\alpha = \omega_\pm$ are defined in Eq. (21). Inserting (44) and (45) in Eq. (64), one obtains the density (56) associated with deformed systems.

From the commutation relations of the pair densities (cf. App. G), we may express the matrix elements of $\delta\rho$ between the collective pair vibrational states in terms of the non-local ground-state densities $\rho_0(A)$, $\rho_0(A+2)$, and $\rho_0(A-2)$ of the nuclei with nucleon number A and $A\pm2$.

We find

$$\langle 1_{\alpha=2} \, | \, \delta\rho \, | \, 0 \rangle = \beta \left(\frac{\partial \rho}{\partial A} + \omega_+ \frac{\partial \rho}{\partial \chi} \right)$$

$$= \langle 1_{\alpha=2} \, | \, M_2 \, | \, 0 \rangle \frac{\rho_0(A+2) - \rho_0(A)}{2}$$

$$\langle 0 \, | \, \delta\rho \, | \, 1_{\alpha=-2} \rangle = \beta \left(\frac{\partial \rho}{\partial A} - \omega_- \frac{\partial \rho}{\partial \chi} \right)$$

$$= \langle 0 \, | \, M_2 \, | \, 1_{\alpha=-2} \rangle \frac{\rho_0(A) - \rho_0(A-2)}{2}.$$

(66)

where $M_{\pm 2}$ are the pair moments

$$M_{\pm 2} = \int \delta\rho_{\pm 2}(\vec{r}\,', \vec{r}\,') d^3 r'. \tag{67}$$

Noting that

$$\int \frac{\partial \rho(\vec{r}, \vec{r})}{\partial A} d^3 r = \frac{\partial}{\partial A} A = 1, \tag{68}$$

and defining

$$\int \frac{\partial \rho(r, r)}{\partial \chi} d^3 r = \tau \tag{69}$$

we find

$$\int \langle 1_{\alpha=2} \, | \, \delta\rho(r,r) \, | \, 0 \rangle d^3 r = \langle 1_{\alpha=2} \, | \, M_2 \, | \, 0 \rangle = (1 + \omega_+ \tau)\beta$$

$$\int \langle 0 \, | \, \delta\rho(r,r) \, | \, 1_{\alpha-2} \rangle \, d^3 r = \langle 0 \, | \, M_2 \, | \, 1_{\alpha=-2} \rangle = (1 - \omega_- \tau)\beta$$

(70)

The two pair moments can thus be written (cf. Bohr and Mottelson, 1975)

$$M_2 = [(\alpha_2)_0 \, \Gamma^\dagger(+2) + (\alpha_{-2})_0 \, \Gamma(-2)] e^{-i\phi}$$

and

$$M_{-2} = [(\alpha_{-2})_0 \, \Gamma^\dagger(-2) + (\alpha_2)_0 \, \Gamma(+2)] e^{i\phi},$$

(71)

441

where we have defined the zero-point amplitudes appearing in these equations by

$$
\begin{aligned}
(\alpha_2)_0 &= (1 + \omega_+ \tau)\beta \\
(\alpha_{-2}) &= (1 - \omega_- \tau)\beta
\end{aligned}
\tag{72}
$$

The fact that $(\alpha_{+2})_0$ and $(\alpha_{-2})_0$ in actual nuclei are different, shows the necessity of introducing the term proportional to $\dot{\alpha}$ in Eq. (64).

While the difference in frequencies for pair addition and pair removal modes can be eliminated by a suitable choice of the Fermi energy, the difference in the corresponding zero-point fluctuations will remain, that is

$$
(\alpha_2)_0 - (\alpha_{-2})_0 = (\omega_+ + \omega_-)\tau\beta = 2\sqrt{\frac{C}{D}}\,\tau\beta.
\tag{73}
$$

Even if the zero-point fluctuations for addition and removal modes are the same, which implies $\tau = 0$, the quantity $\partial\rho/\partial\chi$ is, in general, non-vanishing. This is because the radial density distribution in the nuclei $A + 2$ and $A - 2$ will, in general, be different. In this case, we find from (66)

$$
\frac{\partial\rho}{\partial A} = \frac{1}{2}\left[\left(\frac{\partial\rho}{\partial A}\right)_+ + \left(\frac{\partial\rho}{\partial A}\right)_-\right]
\tag{74}
$$

and

$$
\frac{\partial\rho}{\partial\chi} = \frac{1}{\sqrt{C/D}}\frac{\partial^2\rho}{\partial A^2}.
\tag{75}
$$

where

$$
\left(\frac{\partial\rho}{\partial A}\right)_+ = \frac{\rho(A+2) - \rho(A)}{2}
\tag{76}
$$

and

$$
\left(\frac{\partial\rho}{\partial A}\right)_- = \frac{\rho(A) - \rho(A-2)}{2}.
\tag{77}
$$

while

$$
\frac{\partial^2\rho}{\partial A^2} = \frac{1}{2}\left[\left(\frac{\partial\rho}{\partial A}\right)_+ - \left(\frac{\partial\rho}{\partial A}\right)_-\right].
\tag{78}
$$

From Eq. (66) we may also conclude that in the Fermi gas model the derivatives of ρ, i.e. $\partial\rho/\partial A, \partial\rho/\partial\chi$ or $(\partial\rho/\partial A)_\pm$ all have the same nonlocality, i.e.

$$
\delta\rho(\vec{r}, \vec{r}') = \delta\rho(\vec{q}) \cdot \frac{\sin(k_F\xi)}{k_F\xi}.
\tag{79}
$$

13 MACROSCOPIC DESCRIPTION OF PAIR TRANSFER

In a macroscopic description, one considers the transfer of paired nucleons to or from a nucleus as due to the action of an external field. The external time-dependent field is due to the passing by of a projectile that can donate or absorb pairs, and may have the form

$$\Delta = \Delta_2 \, e^{-i\phi} + \Delta_{-2} \, e^{i\phi}. \tag{1}$$

The fields Δ_2 and Δ_{-2} depend on the degrees of freedom of the projectile, and Δ has no diagonal part since the total number of particles has to be conserved. For a superfluid projectile, the fields $\Delta_{\pm 2}$, e.g., have the form

$$\Delta_{\pm 2} = f(\vec{r}\sigma, \vec{r}'\sigma', t)e^{\pm i\phi'_a}, \tag{2}$$

where ϕ'_a indicates the orientation of the deformation of the projectile in gauge space.

The time-dependent interaction with the target is

$$V(t) = \sum_{\sigma\sigma'} \int d^3r \, d^3r' \, \delta\rho \cdot \Delta \, \frac{d\phi}{2\pi}. \tag{3}$$

Utilizing Eq. (12.79), we may write

$$V(t) = \int \delta\rho(\vec{q}) \, \Delta(\vec{q}, t) \, d^3q \, \frac{d\phi}{2\pi}, \tag{4}$$

where

$$\Delta(\vec{q}, t) = \sum_{\sigma} \int d^3\xi \, \frac{\sin k_F \xi}{k_F \xi} \, \Delta\left(\vec{q} + \frac{\vec{\xi}}{2}, \sigma, \vec{q} - \frac{\vec{\xi}}{2}, \sigma, t\right). \tag{5}$$

We have here also assumed that only the singlet part of the density contributes. The field $\Delta(\vec{q})$ thus mainly acts as a local field on the center of mass of the dineutron or diproton.

The field that causes a pair transfer in actual nuclei has a very complicated structure. This is due to the fact that the pair transfer process is mainly induced by the mean single-particle field in a second-order process and only to a lesser degree, in first order, by the pair field created by the paired nucleons.

One may, however, introduce an effective pair field which in first-order perturbation theory causes the successive transfer of a nucleon pair.

443

This can be obtained from the expression (11.52) by using the parametrization (5.40) of the single-particle form factor. We find

$$
(a)_{\text{succ}} = \frac{1}{i\hbar} \sum_{a_1 a'_1} B^{(A)}(a_1 a_1; 0)\, B^{(b)}(a'_1 a'_1; 0)\,(-1)^{\ell_1 + \ell'_1}
$$
$$
\times\, J^2 \pi (2j_1 + 1)^{1/2}\, (2j'_1 + 1)^{1/2}
$$

$$
\times\, \bar{N}^2_{a_1}\, \bar{N}^2_{a'_1}\, \frac{\hbar^2}{M^2}\, \frac{1}{4\kappa \ddot{r}_o}\, \left(\frac{R_a R_A}{r_o} \right)^2 e^{-2\kappa(r_o - R_A - R_a)}
$$

$$
\times\, R^2_\lambda(\ell_1 \ell'_1) \sum_\gamma \left(h(a_\gamma b_\gamma) - ig(a_\gamma, b_\gamma) \right) |\, C^{(A)}(0\, a_1; I_F)\,|^2\, |\, C^{(b)}(0\, a'_1; I_f)\,|^2,
$$

(6)

where we have used the asymptotic expression (cf. (5.49)) for the spherical Hankel functions. The ratio between this expression and the expression in terms of Hankel functions is denoted $R^2_\lambda(\ell_1 \ell'_1)$. We have furthermore performed the summation over μ introducing the functions $g(a, b)$ and $h(a, b)$ through Eqs. (6.37) and (10.17) with a_γ and b_γ defined by Eqs. (11.55). These functions are depicted in Fig. 27. From this figure, we notice that the modulus of the function $h - ig$ is rather independent of a, i.e., the Q-value for the intermediate state. Neglecting the moderate change in phase, we may, to a rather good approximation, perform the summation over the single-particle strength distribution $|\, C^{(A)}\,|^2$ and $|\, C^{(b)}\,|^2$, i.e. over γ. Finally, we utilize the approximate independence of $\langle R^2_\lambda(\ell_1, \ell'_1)(h - ig) \rangle$ on the intermediate state and utilize that the sum

$$
\sum_{a_1} B^{(A)}(a_1 a_1; 0)\,(-1)^{\ell_1}\, \sqrt{j_1 + \frac{1}{2}}\, \bar{N}^2_{a_1}
$$

$$
= 4\pi \, \langle B \,|\, \rho^{(A)}_{+2}(R_A)\,|\, A \rangle
$$

(7)

is the matrix element of the pair density in the target since $\bar{N}_{a_1}$ is approximately equal to the radial wavefunction at $r = R_A$ (cf. Eq. (5.39)).

The result may therefore (Lotti et al., 1990) be written

$$
(a)_{\text{succ}} = \frac{1}{i\hbar}\, \sqrt{\frac{\pi}{\kappa \ddot{r}_o}}\, F(r_o)\, e^{-q^2}
$$

(8)

with

$$F(r) = \langle B \,|\, \rho_{+2}^{(A)}(R_A) \,|\, A \rangle \, \langle b \,|\, \rho_{-2}^{(a)}(R_a) \,|\, a \rangle$$

$$\times \left(\frac{R_A R_a}{R_a + R_A} \right)^2 e^{-2\kappa(r - R_a - R_A)} L(\tau). \tag{8a}$$

We have furthermore introduced the function

$$L(\tau) = 32\pi^2 \, \frac{\hbar^3}{M^2} \, \sqrt{\frac{\pi}{\kappa \ddot{r}_o}} \, \langle \sum_\lambda R_\lambda^2 (\ell_1 \ell_1') \left(h(a, b) - i g(a, b) \right) \rangle, \tag{9}$$

which depends on the bombarding energy mainly through the collision time $\tau = (2\kappa \ddot{r}_0)^{-1/2}$. The function may be estimated by using Fig. 27 inserting for a and b the values corresponding to a typical intermediate state. The function $F(r)$ acts as an effective form factor for simultaneous pair transfer. This is because the expression (8) has exactly the same form as a first-order transition amplitude evaluated in the parabolic approximation discussed in §6.b. The pair transfer is here evaluated in the prior representation, $2\kappa = \kappa_1 + \kappa_2$ being the slope of the product of the tails of the wave functions of the two neutrons in the projectile. The function $\exp(-q^2)$ correspondingly describes the adiabatic cutoff appropriate for a dineutron transfer in the prior representation. The same formula applies in the post representation when κ is substituted with the slope of the product wavefunction in the target and q is changed correspondingly $(\Delta^{(\alpha)} \to \Delta^{(\beta)})$ (cf. Eq. (6.34)).

From the above discussion, we may identify $F(r)$ with the matrix element of the effective pair interaction (3) in the post representation. We may thus write

$$F(r) = \langle \beta \,|\, V \,|\, \alpha \rangle$$

with

$$\tag{10}$$

$$V(r) = \delta \rho^{(A)}(r - R_a) \left(\frac{R_a R_A}{R_a + R_A} \right)^2 \delta \rho^{(a)}(R_a) L(\tau),$$

where we used that the exponential function (in the post representation) is just the factor that would extrapolate the pair density in the target to the surface of the projectile. Substituting $\delta \rho^{(A)}$ with the macroscopic expression, we find for neutron pair stripping

$$V(r) = \alpha_{+2}(A) \frac{\partial \rho_{ov}^{(A)}(r - R_a)}{\partial N} \left(\frac{R_a R_A}{R_a + R_A} \right)^2 \alpha_{-2}(a) \frac{\partial \rho_{ov}^{(a)}(R_a)}{\partial N} L(\tau),$$

$$\tag{10a}$$

where we have specified explicitly that in this case the equilibrium density $\rho_{o\nu}$ for neutrons given in Eq. (III.A.16) should be used. Similar expressions are obtained for neutron pair pickup and for proton pair transfer. Comparing with the expression (4), we see that the effective pair field $\Delta(\vec{q})$ associated with the projectile is strongly concentrated on the inside of this nucleus. It might be suggestive to write it as

$$\Delta(\vec{q}) \sim \left(U_{1a}(\vec{q})\right)^2$$

where U_{1a} is the mean field of the projectile. This expression differs significantly from the field which was proposed by Dasso and Pollarolo, 1986, and which would lead to an interaction of the form

$$V \sim \alpha_{+2}(A)\,\frac{\partial U}{\partial N}$$

where U is the ion-ion potential.

It should be emphasized that the effective interaction (and the effective pair field) introduced above are of the same type as the absorptive potential. They are fundamentally non-local in space and time and are correspondingly velocity dependent. This also implies that they cannot, in a straighforward way, be used in higher order to describe multi-pair transfer.

Besides this main reaction mechanism, there is also the possibility of pair excitation through the pair field of the projectile. The corresponding interaction is

$$V(t) = \sum_{\alpha\alpha'} \int \delta\rho_\alpha^{(A)} \Delta_{\alpha'}^{(a)}\, d^3r\, d^3r'\,\frac{d\phi}{2\pi}$$
$$= \sum_\alpha \int \delta\rho_\alpha^{(A)} V_{12}\, \delta\rho_{-\alpha}^{(a)}\, d^3r\, d^3r'\, d^3r''\, d^3r''' \tag{11}$$

where V_{12} is the two-body interaction. The interaction (1) can, in first-order perturbation theory, excite a pair in target and projectile simultaneously, and is very much of the same structure as the multipole-multipole interaction for surface deformations. In contrast to other transfer reactions, there is no ambiguity of post and prior representation, and the recoil phase is the one given in (11.46).

We can rather easily estimate the magnitude of the transfer by the pair field. Using (11), we find for pair vibrations

$$V(t) = (\alpha_2^{(A)}\alpha_{-2}^{(a)} + \alpha_{-2}^{(A)}\alpha_2^{(a)})K \cdot \int d^3q\, \frac{\partial\rho^{(a)}(\vec{q})}{\partial A} \cdot \frac{\partial\rho^{(A)}(\vec{q}-\vec{r}(t))}{\partial A}, \tag{12}$$

where

$$K = \int d^3\zeta \left| \frac{\sin(k_F\zeta)}{k_F\zeta} \right|^2 V_{12}(\zeta).$$ (13)

Estimating K by setting $V_{12} = 200\ \delta(\xi)$ MeV/fm^3, we find that the interaction (12) is usually several orders of magnitude smaller than the contribution to the transfer from the mean field (10).

We shall conclude this chapter by performing a purely classical calculation of the probability for pair transfer. As an example, we consider the case of a superfluid projectile where the field is of the form (2).

The total Hamiltonian for nuclei exhibiting pair vibrations is then, according to Eqs. (2)–(4) and (12.17),

$$H(t) = \frac{1}{D}\pi_{-2}\pi_2 + C\,\alpha_{-2}\alpha_2 - i\omega(\alpha_2\pi_2 - \alpha_{-2}\pi_{-2})$$
$$+ \sum_\alpha \int \left(\frac{\partial\rho}{\partial A}\alpha_\alpha - i\frac{\alpha}{2}\frac{\partial\rho}{\partial\chi}\dot\alpha_\alpha \right) f(\vec r,\vec r',t)e^{-i\frac{\alpha}{2}\phi'_a}\,d^3r\,d^3r'$$ (14)

Introducing the variables $A_\pm$ and $\Pi_\pm$ (cf. (12.25) and (12.27)), we find

$$H(t) = i\,\omega_+A_+\Pi_+ + i\,\omega_-A_-\Pi_-$$
$$+ \left(g(t) + \omega_+h(t)\right)\left(e^{-i\phi'_a}A_+ + e^{i\phi'_a}\frac{i}{2\sqrt{CD}}\Pi_+ \right)$$
$$+ \left(g(t) - \omega_-h(t)\right)\left(e^{i\phi'_a}A_- + e^{-i\phi'_a}\frac{i}{2\sqrt{CD}}\Pi_- \right),$$ (15)

where

$$g(t) = \int d^3r\,d^3r'\,\frac{\partial\rho}{\partial A}\,f(\vec r,\vec r',t)$$
$$h(t) = \int d^3r\,d^3r'\,\frac{\partial\rho}{\partial\chi}\,f(\vec r,\vec r',t).$$ (16)

While in Section 12 the value of the frequency ω could be changed by adjusting the Fermi level, it is here fixed by the fact that energies should be measured from the same level in target and projectile.

The classical equations of motion corresponding to the Hamiltonian (15) can be solved analytically, and one may calculate the amplitude after the collision, as was done for the case of surface vibrations in Section II.4 To obtain from here the probability on the various members of the pair removal and pair addition modes is more complicated because, in general, one starts from a state which has a finite number of quanta.

We may also consider the excitation of pair rotations. In this case the Hamiltonian is

$$
\begin{aligned}
H(t) &= \frac{(N'_A)^2}{4D\alpha_o^2} + \frac{2\lambda_A}{\hbar} N'_A \\
&+ 2\alpha_o \int \left(\frac{\partial \rho}{\partial A} + \dot{\phi}'_A \frac{\partial \rho}{\partial \chi} \right) f(\vec{r}, \vec{r}', t) d^3r\, d^3r'\, \cos(\phi'_A - \phi'_a)
\end{aligned}
\tag{17}
$$

$$
\begin{aligned}
&= \frac{(N'_A)^2}{4D\alpha_o^2} + \frac{2\lambda_A}{\hbar} N'_A \\
&+ 2\alpha_o \left(g(t) + \left(\frac{N'_A}{2D\alpha_o^2} + \frac{2\lambda_A}{\hbar} \right) h(t) \right) \cos(\phi'_A - \phi'_a).
\end{aligned}
\tag{18}
$$

The equations of motion are

$$
\dot{\phi}'_A = \frac{\partial H}{\partial N'_A} = \frac{N'_A}{2D\alpha_o^2} + \frac{2\lambda_A}{\hbar} + \frac{h(t)}{2D\alpha_o} \cos(\phi'_A - \phi'_a)
\tag{19}
$$

$$
\dot{N}'_A = -\frac{\partial H}{\partial \phi'_A} = 2\alpha_o \left(g(t) + \left(\frac{N'_A}{2D\alpha_o^2} + \frac{2\lambda_A}{\hbar} \right) h(t) \right) \sin(\phi'_A - \phi'_a).
\tag{20}
$$

The dynamical variables for the projectile satisfy similar equations. We note that since $H(t)$ only depends on ϕ' through the difference $\phi'_A - \phi'_a$, we would find $\dot{N}'_a + \dot{N}'_A = 0$, i.e. conservation of particle number. Equations (19) can be solved easily when we neglect the term proportional to $N'_A/2D\alpha_o^2$ and to $h(t)$. We then find from (19) and the corresponding equation for $\dot{\phi}'_a$ that

$$
\phi'_A - \phi'_a = \frac{2}{\hbar}(\lambda_A - \lambda_a)t + \delta
\tag{21}
$$

and

$$
\Delta N_A = N'_A(\infty) - N'_A(-\infty) = 2\alpha_o \int_{-\infty}^{\infty} g(t) \sin\left(\frac{2}{\hbar}(\lambda_A - \lambda_a)t + \delta \right) dt
\tag{22}
$$

For a time-symmetric interaction, we find

$$\Delta N_A = 2\alpha_o \sin \delta \int_{-\infty}^{\infty} dt\, g(t)\, \cos\left(\frac{2}{\hbar}(\lambda_A - \lambda_a)t\right). \qquad (23)$$

The number of transferred pairs depends on the orientation of the system in gauge space at $t = 0$. In fact, there is a maximum number $\Delta n_m = (\Delta N_A)_{\max}/\hbar$ of transferred pairs which is equal to

$$\Delta n_m = \left| 2\alpha_o \int_{-\infty}^{\infty} g(t)\, \cos\left(\frac{2}{\hbar}(\lambda_A - \lambda_a)t\right) dt \right| /\hbar \qquad (24)$$

Since we must assume that the phase δ at $t = 0$ has an isotropic distribution, we find the probability distribution in Δn to be

$$P\left(\frac{\Delta N}{\hbar}\right) d\left(\frac{\Delta N}{\hbar}\right) = \frac{1}{2\pi} d\delta,$$

or

$$P(\Delta n) = 2 \cdot \frac{\hbar}{2\pi} \left(\frac{d(\Delta N)}{d\delta}\right)^{-1} = \frac{1}{\pi \Delta n_m \sqrt{1 - f^2}}, \qquad (25)$$

where

$$f = \frac{\Delta n}{\Delta n_m}, \qquad (26)$$

with

$$|\Delta n| < \Delta n_m.$$

In (25) we have added the two equal contributions from

$$\delta_1 = \arcsin f \qquad (27)$$

and

$$\delta_2 = \pi - \arcsin f. \qquad (28)$$

The situation is quite similar to the classical description of the Coulomb excitation discussed in Section II.7. In fact, the two contributions, (27) and (28), interfere and the probability distribution is

$$P(\Delta n) = |\sqrt{P_1}\, e^{i\phi_1} - i\sqrt{P_2}\, e^{i\phi_2}|^2 \qquad (29)$$

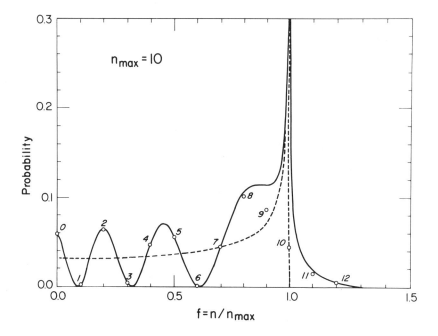

Fig. 33 Probability for transferring different number of pairs in a hypothetical heavy-ion collision. In the abcissa the number of pairs relative to the maximum number allowed classically is shown, the classical probability distribution being given by the dotted curve. The results of the quantal calculation are displayed by circles for the case of $n_{\max} = 10$. The semiclassical result (13.32) is shown as a continuous curve for the same situation. The figure is due to P. Lotti.

where ϕ_1 and ϕ_2 are the action integrals

$$\phi = -\frac{1}{\hbar} \int_{-\infty}^{\infty} (\phi_A' \dot{N}_A' + \phi_a' \dot{N}_a')dt = -\frac{1}{\hbar} \int_{-\infty}^{\infty} (\phi_A' - \phi_a') \dot{N}_A' \, dt \qquad (30)$$

for the two orientations. In (30) we have neglected the reaction on the trajectory. In the sudden approximation we find

$$\phi_1 = -\Delta n \arcsin f$$
$$\phi_2 = -\Delta n(\pi - \arcsin f) \qquad (31)$$

i.e.,

$$
\begin{aligned}
P(\Delta n) &= \frac{1}{2\pi\,\Delta n_m\sqrt{1-f^2}}\,|\,e^{-i\Delta n\ \text{arcsin}\ f}-i\,e^{i\Delta n(\text{arcsin}\ f-\pi)}\,|^2 \\
&= \frac{1}{\pi\,\Delta n_m\sqrt{1-f^2}}\bigl(1+\sin(2\Delta n\ \text{arccos}f)\bigr).
\end{aligned}
\tag{32}
$$

This probability distribution is shown in Fig. 33. The distribution may be supplemented above the classical limit by evaluating the complex roots of Eq. (27) for $f > 1$. These probabilities have been evaluated also through a coupled-channel calculation using the quasispin formalism (Dietrich et al., 1971). The oscillations that were present in these results were taken as a signature of Josephson effect (Josephson (1962)), i.e. of an alternating supercurrent between the two nuclei during contact. As we have seen, the oscillations are, however, not associated with the difference in the Fermi levels of the two nuclei but with an interference between different initial orientations in gauge space. The difference in Fermi levels merely gives rise to an adiabatic cutoff in the maximum number of transferred pairs (cf. Eq. (24)).

APPENDIX A.
COORDINATES OF THE TRANSFER PROCESS

In evaluating the matrix elements for transfer processes, the integration over the coordinate of the transferred particles cannot be performed independently of the integration over the relative center-of-mass coordinate of the colliding nuclei. A separation between a form factor characteristic for the nuclear transition and the relative motion can only be achieved by functions of the relative coordinate, which also contain a momentum dependence that implies a shift of these coordinates. For the coordinate $\vec{r}$ which appears in the form factor, one has a wide choice as e.g., $\vec{r}_{aA}, \vec{r}_{bB}, \vec{r}_{bA}$, etc. A general discussion is given in Broglia et al., 1977. In the semiclassical treatment it is natural to use $\vec{r} = (\vec{r}_{aA} + \vec{r}_{bB})/2$. The geometrical relation between the coordinate of the transferred "cluster" r_d and the coordinates $\vec{r}_a, \vec{r}_b, \vec{r}_A$, and $\vec{r}_B$ is shown in Fig. 4. In the matrix elements (8.3) or (8.5) the integration is often performed over $\vec{r}_{aA}$ and $\vec{r}_{db}$, or $\vec{r}_{bB}$ and $\vec{r}_{dA}$. These are both natural (Jacobian) coordinates, with Jacobian unity, as would be also the coordinates $\vec{r}_{bA}$ and $\vec{r}_{db}$.

In choosing $\vec{r} = (\vec{r}_{aA} + \vec{r}_{bB})/2$ as one of the coordinates, it is natural to choose $\vec{r}_{d\delta} = q(\vec{r}_{aA} - \vec{r}_{bB})/2$ as a second coordinate where q is to be determined such that δ is on the $\vec{r}$ vector. From the figure, we see that

$$\vec{r} = \vec{r}_{PT} \tag{1}$$

and

$$\vec{r}_{d\delta} = q \frac{m_d}{2m_B} \vec{r}_{dA} + q \frac{m_d}{2m_a} \vec{r}_{db}$$

$$= q \frac{m_d}{m_A + m_B} \vec{r}_{dT} + q \frac{m_d}{m_a + m_b} \vec{r}_{dP} \tag{2}$$

From the equilateral triangles shown in the figure, we find

$$\frac{r_{P\delta}}{r_{\delta T}} = \frac{r_{dP} - \dfrac{q m_d}{m_a + m_b} r_{dP}}{q \dfrac{m_d}{m_a + m_b} r_{dP}} = \frac{\dfrac{q m_d}{m_A + m_B} r_{dT}}{r_{dT} - \dfrac{q m_d}{m_A + m_B} r_{dT}} \tag{3}$$

from which

$$q = \frac{(m_a + m_b)(m_A + m_B)}{2m_d(m_a + m_A)}, \tag{4}$$

and

$$\frac{r_{P\delta}}{r_{\delta T}} = \frac{m_a + m_b}{m_A + m_B}. \tag{5}$$

From a similar argument, we find

$$r_{d\beta} = \frac{m_a m_B}{(m_a + m_B) m_d} (\vec{r}_{aA} - \vec{r}_{bB}), \tag{6}$$

and

$$\frac{r_{b\beta}}{r_{\beta A}} = \frac{m_a}{m_B}. \tag{7}$$

The scaling parameter β (cf. Broglia et al. (1977), Eq. (4.12)) corresponding to the variable $r_{d\delta}$ is therefore

$$\beta = \frac{m_B}{m_a + m_B}. \tag{8}$$

The coordinate transformation

$$\vec{r}_{aA} = \vec{r} + \frac{1}{q} \vec{r}_{d\delta}, \tag{9}$$

$$\vec{r}_{db} = \frac{2m_a}{m_a + m_b} \vec{r}_{d\delta} - \frac{m_a}{m_a + m_A} \vec{r}, \tag{10}$$

leads to the Jacobian

$$J_\rho = \left| \frac{\partial(\vec{r}_{aA}, \vec{r}_{db})}{\partial(\vec{r}, \vec{r}_{d\delta})} \right| = \left(\frac{4m_a m_B}{(m_a + m_b)(m_A + m_B)} \right)^3. \tag{11}$$

We find, furthermore, the relations

$$\vec{r}_{dA} = \frac{m_B}{m_a + m_A} \vec{r} + \frac{2m_B}{m_A + m_B} \vec{r}_{d\delta}, \tag{12}$$

and

$$\vec{r}_{bA} = \frac{m_a + m_B}{m_a + m_A} \vec{r} - \frac{2m_d(m_A - m_b)}{(m_a + m_b)(m_A + m_B)} \vec{r}_{d\delta}. \tag{13}$$

APPENDIX B.
ANTISYMMETRIZATION

In the derivation of the coupled equations for transfer reactions, we have neglected the fact that the channel wavefunctions should be anti-symmetric between all protons and all neutrons and not only between the nucleons within each of the two colliding nuclei. This means that the proper channel wavefunction $\tilde{\psi}_\beta$ should be (cf. Eqs. (1.6) and (1.12))

$$\tilde{\psi}_\beta = \frac{1}{\sqrt{N_{bB}}} \sum_{P_\beta} (-1)^{P_\beta} \, \psi_b(\zeta_b) \, \psi_B(\zeta_B) e^{i\delta_\beta} \chi(\vec{r}_\beta - \vec{R}_\beta(t)), \qquad (1)$$

where P_β indicates the N_{bB} permutations between the nucleons in b and in B. To simplify the arguments, we consider that the nuclei only consist of one kind of particles, i.e. n_b in b and n_B in B. We then find

$$N_{bB} = \frac{(n_b + n_B)!}{n_b! n_B!} \qquad (2)$$

In (1) we have indicated also the wavefunction for the wavepacket χ. It depends on a different center-of-mass coordinate for each permutation. In calculating the flux in channel β, these different coordinates do not interfere, and we obtain N_{bB} incoherent contributions which in the probability current would cancel the factor $(N_{bB})^{-1/2}$ in (1) to give unit flux to the channel wavefunction.

The change in the coupled equations which is introduced by utilizing the channel wavefunctions (1) instead of ψ_β is that the matrix elements in (1.17) should be substituted by

$$\langle \tilde{\psi}_\gamma \,|\, V_\beta - U_\beta \,|\, \tilde{\psi}_\beta \rangle$$

$$= \frac{1}{\sqrt{N_{fF} N_{bB}}} \sum_{P_\beta P_\gamma} (-1)^{P_\beta + P_\gamma} \langle \psi_\gamma \,|\, V_\beta - U_\beta \,|\, \psi_\beta \rangle \qquad (3)$$

$$(\beta \equiv (b, B); \gamma \equiv (f, F))$$

The different matrix elements appearing in this sum can be classified into: 1) a simple stripping of

$$n_s = n_b - n_f = n_F - n_B \qquad (4)$$

nucleons from the projectile b to target F when $b > f$, 2) a stripping plus an exchange of one nucleon, 3) a stripping plus exchange of two nucleons, etc. until all nucleons of f have been exchanged.

Each matrix element can be evaluated by the fractional parentage expansion of the type shown in Eq. (4.17). In this expansion the identification of the nucleons taking part in the reaction is explicitly specified. Many of the $N_{fF} \cdot N_{bB}$ matrix elements appearing in the sum (3) are identical except for a renumbering. In fact, for the simple stripping matrix elements, there are

$$N_{n_s}(0) = \frac{(n_b + n_B)!}{n_s! \, n_f! \, n_B!} \tag{5}$$

identical matrix elements. This is because the cluster of n_s nucleons, as well as the cores f and B, are not changed by the permutations. If, in addition, n_e nucleons are exchanged, there will similarly be

$$N_{n_s}(n_e) = \frac{(n_b + n_B)!}{(n_s + n_e)! \, n_e! \, (n_f - n_e)! \, (n_B - n_e)!} \tag{6}$$

identical form factors.

We may thus write the matrix element

$$\langle \tilde{\psi}_\gamma \,|\, V_\beta - U_\beta \,|\, \tilde{\psi}_\beta \rangle = \sum_{n_e} \frac{N_{n_s}(n_e)}{\sqrt{N_{bB} N_{fF}}} \, f_{n_s, n_e}(\vec{r}). \tag{7}$$

The quantity $f_{n_s, n_e}(\vec{r})$ is the form factor as calculated making use of the fractional parentage expansion corresponding to a reaction in which n_s particles are stripped and, in addition, n_e particles exchanged. The relative sign of the different form factors $f_{n_s, n_e}(\vec{r})$ which contribute coherently is directly determined from the parity of the number of exchanges.

Since the matrix elements decrease progressively with the number of nucleons exchanged, it is only necessary to specify a few of the active nucleons in each of the interacting ions. The same expressions (1)–(7) are applicable to this situation except that n_b, n_B, n_f, and n_F now indicate the number of specified nucleons in the respective nuclei, i.e. the nucleons specified in the nuclear wavefunctions.

It is noted that the numerical factor in (7) can be written as

$$\frac{N_B(n_e)}{\sqrt{N_{bB} N_{fF}}} = \left(\frac{n_b!}{(n_s + n_e)!(n_f - n_e)!} \right)^{1/2} \left(\frac{n_B!}{n_e!(n_B - n_e)!} \right)^{1/2}$$

$$\times \left(\frac{n_f!}{n_e!(n_f - n_e)!} \right)^{1/2} \left(\frac{n_F!}{(n_s + n_e)!(n_B - n_e)!} \right)^{1/2}. \tag{8}$$

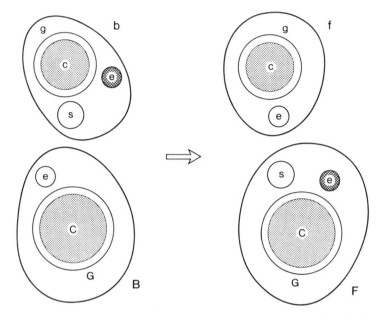

Fig. 34 Schematic representation of the reaction $b + B \rightarrow f + F$ where a number n_s of nucleons are stripped, and a group n_e of particles exchanged. In a description where the degrees of freedom of all nucleons are explicitly taken into account, the wavefunctions of the different nuclei b and B participating in the reaction will be written as

$$\psi^{(b)}(1, 2, \ldots, n_g; \; n_g + 1, \ldots, \; n_g + n_s; n_g + n_s + 1, \ldots, n_b),$$

and

$$\psi^{(B)}\big((n_b + 1), \ldots, (n_b + 1) + n_G; \; (n_b + 1) + n_G + 1, \ldots, n_b + n_B\big).$$

This representation may, however, be too detailed for the description of the nuclear states connected in the transfer process. In fact, in most cases, only a reduced number of nucleons will be explicitly taken into account in the nuclear structure calculations. A large fraction of the nucleons in both target and projectile are usually considered as forming part of a core denoted c and C respectively. The system c and C may correspond to closed-shell systems in many situations. However, other less justified choices of the cores are made to simplify nuclear structure calculations. The wavefunctions describing the intrinsic degrees of freedom of nuclei b and B are in the present example

This is because (cf., Fig. 34)

$$n_b + n_B = n_f + n_F, \tag{9}$$

$$n_b - n_s - n_e = n_f - n_e = n_g, \tag{10}$$

and

$$n_F - n_s - n_e = n_B - n_e = n_G, \tag{11}$$

n_g and n_G being the number of specified nucleons which do not take part in the process $n_b + n_B \to n_f + n_F$. In order to calculate the form factor $f_{n_s n_e}(\vec{r})$, one should use the parentage functions $\phi^{b(g)}, \phi^{B(G)}, \phi^{f(g)}$, and $\phi^{F(G)}$. Associating each of the four factors in (8) to the corresponding parentage function through the substitution

$$\phi^{b(g)} \to \tilde{\phi}^{b(g)} = \left(\frac{n_b}{n_s + n_e} \right)^{1/2} \phi^{b(g)}, \tag{12}$$

one can include the effect of antisymmetrization in the coupled equations. The binomial coefficient indicates the number of ways the active transferred particles can be selected among the specified particles in b. With a more general notation, we thus find

$$\tilde{\phi}^{b(g)} = \left[\binom{n_b}{n_g} \binom{z_b}{z_g} \right]^{1/2} \phi^{b(g)}, \tag{13}$$

where n_i now indicates the number of specified neutrons and z_i the number of specified protons in the system i.

The relative sign of the form factors is relevant in evaluating (7) but is also important in the coupled equations when one can arrive to a given channel via different intermediate channels. We note also that the relative signs become acute when interference is observed between different reactions with nearly identical target and projectile (cf. Appendix D). The relative sign is determined from the parentage functions from the condition of antisymmetry by keeping track of the numbering of the specified nucleons. For the calculation of the cross section for nearly symmetric systems, one needs, in addition, to take into account the exchange property of the inert cores.

---

formally obtained by making the replacements $n_g \to (n_g - n_c)$, $n_b \to n_b - n_c$, $n_G \to n_G - n_c$, and $n_B \to n_C$ in the wavefunctions $\psi^{(b)}(1, \ldots, n_b)$ and $\psi^{(B)}(n_b + 1, \ldots, n_b + n_B)$ written above. Similarly, for $\psi^{(f)}$ and $\psi^{(F)}$.

APPENDIX C.
MOMENTUM REPRESENTATION

The single-particle state (4.41) can be written in momentum representation

$$\phi_{j_1 m_1}^{(A)}(a_1; \vec{r}_{1A}\zeta_1) = (2\pi)^{-3/2} \int d^3k \, \tilde{\phi}_{j_1 m_1}^{(A)}(a_1; \vec{k}\zeta_1)e^{i\vec{k}\cdot\vec{r}_{1A}}, \qquad (1)$$

where the single-particle wavefunction in $\vec{k}$-space is given by

$$\tilde{\phi}_{j_1 m_1}^{(A)}(a_1; \vec{k}\,\zeta_1) = (2\pi)^{-3/2} \int d^3 r_{1A} \, \phi_{j_1 m_1}^{(A)}(a_1; \vec{r}_{1A}\zeta_1)e^{-i\vec{k}\cdot\vec{r}_{1A}}$$
$$= i^{-\ell_1} \tilde{R}_{a_1}^{(A)}(k)[Y_{\ell_1}(\hat{k})\chi]_{j_1 m_1}. \qquad (2)$$

The radial function in k-space

$$\tilde{R}_{a_1}^{(A)}(k) = \sqrt{\frac{2}{\pi}} \int_o^\infty j_{\ell_1}(kr) \, R_{a_1}^{(A)}(r) r^2 dr \qquad (3)$$

is given as the Hankel transform of the radial wavefunction in r-space.
The reciprocal transformation is

$$R_{a_1}^{(A)}(r) = \sqrt{\frac{2}{\pi}} \int_o^\infty j_{\ell_1}(kr) \, \tilde{R}_{a_1}^{(A)}(k) k^2 dk. \qquad (4)$$

Note that the index ℓ_1 on the spherical Bessel function must be identical
to the quantum number ℓ_1 of the state.

   It may sometimes be useful to use the helicity representation where
we specify the component h of the spin along the momentum, i.e.

$$\chi_{m_s} = \sum_h D_{m_s h}^{1/2}(\hat{k})\chi_h. \qquad (5)$$

One finds (cf. Bohr and Mottelson (1969))

$$\tilde{\phi}_{j_1 m_1}(a_1; \vec{k}\zeta_1) = \left(\frac{2j_1+1}{16\pi^2}\right)^{1/2} i^{-\ell_1} \tilde{R}_{a_1}^{(A)}(k)$$

$$\times \left\{ D_{m_1 -1/2}^{j_1}(\hat{k})\chi_{h=-1/2}(\zeta_1) + (-1)^{j_1-\ell_1-1/2} D_{m_1 1/2}^{j_1}(\hat{k})\chi_{h=1/2}(\zeta_1) \right\}, \qquad (6)$$

which is normalized to be integrated over all three Eulerian angles besides over $k = |\vec{k}|$.

The tail of the radial wavefunction for neutrons is given by

$$R_{a_1}^{(A)}(r) = N_{a_1}^{(A)} k_{\ell_1}(\kappa_{a_1} r) \text{ for } r \to \infty, \tag{7}$$

where $k_\ell(z)$ is the modified spherical Bessel function

$$k_\ell(z) = \sqrt{\frac{\pi}{2z}} K_{\ell+1/2}(z) = -\frac{\pi}{2} e^{i\ell\pi/2} h_\ell^{(1)}(e^{i\pi/2} z)$$

$$= \frac{\pi}{2z} e^{-z} \sum_{n=o}^{\ell} \frac{(\ell+n)!}{n!(\ell-n)!} \frac{1}{(2z)^n}. \tag{8}$$

The Hankel transform of (7) is (cf. Erdelyi (1954) Vol. II, p. 63)

$$\tilde{R}_{a_1}^{(A)}(k) = N_{a_1}^{(A)} \sqrt{\frac{\pi}{2}} \kappa_{a_1}^{-\ell_1-1} \frac{k^{\ell_1}}{k^2 + (\kappa_{a_1})^2}. \tag{9}$$

A useful relation is (cf. Erdelyi (1954) Vol. II, i 49)

$$\int_o^\infty j_\lambda(kr) j_{\ell'}(kr') \kappa^{-\ell-1} \frac{k^\ell}{k^2 + \kappa^2} k^2 \, dk$$

$$= (-1)^{\frac{\ell+\ell'-\lambda}{2}} i_{\ell'}(\kappa r') k_\lambda(\kappa r), \quad (\text{for } r > r') \tag{10}$$

where the function $i_\ell(z)$ is the modified spherical Bessel function

$$i_\ell(z) = \sqrt{\frac{\pi}{2z}} I_{\ell+1/2}(z) = e^{-i\ell\pi/2} j_\ell(e^{i\pi/2} z). \tag{11}$$

The functions k_ℓ and i_ℓ have the following asymptotic behavior

$$i_\ell(z) = \frac{z^\ell}{(2\ell+1)!!} \left(1 + \frac{1/2\, z^2}{1!(2\ell+3)} + \frac{(1/2\, z^2)^2}{2!(2\ell+3)(2\ell+5)} + \cdots\right),$$

$$k_\ell(z) = \frac{\pi(2\ell-1)!!}{2z^{\ell+1}}, \tag{12}$$

for $z \to 0^+$ and

$$i_\ell(z) \approx \frac{e^z}{2z},$$

$$k_\ell(z) \approx \pi \frac{e^{-z}}{2z}, \tag{13}$$

for $z \to \infty$.

The functions i_ℓ and k_ℓ both satisfy the radial Schrödinger equation

$$\left(\frac{d^2}{dr^2} - \frac{\ell(\ell+1)}{r^2} - \kappa^2\right)\left(r\, i_\ell(\kappa r)\right) = 0 \tag{14}$$

corresponding to a constant negative energy.

APPENDIX D.
NEARLY SYMMETRIC SYSTEMS

The effect of antisymmetrization manifests itself strikingly in collisions of identical or nearly identical heavy ions. In such collisions, one may observe the interference between a scattering through an angle θ where $a \to b$ (and $A \to B$) and a scattering where $a \to B$ (and $A \to b$) through an angle $\pi - \theta$ (cf. Fig. 35).

The interference pattern which is observed can be understood in terms of the two indistinguishable contributions to the scattering amplitude which should be added before the cross section is evaluated, i.e.

$$\frac{d\sigma}{d\Omega}_{aA \to bB} = \left| f_{aA \to bB}(\theta) \pm f_{aA \to Bb}(\pi - \theta) \right|^2. \tag{1}$$

The relative sign of the two contributions is determined from the condition of antisymmetrization not only of the specified nucleons but also of the inert cores. This is because the substitution $\vec{r}_\beta \to -\vec{r}_\beta$ implies the exchange of all nucleons.

For grazing reactions, this intereference is only seen if both reactions $aA \to bB$ and $aA \to Bb$ imply the transfer of a few nucleons, i.e. if a and A are very similar in mass number. In the case where first-order perturbation theory applies for both reactions, we find that the cross section (cf. §8) for detecting the particle b in the direction $\hat{k}_\beta$ is given by

$$d\sigma_{\alpha \to \beta}(k_\beta) = \frac{k_\beta}{k_\alpha} \frac{1}{E(\alpha)\,E(\beta)} \frac{1}{(2I_a+1)(2I_A+1)}$$

$$\times \sum_{M_a M_A M_b M_B} \left| \sum_{\lambda J J' m} \{ \langle I_A M_A J M \mid I_B M_B \rangle \langle I_b M_b J' M' \mid I_a M_a \rangle \right.$$

$$\times \langle \lambda m J M \mid J' M' \rangle\, t_{aA \to bB}(\lambda J J' m, \hat{k}_\beta) \tag{2}$$

$$+ \langle I_A M_A J M \mid I_b M_b \rangle \langle I_B M_B J' M' \mid I_a M_a \rangle$$

$$\left. \times \langle \lambda m J M \mid J' M' \rangle\, t_{aA \to Bb}(\lambda J J' m, -\hat{k}_\beta) \right|^2 d\Omega$$

$$= d\sigma_{aA \to bB}(\hat{k}_\beta) + d\sigma_{aA \to Bb}(-\hat{k}_\beta) + d\sigma_{\text{int}},$$

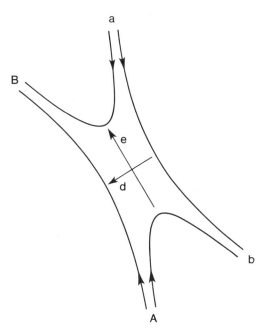

Fig. 35 Schematic representation of the center-of-mass trajectories of the scattering of nearly identical particles.

where $d\sigma_{aA\to bB}$ is given in Eq. (8.25), while $d\sigma_{aA\to Bb}$ stands for the cross sections for the reactions $aA \to Bb$, and $d\sigma_{\text{int}}$ is the interference term. It is noted that the total cross section is equal to the sum of the cross sections $\sigma_{aA\to bB}$ and $\sigma_{aA\to Bb}$. For the scattering of identical nuclei, it is therefore equal to twice the cross section associated with the reaction $a + A \to b + B$, as it would be in a classical description.

In a partial-wave expansion we may use that

$$Y_{\ell_\beta m_\beta}(-\hat{k}_\beta) = (-1)^{\ell_\beta} Y_{\ell_\beta m_\beta}(\hat{k}_\beta). \tag{3}$$

This means that in the scattering amplitude even and odd partial waves give quite different contributions. For reactions with identical target and projectile, it means that only even or only odd partial waves appear.

In the classical limit described in Sections IV.9 and V.8 it is seen that the cross section can be calculated from the classical picture of Fig. 35 taking into account the interference with the exchange reaction

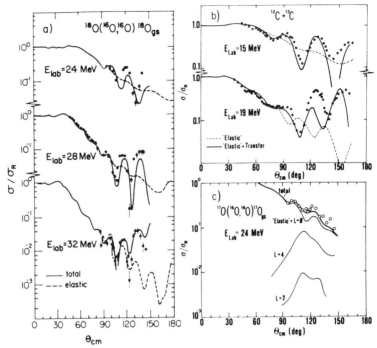

Fig. 36 Angular distribution for elastic scattering of nearly identical systems. The experimental elastic scattering data was taken from Gelbke *et al.*
(1972) (a), Bohlen and von Oertzen (1971) (b), and Gelbke *et al.* (1972) (c).
The full drawn curve is the result of a semiclassical evaluation of Eq. (D2).
The optical model parameters are those given in (III.1.44–45) for the real part,
while the imaginary part was calculated microscopically according to Sect. 7.
The two-particle transfer process in (a) was calculated including both successive and simultaneous transfer, the ground state of ^{18}O being described by
the wavefunction $0.9(0d_{5/2})^2 + 0.4(1s_{1/2})^2 + 0.2(0d_{3/2})^2$ of Glendenning and
Wolschin (1977). For the single-particle elastic transfer ^{16}O $+^{17}$O three transferred angular momenta $\lambda = 0, 2$, and 4 contribute, of which only the former
interfere with elastic scattering.

in the same way as one accounts in this picture for the interference
between different values of ℓ.

An example of a reaction between nearly identical nuclei is given in
Fig. 36 (cf. Maglione *et al.* (1987)).

APPENDIX E.
TRANSFER REACTIONS
WELL BELOW THE COULOMB BARRIER

For bombarding energies well below the Coulomb barrier, the quantal penetrability may be important for the transfer processes invalidating the semiclassical expressions. In the derivation of these expressions, it was assumed that the main contribution to the radial integral (8.24) came from distances close to the classical turning point. For very low bombarding energies and not too heavy ions, it may happen that the exponential decay of the radial wavefunction is slower than the exponential increase in the form factor in which case the radial integral receives the main contribution from a region inside the barrier.

We assume that the form factor has an exponential shape

$$f_\lambda \equiv \langle \ell_\beta \, || \, f_\lambda \, || \, \ell_\alpha \rangle \sim e^{-\kappa r}. \tag{1}$$

In the WKB approximation the product of the radial wavefunction and f_λ is therefore

$$f_\lambda(r)\chi_{\ell_\alpha}(s_A r) \sim \exp\left\{ -\kappa r + \int_{r_{o\alpha}}^{s_A r} \kappa_\alpha(r')dr' \right\}, \tag{2}$$

where

$$s_A r = \frac{2m_b}{m_a + m_b} r + \frac{m_d}{2m_o} R_A \tag{3}$$

Furthermore

$$\kappa_\alpha(r) = \left(\frac{2m_\alpha}{\hbar^2} \left(\frac{Z_a Z_A e^2}{r} + \frac{\hbar^2(\ell_\alpha + 1/2)^2}{2m_\alpha r^2} - E(\alpha) \right) \right)^{1/2} \tag{4}$$

since at these low energies we may neglect the nuclear field.

The maximum of the product occurs at a distance r_m determined by

$$\kappa = \frac{2m_b}{m_a + m_b} \kappa_\alpha(s_A r_m) \tag{5}$$

i.e.

$$r_m = \frac{1}{q^2} \frac{\eta_\alpha}{k_a} \left(1 + \sqrt{1 + q^2 \frac{(\ell_\alpha + 1/2)^2}{\eta_\alpha^2}} \right) - \frac{m_d}{2m_o} R_A, \tag{6}$$

463

with

$$k_a = \frac{2m_b}{m_a + m_b} k_\alpha, \tag{7}$$

and

$$q^2 = 1 + \frac{\kappa^2}{k_a^2}. \tag{8}$$

Usually the ratio κ/k_a is much smaller than unity, and the distance r_m is then equal to the classical turning point in the Rutherford trajectory (cf. (III.B.17)). We shall here especially consider the situation where $\kappa/k_a \gtrsim 1$ and where the maximum is well inside this point.

It is thus seen that as the energy is decreased and q increases the point r_m moves outwards slower than the classical turning point, i.e. further and further under the barrier until it finally reaches the point

$$(r_m)_{\min} = \kappa^{-1} \frac{k_a \eta_\alpha}{\kappa} \left(1 + \sqrt{1 + \left(\frac{(\ell_\alpha + 1/2)\kappa}{k_a \eta_\alpha} \right)^2} \right) - \frac{m_d}{2m_o} R_A. \tag{9}$$

The dimensionless quantity

$$\frac{k_a \eta_\alpha}{\kappa} = \frac{2m_b}{m_a + m_b} \frac{A_a A_A Z_a Z_A}{28(A_a + A_A)\kappa_{\mathrm{fm}^{-1}}}, \tag{10}$$

is energy independent and is usually a large number. Obviously the present considerations can only be used if

$$r_m > r_B, \tag{11}$$

where r_B is the radius of the Coulomb barrier (cf. eq. (III.3.26)).

We can estimate the product (2) by expanding around $r = r_m$, i.e.

$$f_\lambda(r)\chi_{\ell_\alpha}(s_A r) \sim \exp\left(-\kappa r_m + \int_{r_{o\alpha}}^{s_A r_m} \kappa_\alpha(r')dr' \right) e^{-\frac{(r - r_m)^2}{2(\sigma r_m)^2}} \tag{12}$$

with

$$\sigma = \left[\frac{k_a \eta_\alpha}{\kappa} \left(1 + \frac{(\ell_\alpha + 1/2)^2}{\eta_\alpha k_a r_m} \right) \right]^{-1/2}. \tag{13}$$

According to (10), the maximum of (2) around r_m is rather well defined.

The maximum cross section for a transfer reaction to a channel β is obtained if the radial wavefunction in this channel has its turning point close to r_m, i.e. if

$$s_b r_o = \frac{\eta_\beta}{k_\beta} \left(1 + \sqrt{1 + \frac{(\ell_\beta + 1/2)^2}{\eta_\beta^2}} \right) = s_b r_m, \qquad (14)$$

with

$$s_b r_m = \frac{2m_a}{m_a + m_b} r_m - \frac{m_d}{2m_o} R_A. \qquad (15)$$

For $\ell_\beta \approx \ell_\alpha \ll \eta_\alpha \approx \eta_\beta$, we find that the condition (14) is equivalent to

$$\frac{\eta_\beta}{k_\beta} = \frac{2m_a}{m_a + m_b} \left(\frac{\eta_\alpha}{k_a q^2} - \frac{m_d}{2m_o} R_A \right) - \frac{m_d}{2m_o} R_A = \frac{\eta_\alpha}{k_a q^2} - \frac{m_d}{m_o} R_A. \qquad (16)$$

In most cases the last term is of minor significance, and we may formulate the equation as a condition for the optimum Q-value for stripping reactions $E(\beta) = E(\alpha) + Q_{\text{opt}}$ leading to

$$Q_{\text{opt}} = \frac{Z_b Z_B}{Z_a Z_A} \left(E(\alpha) + \frac{\hbar^2 \kappa^2}{2m_{aA}} \right) - E(\alpha). \qquad (17)$$

This equation is to be compared to the expressions (6.4) and (6.18). It differs from them by the quantal correction term, which should be applied for reactions well below the barrier.

We may also use the present results to estimate the energy dependence of the stripping cross section in the low-energy limit. According to (8.46), the total stripping cross section is given by

$$\sigma_\beta = \frac{2I_B + 1}{2I_A + 1} \frac{4\pi}{E(\alpha)E(\beta)} \frac{k_\beta}{k_\alpha} \sum_{\substack{\ell_\alpha \ell_\beta \\ J \lambda J'}} \frac{2\ell_\beta + 1}{(2J+1)(2\lambda+1)} \left| I_{\beta\alpha} \right|^2, \qquad (18)$$

with

$$I_{\beta\alpha} = (2\ell_\alpha + 1)^{1/2} (-1)^{\ell_\beta} \begin{pmatrix} \ell_\beta & \lambda & \ell_\alpha \\ 0 & 0 & 0 \end{pmatrix}$$

$$\times \frac{1}{2} f_{\lambda 0}^{JJ'} (0, b) e^{\kappa b} \int \chi_{\ell_\beta}(s_b r) \sqrt{\frac{k_\alpha}{\kappa_\alpha(r)}} \qquad (19)$$

$$\times \exp\left(-\kappa \, r_m + \int_{r_{o\alpha}}^{s_A r_m} \kappa_\alpha(r') dr' \right) e^{-\frac{(r - r_m)^2}{2(\sigma r_m)^2}} dr,$$

where we used (12) and leave b undetermined. The width of the maximum σr_m is so small that we may consider χ_{ℓ_β} constant especially for optimum Q-value where the maximum χ_{ℓ_β} coincides with r_m. We thus find

$$I_{\beta\alpha} = (2\ell_\alpha + 1)^{1/2}(-1)^{\ell_\beta} \begin{pmatrix} \ell_\beta & \lambda & \ell_\alpha \\ 0 & 0 & 0 \end{pmatrix} f_\lambda^{JJ'}(b) e^{\kappa b}$$

$$\sqrt{\frac{2\pi\eta_\alpha}{k_a^2}} \frac{1}{q^2} \chi_{\ell_\beta}(s_b r_m) \exp\left\{ -2\eta_\alpha \arctan \frac{\kappa}{k_a} - \frac{(\ell_\alpha + 1/2)^2}{2\eta_\alpha k_a/\kappa} \right\}.$$

$$(20)$$

We have here introduced the analytic expression for the action integral $\int \kappa_\alpha(r)dr$ as derived from (4) and have performed a series expansion of the result in the quantity ℓ_α/η_α. The cutoff in ℓ_α at $\ell_{\alpha\max} \approx \sqrt{\eta_\alpha k_a/\kappa}$, which follows from (20), implies that we may rather safely neglect the dependence of χ_{ℓ_β} on ℓ_β. This is because ℓ_β can only differ from ℓ_α by λ and because $\ell_{\alpha\max}$ is smaller than η_β which indicates the value of ℓ_β where χ_{ℓ_β} is changed significantly. Inserting (20) into (18), we may therefore perform the summation over ℓ_β. The summation over ℓ_α is then easily performed by changing the sum into an integral over $(\ell_\alpha + 1/2)^2$.

The result is

$$\sigma = \frac{2I_B + 1}{2I_A + 1} \frac{8\pi^2}{E(\alpha)E(\beta)} \frac{k_\beta(\eta_\alpha k_a)^2}{\kappa(k_a^2 + \kappa^2)^2} |\chi_o(s_b r_m)|^2$$

$$\times \sum_{\lambda JJ'} \frac{|f_\lambda^{JJ'}(b)|^2}{(2J+1)(2\lambda+1)} e^{-4\eta_\alpha \arctan \frac{\kappa}{k_a} + 2\kappa b}.$$

$$(21)$$

It is seen that if we choose

$$b = \frac{2\eta_\alpha}{\kappa} \arctan \frac{\kappa}{k_a} \qquad (22)$$

the last factor disappears, and the result can then be directly compared to the result (8.53) with the expression (6.36) for p_β. For the normal case where $\kappa/k_a \ll 1$, both expressions are proportional to the square of the form factor at the distance of closest approach for the Coulomb scattering $2\eta_\alpha/k_a$. For low energies b is smaller than $2\eta_\alpha/k_a$. In the

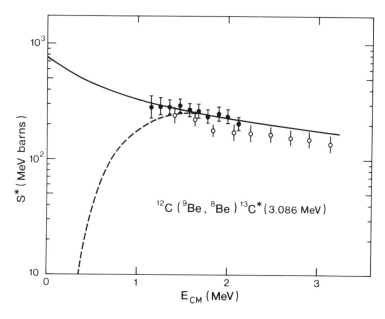

Fig. 37 The neutron stripping of ^{9}Be on ^{12}C at low bombarding energies. The quantity

$$S^* = E(\alpha)\,e^{4\eta_\alpha\ \mathrm{arctan}\ \kappa/k_a}\,\sigma,$$

where σ is the total transfer cross section is plotted as a function of the center-of-mass energy $E_{CM} = E(\alpha)$. The experimental data are taken from Caltech (open circles) and Melbourne (filled circles). The full drawn curve for $E_{CM} < 1.5$ MeV indicates the theoretical low-energy limit (E.21), while the full drawn curve for $E > 1.5$ MeV shows the semiclassical result (8.53) as calculated with a spectroscopic factor $\mathcal{S}(^9\mathrm{Be}(1p_{3/2})) \cdot \mathcal{S}(^{13}\mathrm{C}(1s_{1/2})) = 0.45$. The dashed curve indicates the prolongation of this result into the region below 1.5 MeV where it is not expected to apply. The classical distance of closest approach in a head-on collision is 35 fm at 1 MeV. The form factor was calculated from Eq. (5.19) and was parametrized using the standard shell model (5.7) adjusting the well depth V_o to give the correct binding energies of the $1p_{3/2}$ state in ^{9}Be and the 3.086 MeV excited $2s_{1/2}$ state of ^{13}C. The value of the decay constant $\kappa = 0.348$ fm^{-1} of the form factor is slightly larger than would be expected from Eq. (5.40) based on the binding energy of the $p_{3/2}$ state. This is because the contribution from $\langle U_{1A}\rangle$ in the form factor is significant for the low binding energies which occur in this reaction. The experimental data are from Switkowski et al., 1977 and Cheung et al., 1978.

467

limit of small energies we may write the energy dependence of the cross section (21) as

$$
\sigma \sim \left(E(\alpha)\right)^{-1} e^{-4\eta_\alpha \, \mathrm{arctan} \, \frac{\kappa}{k_a}}
$$
$$
\rightarrow \left(E(\alpha)\right)^{-1} e^{-2\pi\eta_\alpha} \quad \mathrm{for} \ \frac{k_a}{\kappa} \rightarrow 0.
\tag{23}
$$

It is interesting that the energy dependence in this limit is the same as that expected for fusion reactions (cf. Chapter VIII).

The expression (21) has a very similar energy dependence as the formulae derived by Breit et al. (1964) and Buttle and Goldfarb (1966) (cf. Eq. (4.11) in the latter reference). In the very low-energy limit they differ significantly because the earlier derivations relied on η_α being approximately equal to η_β.

An example of a low-energy stripping reaction is given in Fig. 37.

APPENDIX F.
ADIABATIC DESCRIPTION

The problems associated with the non-orthogonality of channel wave-functions as described in § 2 can be avoided by using as a set of basis states the eigenstates of both nuclei at a fixed separation. This method, which is especially appropriate when the relative velocity of the two collision partners is very small compared to the velocity of the particles, i.e. in the adiabatic limit, is widely used in the description of atomic collisions (see, e.g., Delos, 1981). It has also been used in the study of nuclear collisions by several groups (see, e.g., Voit et al., 1988).

Here we shall only give a short summary of this method for the simple case considered in § 1 of a particle moving in the time-dependent field of two potentials that move along classical trajectories. The Hamiltonian is

$$
H = T + U_a\left(\vec{r} - \vec{R}_a(t)\right) + U_A\left(\vec{r} - \vec{R}_A(t)\right)
\tag{1}
$$

where T is the kinetic energy. Before the collision, the particle is bound in the ground state of system a. To solve the associated time-dependent Schrödinger equation

$$
i\hbar \frac{\partial \psi}{\partial t} = H(t)\psi,
\tag{2}
$$

we introduce the complete set of normalized states $|n(t)\rangle$ which are the eigenstates of the Hamiltonian H at time t, i.e.

$$H(t)|n(t)\rangle = E_n(t)|n(t)\rangle. \tag{3}$$

We thus use the ansatz

$$|\psi\rangle = \sum_n a_n(t)|n(t)\rangle e^{-i\int_o^t E_n(t)dt/\hbar}, \tag{4}$$

and find that the coefficients $a_n(t)$ must satisfy the coupled equations

$$\dot{a}_n = -\sum_m \langle n(t)|\frac{\partial}{\partial t}|m(t)\rangle e^{i\int_o^t (E_n(t')-E_m(t'))dt'/\hbar} a_m(t). \tag{5}$$

It is often convenient to choose the phase of the solutions $|n(t)\rangle$ such that the diagonal matrix elements of (5) vanish, i.e.

$$\langle n(t)|\frac{\partial}{\partial t}|n(t)\rangle = 0. \tag{6}$$

In this case the equations (5) can be written

$$\dot{a}_n = \sum_{m\neq n} \frac{\langle n(t)|\partial H/\partial t|m(t)\rangle}{E_n(t) - E_m(t)}$$

$$\times \exp\left\{\frac{i}{\hbar}\int_o^t (E_n(t') - E_m(t'))dt'\right\} a_m(t), \tag{7}$$

where we expressed $\partial|n\rangle/\partial t$ in terms of $\partial H/\partial t$ by means of (3).

If the two potentials U_a and U_A are spherically symmetric, the states $|n(t)\rangle$ may be chosen to have a given total angular momentum Ω along the intrinsic three-axis R_{aA}. They may, furthermore, be characterized by the asymptotic quantum numbers in the two nuclei since at large distances they are identical to the states of the isolated nuclei. At any distance, there is a degeneracy of states with $I_3 = \pm\Omega$. As is customary in atomic physics, one may thus, instead of $|\Omega\rangle$, use the real linear combinations of the eigenfunctions $|\Omega\rangle$ and $|-\Omega\rangle$. For the collision of identical nuclei, a further degeneracy appears because of the symmetry under exchange.

For grazing collisions the adiabatic eigenstates may be described sufficiently well by perturbation theory based on the eigenstates in the two potentials U_a and U_A, i.e.

$$(T + U_a)\psi_n^{(a)} = E_n^{(a)}\psi_n^{(a)}$$

and

$$(T + U_A)\psi_m^{(A)} = E_m^{(A)}\psi_m^{(A)}. \tag{8}$$

This set of basis states is non-orthogonal, which poses no serious problems as long as the set is truncated appropriately. It leads, however, to a modification in the result of the perturbation expansion since there are now two expansion parameters: the matrix elements of the interaction, and the overlaps. The adiabatic eigenstates $|N\rangle$ satisfying

$$(T + U_a + U_A)|N,i,\Omega\rangle = E_{N,i,\Omega}|N,i,\Omega\rangle \tag{9}$$

have asymptotically the quantum numbers of either nucleus ($i = a$) or ($i = A$) and have a well-defined projection Ω of angular momentum along $\vec{R}_{aA}$. To first order in the two expansion parameters, one finds

$$E_{Ni\Omega} = \begin{cases} E_N^{(a)} & + & \langle U_A \rangle & i = a \\ E_N^{(A)} & + & \langle U_a \rangle & i = A \end{cases} \tag{10}$$

with

$$\langle U_A \rangle = \langle \psi_{N\Omega}^{(a)} | U_A | \psi_{N\Omega}^{(a)} \rangle$$
$$\langle U_a \rangle = \langle \psi_{N\Omega}^{(A)} | U_a | \psi_{N\Omega}^{(A)} \rangle \tag{11}$$

and

$$\langle \vec{r}\xi | N i\Omega \rangle = \begin{cases} \psi_{N\Omega}^{(a)} + \sum_{nj} \dfrac{\langle \psi_{n\Omega}^{(j)} | U_A - \langle U_A \rangle | \psi_{N\Omega}^{(a)} \rangle}{E_{N\Omega}^{(a)} - E_{n\Omega}^{(j)}} \psi_{n\Omega}^{(j)}, & i = a \\[4mm] \psi_{N\Omega}^{(A)} + \sum_{nj} \dfrac{\langle \psi_{n\Omega}^{(j)} | U_a - \langle U_a \rangle | \psi_{N\Omega}^{(A)} \rangle}{E_{N\Omega}^{(A)} - E_{n\Omega}^{(j)}} \psi_{n\Omega}^{(j)}, & i = A \end{cases} \tag{12}$$

The matrix elements appearing here are identical to the form factors in the intrinsic frame for single-particle transfer and inelastic scattering neglecting recoil. The appearance of the expectation values of $\langle U_A \rangle$ and $\langle U_a \rangle$ in the matrix element is associated with the non-orthogonality and is only important for proton transfer. The matrix elements for transfer are, according to (4.31), (4.52), and (5.22), explicitly given by

$$\langle \psi^{(A)}_{a_1\Omega} \, | \, U_A - \langle U_A \rangle \, | \, \psi^{(a)}_{a'_1\Omega} \rangle = \sum_\lambda \langle \lambda 0 j_1 \Omega \, | \, j'_1 \Omega \rangle \, f^{a_1 a'_1}_\lambda (r) \sqrt{\frac{2\lambda + 1}{4\pi}} \quad (13)$$

in the intrinsic frame. It follows from the parametrization (5.40) that the largest form factors are those for which $\Omega = \pm 1/2$.

The transfer form factor appears in (12) as giving rise to the small amplitude of the state $\psi_{N, i=a}$ in nucleus A. When one of the energy denominators in (12) vanishes, i.e. when there is a crossing of levels with the same value of Ω, this amplitude is not small, and the perturbation theory breaks down.

We may estimate the energy shift (10) from the single-particle inelastic form factors given in Fig. IV.2 to be of the order of a few tenths of one MeV at the distance r_B. Since all shifts are negative, the relative shift is typically less than 100 keV for grazing collisions. Level crossing in the grazing region occurs only for states which are accidentally close in energy, i.e. for transitions with energies very close to the optimum Q-value. In this case, the amplitude of the adiabatic eigenstates is the same in both nuclei, independent of the form factor.

For transitions to a continuous spectrum in A, perturbation theory leads to the "golden rule" for the transfer probability per unit time, i.e. the tunneling through the barrier between the two nuclei is proportional to the square of the matrix element. This is related to the fact that for optimal Q-value, the time-dependent description shows that the matrix element, i.e. the form factor, is proportional to the transfer amplitude per unit time. In this case, it is noted that although the perturbation theory (12) breaks down, time-dependent perturbation theory may still be valid.

For the nuclear collisions we consider, we must take into account that the equations (5) or (7) cannot be used as they stand because the matrix elements are non-vanishing even at time $t = \pm \infty$, i.e.

$$\langle n(t) \, | \, \frac{\partial}{\partial t} \, | \, m(t) \rangle = \vec{v}(t) \cdot \langle n(t) \, | \, \vec{\nabla} \, | \, m(t) \rangle.$$

Correspondingly, the states $| \, n(t) \rangle$ do not have the correct asymptotic behavior which is given by $G \, | \, n(t) \rangle$, i.e. a Galilean transformation of the asymptotic state $|n(t)\rangle$ at large separation between the centers (cf. eq. (1.2)). In atomic physics, this problem is often dealt with by using as basis states in the expansion (4) the states $|n(t)\rangle$ multiplied by the function

$$G_a(r_1)f_a(r_1) + G_A(r_1)f_A(r_1),$$

which behaves like the Galilean transformation G_a over the region of the projectile a, and like G_A over the target A, the so-called switching functions f being similar to step functions. Another possibility might be to expand the eigenstates $|n(t)\rangle$ in terms of Galilean-transformed eigenstates. In this description, which to our knowledge has not been used in practice, the expansion coefficients are determined by the overlaps and form factors including recoil.

Finally, it should be observed that the relative motion implies not only a translation but also a rotation. The eigenstates $|n(t)\rangle$ are thus rather to be compared to the wavefunction in the intrinsic system which, in the absence of couplings, are given by rotation matrices acting on the proper eigenstates in a space-fixed system.

In spite of all these complications which make the adiabatic description rather impractical for quantitative estimates in grazing heavy-ion collisions, it seems the natural scheme for the description of nuclear structure effects in fusion processes. In this case, the final state of high spin is expressed in terms of the cranking eigenstates of a classically rotating deformed potential.

APPENDIX G.
SINGLE-PARTICLE DENSITIES

The nonlocal single-particle density operator is defined by

$$\hat{\rho}(\vec{r}\sigma; \vec{r}'\sigma') = a^\dagger(\vec{r}'\sigma')\, a(\vec{r}\sigma). \tag{1}$$

It annihilates a particle of spin quantum number σ in point $\vec{r}$ and creates it in point $\vec{r}'$ with spin σ'. The single-particle density in the state ψ

$$\rho(\vec{r}\sigma, \vec{r}'\sigma') = \langle \psi \,|\, \hat{\rho}(\vec{r}\sigma; \vec{r}'\sigma') \,|\, \psi \rangle \tag{2}$$

can be interpreted as the $\vec{r}, \sigma$ representation of the density matrix $\hat{\rho}^{(\psi)}$ for the state ψ integrated over all particles except one, i.e.

$$\rho(\vec{r}\sigma, \vec{r}'\sigma') = \langle \vec{r}\sigma \,|\, \hat{\rho}^{(\psi)} \,|\, \vec{r}'\sigma' \rangle. \tag{3}$$

It is noted that

$$\rho^*(r\sigma, r'\sigma') = \rho(r'\sigma', r\sigma). \tag{3a}$$

By expanding $a(\vec{r}, \sigma) = \sum_k a_k \psi_k(\vec{r}, \sigma)$ on a (complete) set of single-particle states, we may obtain other representations of the density matrix

$$\rho(\vec{r}\sigma, \vec{r}'\sigma') = \sum_{ik} \rho_{ik} \, \psi_k^*(\vec{r}', \sigma')\psi_i(\vec{r}, \sigma), \tag{4}$$

with

$$\begin{aligned}
\rho_{ik} &= \langle i \,|\, \hat{\rho}^{(\psi)} \,|\, k\rangle \\
&= \langle \psi \,|\, a_k^\dagger a_i \,|\, \psi\rangle \\
&= \sum_{\sigma\sigma'} \int d^3r\, d^3r' \, \rho(\vec{r}\sigma, \vec{r}'\sigma') \, \psi_i^*(\vec{r}, \sigma) \, \psi_k(\vec{r}', \sigma').
\end{aligned} \tag{5}$$

Since we shall also consider transition densities $\langle \psi \,|\, \hat{\rho}(\vec{r}\sigma, \vec{r}'\sigma') \,|\, \phi\rangle$ we shall usually leave out the index ψ on the density matrix.

The density matrix is especially simple if the state is described by a Slater determinant. In this case, the density matrix is a projection operator, i.e.,

$$\begin{aligned}
\langle \vec{r}\sigma \,|\, \rho^2 \,|\, \vec{r}'\sigma'\rangle &= \sum_{\sigma''} \int d^3r'' \langle \vec{r}\sigma \,|\, \rho \,|\, \vec{r}''\sigma''\rangle \, \langle \vec{r}''\sigma'' \,|\, \rho \,|\, \vec{r}'\sigma'\rangle \\
&= \langle \vec{r}\sigma \,|\, \rho \,|\, \vec{r}'\sigma'\rangle,
\end{aligned} \tag{6}$$

which contains all information about the system. In general, the density matrix only contains the information relevant for the calculation of matrix elements of single-particle operators since

$$\langle \psi \,|\, \hat{F} \,|\, \phi\rangle = \sum_{\ell\ell'} \langle \ell \,|\, F \,|\, \ell'\rangle \, \langle \psi \,|\, a_\ell^\dagger a_{\ell'} \,|\, \phi\rangle = tr(F\rho). \tag{7}$$

(a) Wigner transformation

While the local density $\rho(\vec{r}\sigma; \vec{r}\sigma)$ describe the density of particles at the point $\vec{r}$ with spin σ, the nonlocal density $\rho(\vec{r}\sigma, \vec{r}'\sigma)$ corresponds to the

473

density in phase space of particles with spin component σ. This is seen by using the Wigner transformation defining the quantity

$$\bar{\rho}(\vec{p}, \vec{q}, \sigma) = \frac{1}{(2\pi\hbar)^3} \int e^{-i\vec{p}\vec{\xi}/\hbar} \langle q + \frac{\vec{\xi}}{2}, \sigma|\hat{\rho}|\vec{q} - \frac{\vec{\xi}}{2}, \sigma\rangle, d^3\xi. \qquad (8)$$

The nondiagonal matrix elements in σ are important if one wishes to obtain the density of particles with spin direction different from that of the z-direction.

The real quantity $\bar{\rho}(\vec{p}, \vec{q}, \sigma)$ has many properties in common with the phase-space density. Thus the integral over $\vec{p}$ leads to

$$\int \bar{\rho}(\vec{p}, \vec{q}, \sigma)d^3p = \rho(\vec{q}\sigma, \vec{q}\sigma) = \rho(\vec{q}, \sigma) \qquad (9)$$

i.e., to the local density of particles. Similarly,

$$\int \bar{\rho}(\vec{p}, \vec{q}, \sigma)d^3q = \tilde{\rho}(\vec{p}\sigma, \vec{p}\sigma) = \tilde{\rho}(\vec{p}, \sigma) \qquad (10)$$

is the (local) momentum distribution in the nucleus as can be seen from the momentum representation of the density matrix defined by

$$\rho(\vec{r}\sigma, \vec{r}'\sigma) = \int \frac{d^3p d^3p'}{(2\pi\hbar)^3} e^{i\vec{p}\cdot\vec{r}/\hbar} \langle \vec{p}\sigma \,|\, \rho \,|\, \vec{p}'\sigma\rangle \, e^{-i\vec{p}'\vec{r}'/\hbar} \qquad (11)$$

and the corresponding connection

$$\bar{\rho}(\vec{p}, \vec{q}, \sigma) = \frac{1}{(2\pi\hbar)^3} \int d^3\zeta \, \langle \vec{p} + \frac{\vec{\zeta}}{2}, \sigma \,|\, \rho \,|\, \vec{p} - \frac{\vec{\zeta}}{2}, \sigma\rangle \, e^{-iq\zeta/\hbar}. \qquad (12)$$

Expectation values of single-particle operators $\widehat{F}$ can be calculated by the familiar relation

$$\langle F\rangle = \sum_\sigma \int d^3p \, d^3q \, F(\vec{p}, \vec{q})\rho(\vec{p}, \vec{q}\sigma), \qquad (13)$$

where we have defined

$$F(\vec{p}, \vec{q}) = \int e^{-i\vec{p}\vec{\xi}/\hbar} \langle \vec{q} + \frac{\vec{\xi}}{2} \,|\, F \,|\, \vec{q} - \frac{\vec{\xi}}{2}\rangle \, d^3\xi. \qquad (13a)$$

Relation (13) holds even though the quantity $\rho(\vec{p}, \vec{q}, \sigma)$ is not necessarily positive everywhere, as the classical phase-space density would be.

For a two-body operator, the expectation value cannot, in general, be expressed in terms of the single-particle density. One would need the two-particle density

$$\rho^{(2)}_{k\ell pq} = \langle \psi \,|\, a^\dagger_p a^\dagger_q \, a_\ell a_k \,|\, \psi \rangle, \tag{13b}$$

which in the $\vec{r}$-representation would depend on four coordinates and spins. Extracting out of $\rho^{(2)}$ the uncorrelated pairs through

$$\rho^{(2)}_{k\ell pq} = \rho_{kp}\rho_{\ell q} - \rho_{kq}\rho_{\ell p} + g^{(2)}_{k\ell pq}, \tag{13c}$$

and neglecting the correlation function $g^{(2)}$, one is led to the Hartree-Fock approximation. The expectation value of the two-body interaction

$$v(\vec{r}'_1 \vec{r}'_2 \vec{r}_1 \vec{r}_2) = \langle \vec{r}'_1 \vec{r}'_2 \,|\, V \,|\, \vec{r}_1 \vec{r}_2 \rangle \tag{14}$$

is then given by

$$\langle V \rangle = 2 \int d^3 r_1\, d^3 r_2\, d^3 r'_1\, d^3 r'_2\, \{ v(\vec{r}'_1 \vec{r}'_2 \vec{r}_1 \vec{r}_2) - v(\vec{r}'_1 \vec{r}'_2 \vec{r}_2 \vec{r}_1) \} \\ \times \rho(r_1 r'_1)\, \rho(r_2 r'_2). \tag{15}$$

We have here, for convenience, assumed spin independence and have defined

$$\rho(\vec{r}, \vec{r}') = \frac{1}{2} \sum_\sigma \rho(\vec{r}\sigma, \vec{r}'\sigma). \tag{16}$$

For a local two-body interaction, one finds

$$v(\vec{r}'_1 \vec{r}'_2 \vec{r}_1 \vec{r}_2) = \delta(\vec{r}_1 - \vec{r}'_1)\delta(\vec{r}_2 - \vec{r}'_2)V(\vec{r}_1 - \vec{r}_2) \tag{17}$$

where we correspondingly left out the factor $\delta(\sigma_1\sigma'_1)\,\delta(\sigma_2\sigma'_2)$. Note that (17) has a dimension of MeV fm^{-6}.

The quantity (16) is the singlet density. In general, the density matrix for a given nuclear state is not diagonal in σ. In the spherical shell model it is thus diagonal in an appropriate $j - j$ coupling particle representation $\rho_{ik} = \delta_{ik}$, and it would, according to (5), not be diagonal in σ. It is, however, always possible to separate ρ into a singlet and triplet part, i.e.,

$$\rho(\vec{r}\sigma, \vec{r}'\sigma') = \rho(\vec{r}, \vec{r}') + \sum_i \sigma_i \rho_{1i}(\vec{r}, \vec{r}') \qquad (17a)$$

where $\sigma_i = 2s_i$ are the Pauli-matrices and where the triplet density is

$$\rho_{1i}(r, r') = tr(s_i \rho). \qquad (17b)$$

We have throughout neglected isospin (cf. Bohr and Mottelson, Vol. I, Eq. (2A-41)).

A fundamental quantity in the Hartree-Fock description is the mean field

$$U(\vec{r}'_1, \vec{r}_1) = U_o(\vec{r}'_1, \vec{r}_1) - U_{\text{exch}}(\vec{r}'_1, \vec{r}_1) \qquad (17c)$$

where the first term

$$\begin{aligned}
U_o(\vec{r}'_1 \vec{r}_1) &= \int d^3 r_2\, d^3 r'_2\, v(\vec{r}'_1 \vec{r}'_2 \vec{r}_1 \vec{r}_2) \rho(\vec{r}_2 \vec{r}'_2) \\
&= \delta(\vec{r}'_1 - \vec{r}_1) \int d^3 r_2\, V(\vec{r}_1 - \vec{r}_2) \rho(\vec{r}_2 \vec{r}_2)
\end{aligned} \qquad (18)$$

is local for a local interaction while the exchange interaction

$$\begin{aligned}
U_{\text{exch}}(\vec{r}'_1 \vec{r}_1) &= \int d^3 r_2\, d^3 r'_2\, v(\vec{r}'_1 \vec{r}'_2 \vec{r}_2 \vec{r}_1) \rho(\vec{r}_2 \vec{r}'_2) \\
&= V(\vec{r}'_1 - \vec{r}_1) \rho(\vec{r}'_1, \vec{r}_1)
\end{aligned} \qquad (19)$$

is non-local.

Performing a Wigner transformation of the densities in Eq. (15), we obtain

$$\begin{aligned}
\langle V \rangle = 2 \int d^3 p_1\, d^3 p_2\, d^3 r_1\, d^3 r_2\, \bar{\rho}(\vec{p}_1 \vec{q}_1) \bar{\rho}(\vec{p}_2 \vec{q}_2) \\
\times \{ \tilde{v}_o(\vec{p}_1 \vec{q}_1 \vec{p}_2 \vec{q}_2) - \tilde{v}_{\text{exch}}(\vec{p}_1 \vec{q}_1 \vec{p}_2 \vec{q}_2) \}.
\end{aligned} \qquad (20)$$

The quantity

$$\begin{aligned}
\tilde{v}_o(\vec{p}_1 \vec{q}_1 \vec{p}_2 \vec{q}_2) = \int d^3 \xi_1\, d^3 \xi_2\, e^{i \vec{p}_1 \vec{\xi}_1 / \hbar}\, e^{i \vec{p}_2 \vec{\xi}_2 / \hbar} \\
\times v \left(q_1 + \frac{\vec{\xi}_1}{2}, q_2 + \frac{\vec{\xi}_2}{2}, q_1 - \frac{\vec{\xi}_1}{2}, q_2 - \frac{\vec{\xi}_2}{2} \right)
\end{aligned} \qquad (21)$$

and similar for v_{exch} are the Wigner transform of $\tilde{v}(\vec{r}'_1 \vec{r}'_2 \vec{r}_1 \vec{r}_2)$ and $\tilde{v}(\vec{r}_1 \vec{r}'_2 \vec{r}'_1 \vec{r}_2)$, respectively. For a local interaction (17), we have

$$\tilde{v}_o(\vec{p}_1 \vec{q}_1 \vec{p}_2 \vec{q}_2) = V(\vec{q}_1 - \vec{q}_2) \tag{22}$$

and

$$\tilde{v}_{\text{exch}}(\vec{p}_1 \vec{q}_1 \vec{p}_2 \vec{q}_2) = \delta(\vec{q}_2 - \vec{q}_1) \int d^3\xi \, e^{(i\vec{p}_1 - \vec{p}_2)\vec{\xi}/\hbar} \, V(\xi)$$
$$= \delta(\vec{q}_2 - \vec{q}_1) \, \tilde{V}(\vec{p}_2 - \vec{p}_1), \tag{23}$$

where $\tilde{V}$ is the Fourier transform of the two-body interaction.

In the phase-space representation the mean field is

$$U(\vec{p}, \vec{q}) = U_o(\vec{p}, \vec{q}) - U_{\text{exch}}(\vec{p}, \vec{q}) \tag{24}$$

where

$$U(\vec{p}_1 \vec{q}_1) = \int d^3 p_2 \, d^3 q_2 \, \tilde{v}(\vec{p}_1 \vec{q}_1 \vec{p}_2 \vec{q}_2) \bar{\rho}(\vec{p}_2 \vec{q}_2). \tag{25}$$

For the local interaction, we find the field

$$U_o(\vec{p}_1 \vec{q}_1) = \int V(\vec{q}_1 - \vec{q}_2) \bar{\rho}(\vec{q}_2) d^3 q_2 \tag{26}$$

while the mean exchange field is

$$U_{\text{exch}}(\vec{p}_1 \vec{q}_1) = \int \tilde{V}(\vec{p}_1 - \vec{p}_2) \bar{\rho}(\vec{p}_2 \vec{q}_1) d^3 p_2. \tag{27}$$

(b) Hartree-Fock approximation

The Hartree-Fock approximation may be derived from the general equation of motion of the density matrix

$$i\hbar \, \dot{\rho}_{k\ell} = \langle \psi \,|\, [a^\dagger_\ell a_k, H] \,|\, \psi \rangle \tag{28}$$

by neglecting the correlations $g^{(2)}$ (cf. (13c)) in the two-particle density. We thus obtain

$$i\hbar \dot{\rho}_{k\ell} = [h, \rho]_{k\ell} = 0 \tag{29}$$

where

$$h = h(\rho) = t + U_o - U_{\text{exch}} \tag{30}$$

477

is the Hartree-Fock Hamiltonian. We can reformulate (29) into an eigen-value problem by performing a unitary transformation which simultaneously diagonalizes h and ρ. In the new set of basis states n the ground state density is

$$\rho = \sum_{\text{occ}} a_n^\dagger a_n \tag{31}$$

and we have

$$h_{nn'} = \varepsilon_n\,\delta(n,n'). \tag{32}$$

If we perform a Wigner transformation of Eq. (29), we obtain an equation of motion which in the limit $\hbar \to 0$ leads to the Liouville equation

$$\frac{\partial}{\partial t}\rho(\vec{p},\vec{q},t) = \{h(p,q),\rho(p,q)\} = \frac{\partial h}{\partial p}\frac{\partial \rho}{\partial q} - \frac{\partial h}{\partial q}\frac{\partial \rho}{\partial p} = 0, \tag{33}$$

where, according to (13a), $h(p,q) = \frac{p^2}{2M} + U(p,q)$. Equation (33) describes a collisionless gas of particles, the so-called Vlasov gas. The solution of Eq. (33) is

$$\rho(\vec{p},\vec{q}) = f\big(h(\vec{p},\vec{q})\big) \tag{34}$$

where f is an arbitrary function of the Hamiltonian. In order to satisfy the Pauli principle, we must furthermore enforce the condition $\rho^2 = \rho$ which in the limit where the density does not depend strongly on q takes the form

$$\rho(\vec{p},\vec{q}) = (2\pi\hbar)^3 \left(\rho(\vec{p},\vec{q})\right)^2. \tag{35}$$

This is the Thomas-Fermi approximation where the density is either 0 or 1 in units of $(2\pi\hbar)^3$. Combining with (34), we find in this approximation for the ground state

$$\rho(\vec{p},\vec{q}) = \frac{1}{(2\pi\hbar)^3}\,\Theta\big(\lambda - h(\vec{p},\vec{q})\big), \tag{36}$$

where Θ is the step function and λ is the Fermi energy determined from

$$2\int \rho(\vec{p},\vec{q})d^3p\,d^3q = A, \tag{37}$$

where A is the number of particles.

When the self-consistent Hartree-Fock equation (29) and (30) with U given by (18) and (19) are solved by diagonalization or by a corresponding minimization principle, one most often finds that the solution leads to a non-spherical density ρ and mean field U. This is the spontaneous symmetry breaking phenomenon. Although this is acceptable in a classical description, it does not describe an eigenstate of the total angular momentum. It is quite analogous to the fact that the density does not display translational invariances and is not in an eigenstate of the total momentum. The interpretation is that we obtain, by the mean field approximation, the intrinsic state of the system. In the laboratory system the wavefunction with the total momentum $\vec{P}$ is

$$|\psi\rangle = e^{i\vec{R}\cdot\vec{P}/\hbar} |\psi'\rangle, \tag{38}$$

where $\vec{R}$ is the center-of-mass coordinate, and where all coordinates in ψ' are interpreted as intrinsic coordinates. Similarly, we may obtain the wavefunction of given total angular momentum I, M by

$$|\psi\rangle = D^I_{MK}(\varphi, \theta, \psi) |\psi'\rangle_K \tag{38a}$$

where φ, θ and ψ indicate the Eulerian angles specifying the orientation of the intrinsic frame with respect to the laboratory. We assumed in (38a) axial symmetry of the mean field, and K indicates the total intrinsic angular momentum along this z-axis.

While this would be a good description for $I = 0$ where D^I is a constant, it does not work for high angular momenta, as can be seen from the time-dependent Hartree-Fock equation (29). We thus find for a rotation with constant angular velocity $\vec{\omega}$

$$\dot{\rho}_{k\ell} = -\frac{i}{\hbar}[\vec{\omega} \cdot \vec{J}, \rho]_{k\ell} \tag{39}$$

$\vec{J}$ being the total angular momentum operator. The equation of motion (29) therefore takes the form

$$[h - \vec{\omega} \cdot \vec{J}, \rho]_{k\ell} = 0. \tag{39a}$$

The frequency ω should be determined from the classical equation $\omega = dE/dI$ by calculating the energy E as a function of I.

The result is therefore that for large angular momenta one should reconsider the eigenvalue problem in the intrinsic frame taking into account the centrifugal and Coriolis forces, i.e., one should substitute the Hamiltonian h in (30) by the Hamiltonian in the intrinsic frame

$$h \to h - \vec{\omega} \cdot \vec{J}. \tag{40}$$

Here ω is the frequency of rotation, $\hat{\omega}$ being the axis of rotation, while $\vec{J}$ is the total angular momentum operator, which is the generator of the rotation.

(c) Pairing

As an improved wavefunction was obtained in the Hartree-Fock description by allowing a violation of the conservation of angular momentum, a further improvement can be obtained by letting $|\psi\rangle$ violate the conservation of particle number. This means allowing expectation values of the type

$$\rho_2(\vec{r}\sigma, \vec{r}'\sigma') = \langle \psi \,|\, a^\dagger(\vec{r}', \sigma') a^\dagger(\vec{r}, \tilde{\sigma}) \,|\, \psi \rangle \tag{41}$$

to be nonvanishing, where

$$a^\dagger(\vec{r}, \tilde{\sigma}) = T a^\dagger(\vec{r}, \sigma) T^{-1} = (-1)^{1/2 + \sigma} a^\dagger(\vec{r}, -\sigma),$$

T being the time-reversal operator. The quantity $\hat{\rho}_2 = a^\dagger(\vec{r}', \sigma') a^\dagger(\vec{r}, \tilde{\sigma})$ is also a single-particle operator like $\hat{\rho}$ and the matrix element (41) may be considered as the $\vec{r}$ representation of an operator $\hat{\rho}_2^{(\psi)}$, i.e.,

$$\rho_2(\vec{r}\sigma, \vec{r}'\sigma') = \langle \vec{r}, \sigma \,|\, \hat{\rho}_{+2}^{(\psi)} \,|\, \vec{r}', \sigma' \rangle. \tag{42}$$

Similarly, we define the quantity

$$\rho_{-2}(\vec{r}\sigma, \vec{r}'\sigma') = \langle \psi \,|\, a(\vec{r}', \tilde{\sigma}') a(\vec{r}, \sigma) \,|\, \psi \rangle, \tag{43}$$

which we may consider as the r representation of $\hat{\rho}_{-2}^{(\psi)}$

$$\rho_{-2}(\vec{r}\sigma, \vec{r}'\sigma') = \langle \vec{r}\sigma \,|\, \hat{\rho}_{-2}^{(\psi)} \,|\, \vec{r}'\sigma' \rangle. \tag{44}$$

In fact, the quantities

$$\hat{\rho}(\vec{r}\sigma v, \vec{r}'\sigma'v') = c^\dagger(\vec{r}', \sigma') c(\vec{r}, \sigma)$$

with

$$c(\vec{r},\sigma) = \begin{cases} a(\vec{r},\sigma) & \text{for } v = +\frac{1}{2} \\ b(\vec{r},\sigma) = a^\dagger(\vec{r},\tilde{\sigma}) & \text{for } v = -\frac{1}{2} \end{cases} \tag{45}$$

have similar algebraic properties in the variables v, v' as in the spin variables σ, σ'. Suppressing the spin indices, we may thus write in analogy to (17a)

$$\hat{\rho}(\vec{r}v, \vec{r}'v') = \hat{\rho}_o(\vec{r}, \vec{r}') + \sum_i v_i(v, v')\, \hat{\rho}_i(\vec{r}, \vec{r}'), \tag{46}$$

where v_i are the Pauli spin matrices. We note that the density $\rho(\vec{r}\frac{1}{2}, \vec{r}'\frac{1}{2})$ is given by

$$\rho(\vec{r}\frac{1}{2}, \vec{r}'\frac{1}{2}) = \rho_o(\vec{r}, \vec{r}') + \rho_z(\vec{r}, \vec{r}'), \tag{47}$$

while

$$\rho_{\pm 2}(\vec{r}\,\vec{r}') = \rho_x(\vec{r}\,\vec{r}') \pm i\rho_y(\vec{r}, \vec{r}'). \tag{48}$$

We also introduce the moments, the total quasispins,

$$\widehat{M_i} = \int d^3r\, \hat{\rho}_i(\vec{r}, \vec{r}). \tag{49}$$

These quantities satisfy the commutation relations

$$[\widehat{M_i}, \widehat{M_j}] = i\,\varepsilon_{ijk}\,\widehat{M_k} \tag{50}$$

appropriate for angular momenta, and they generate rotations among the density operators $\hat{\rho}_i$. We thus have

$$[\hat{\rho}_i(\vec{r}, \vec{r}'), \widehat{M_j}] = i\,\varepsilon_{ijk}\,\hat{\rho}_k(\vec{r}, \vec{r}'), \tag{51}$$

which, for instance, implies that

$$[\hat{\rho}_z(\vec{r}, \vec{r}'), \widehat{M}_{\pm,2}] = \pm\,\hat{\rho}_{\pm 2}(\vec{r}, \vec{r}'), \tag{52}$$

where

$$\widehat{M}_{\pm 2} = \widehat{M_x} \pm i\widehat{M_y} = \int d^3r\, \rho_{\pm 2}(\vec{r}, \vec{r}) \tag{53}$$

are the pair moments.

From (47) we see that the moment $\widehat{M_z}$ is given by

$$\widehat{M_z} = \frac{1}{2}\hat{N} - \widehat{M_o}, \tag{54}$$

where $\widehat{N}$ is the particle number operator. The moment $\widehat{M_o}$ is invariant for rotations in quasispin space and depends on the dimension of the configuration space.

A general rotation in the quasispin space through an angle ϕ around the axis $\hat{\phi}$ is given by

$$\mathcal{G}_{\hat{\phi}}(\phi) = e^{-i\vec{M}\cdot\vec{\phi}}, \tag{55}$$

where $\vec{M} = (M_x, M_y, M_z)$. The rotations around the z-axis are the gauge transformations (cf. Eq. (11.5)), i.e.

$$\mathcal{G}_{\hat{z}}(\phi)\hat{\rho}_{\pm 2}\,\mathcal{G}_{\hat{z}}^{-1}(\phi) = e^{\pm i\phi}\hat{\rho}_{\pm 2}, \tag{56}$$

while the rotations around the y-axis are equivalent to the Bogoliubov-Valatin transformation, i.e.

$$\mathcal{G}_{\hat{y}}(\theta)\hat{\rho}_z\,\mathcal{G}_{\hat{y}}^{-1}(\theta) = \hat{\rho}_z\,\cos\theta + \hat{\rho}_x\,\sin\theta. \tag{57}$$

The mean-field equations are obtained from the general equation (28) by extracting from the two-particle density (13c) also the product $\rho_{+2}\rho_{-2} = \rho_x^2 + \rho_y^2$. One thus obtains again Eq. (29) with

$$h = (t + U_z)v_z + \sum_{xy} U_i v_i \tag{58}$$

and with $U_z = U_o - U_{\text{ex}}$. We have here introduced the pair field

$$\begin{aligned} U_i(\vec{r}_1'\vec{r}_1) &= \int d^3 r_2\, d^3 r_2'\, V(r_1' r_1, r_2 r_2')\rho_i(r_2 r_2') \\ &= V(\vec{r}_1 - \vec{r}_1')\rho(\vec{r}_1'\vec{r}_1) \end{aligned} \tag{59}$$

and have specified the dependence on the quasispin variables through the spin matrices v_i.

The mean-field equation (29) leads to a nonvanishing value for ρ_x and ρ_y, but we may without loss of generality through a rotation of angle ϕ' around the z-axis transform to an intrinsic frame in quasispin space where $\rho_y = 0$. We must, however, envisage that this intrinsic frame rotates with frequency $\dot{\phi}'$ with respect to the laboratory frame.

Since ϕ' and $\widehat{N}/2$ are conjugate variables, i.e.

$$\left[\phi', \frac{\widehat{N}}{2}\right] = i\hbar \tag{60}$$

we may write $\phi' = 2i\hbar\,\partial/\partial N$. We therefore find

$$i\hbar\dot{\phi}' = [\phi', H] \tag{61}$$

i.e.

$$\dot{\phi}' = \frac{2\lambda}{\hbar} \tag{62}$$

where $\lambda = \partial H/\partial N$ indicates the Fermi energy.

   This rotation gives rise to a time derivative of ρ in the laboratory frame which is

$$\dot{\rho} = \frac{d}{dt}\{\mathcal{G}_{\hat{z}}(\dot{\phi}'t)\rho\,\mathcal{G}_{\hat{z}}^{-1}(\dot{\phi}t)\}$$
$$= \left[\frac{2\lambda}{i\hbar}\widehat{M}_z v_z, \rho\right]. \tag{63}$$

We therefore find

$$[(t + U_z - \lambda)v_z + U_x v_x, \sum_i \rho_i v_i] = 0, \tag{64}$$

where we used that ρ_o is a constant for an even system.

   We assume the density to correspond to a Slater determinant, i.e.

$$\rho^2 = \rho$$

or

$$\left(\rho_o + \sum_i \rho_i v_i\right)^2 = \rho_o + \sum_i \rho_i v_i.$$

This leads to the relations

$$\rho_o^2 + \sum_i \rho_i^2 = \rho_o \tag{65}$$

and

$$\rho_o\rho_i + \rho_i\rho_o = \rho_i. \tag{66}$$

   These equations also imply that the normal density $\rho\left(v = v' = \frac{1}{2}\right)$ satisfies the relation

$$\rho^2 + \rho_x^2 = \rho, \tag{67}$$

where we assumed $\rho_y = 0$.

The equation (64) ensures that the Hartree-Fock-Bogoliubov Hamiltonian

$$H = (t + U_z - \lambda)v_z + U_x v_x \tag{68}$$

can be diagonalized together with the density

$$\rho = \rho_o + \sum_{xz} \rho_i v_i. \tag{69}$$

The common eigenstates satisfy

$$H \, | \, i \rangle = E_i \, | \, i \rangle$$

or

$$\sum_{v'} \int d^3 r' \, \langle \vec{r}v \, | \, H \, | \, \vec{r}'v' \rangle \, \langle \vec{r}'v' \, | \, i \rangle = E_i \, \langle \vec{r}v \, | \, i \rangle. \tag{70}$$

This is the Hartree-Fock-Bogoliubov equation in the coordinate representation for the two-component eigenstates. The ground state is described by the density matrix

$$\rho = \sum_{E_i < o} \langle \vec{r}v \, | \, i \rangle \langle i \, | \, \vec{r}'v' \rangle. \tag{71}$$

For more details cf. Dobaczewski et al., 1984.

The solution of (70) is especially simple if we may assume U_x to be local and a constant, i.e.

$$U_x(\vec{r}, \vec{r}') = \Delta \, \delta(\vec{r} - \vec{r}'). \tag{72}$$

The solution is then

$$\psi_i(\vec{r}, v) = \langle \vec{r}v \, | \, i \rangle = \begin{pmatrix} v_i \\ u_i \end{pmatrix} \varphi_i(\vec{r}), \tag{73}$$

where φ_i is the solution of the Hartree-Fock equation (32) with eigenvalue ε_i. The coefficients u_i and v_i are determined by noting that the Hamiltonian (68) is made diagonal in v by a rotation of θ around the y-axis in quasispin space where

$$\tan \theta = \frac{\Delta}{\varepsilon_i - \lambda}. \tag{74}$$

Since the relation (51) is linear in the density operators, it also holds in the Wigner representation

$$\bar\rho_i(\vec p, \vec q) = \frac{1}{(2\pi\hbar)^3} \int d^3\xi \, e^{-i\vec p \vec\xi/\hbar} \, \hat\rho_i \left(\vec q + \frac{\vec\xi}{2}, q - \frac{\vec\xi}{2}\right). \tag{75}$$

The moments may be written

$$\widehat{M_i} = \int d^3p \, d^3q \, \bar\rho_i(\vec p, \vec q), \tag{76}$$

and we may thus interpret $\bar\rho_i(\vec p, \vec q)$ as a quasispin associated with the point $\vec p, \vec q$ in phase space. The total quasispin (76) is the generator of rotations in the quasispin space. We note that according to the definition (46), the density $\bar\rho_o(\vec p, \vec q)$ is given by (cf. (12))

$$\bar\rho_o(\vec p, \vec q) = \frac{1}{2} \left[\bar\rho(\vec p, \vec q) + \bar\rho(-\vec p, \vec q) - (2\pi\hbar)^{-3}\right]. \tag{77}$$

The expectation value of this quantity is therefore $1/2(2\pi\hbar)^{-3}$ if the volume elements $\vec p, \vec q$ and the time-reversed element $-\vec p, \vec q$ are occupied and are equal to $-1/2(2\pi\hbar)^{-3}$ if they are empty.

In the Thomas-Fermi approximation, the density $\bar\rho(\vec p, \vec q)$ satisfies the rules (65)–(66) with

$$\bar\rho_o(\vec p, \vec q) = \frac{1}{2} \tag{78}$$

and

$$\sum_i \left(\bar\rho_i(\vec p, \vec q)\right)^2 = \frac{1}{4} \tag{79}$$

when the densities are measured in units of $(2\pi\hbar)^{-3}$.

In the intrinsic system where $\rho_y = 0$, we may therefore write

$$\bar\rho_z(\vec p, \vec q) = \frac{1}{2} \cos\theta(\vec p, \vec q) \tag{80}$$

and

$$\bar\rho_x(\vec p, \vec q) = \frac{1}{2} \sin\theta(\vec p, \vec q). \tag{81}$$

Utilizing the commutation relations of the Pauli matrices v_i, we find that the mean-field equation (64) can be written in terms of the two equations

$$[h - \lambda, \bar\rho_z] + [U_x, \bar\rho_x] = 0 \tag{82}$$

485

and

$$\{h - \lambda, \bar{\rho}_x\} + \{U_x, \bar{\rho}_z\} = 0. \tag{83}$$

In the Thomas-Fermi approximation the latter equation means

$$(h - \lambda)\bar{\rho}_x = U_x \bar{\rho}_z \tag{84}$$

implying that

$$\tan \theta(\vec{p}, \vec{q}) = \frac{U_x(\vec{p}, \vec{q})}{h(\vec{p}, \vec{q}) - \lambda}. \tag{85}$$

Equation (82) is equivalent to a Liouville equation which is identically satisfied with the ansatz (80), (81), and (85). The classical solution which satisfies the Hartree-Fock-Bogoliubov equations (64) is therefore

$$\bar{\rho}_x(\vec{p}, \vec{q}) = \frac{U_x(\vec{p}, \vec{q})}{\sqrt{U_x^2 + (h - \lambda)^2}} \tag{86}$$

$$\bar{\rho}_z(\vec{p}, \vec{q}) = \frac{\lambda - h(\vec{p}, \vec{q})}{\sqrt{U_x^2 + (h - \lambda)^2}}, \tag{87}$$

implying that the normal density is

$$\bar{\rho}\left(\vec{p}, \vec{q}, v = v' = \frac{1}{2}\right) = \frac{1}{2}\left(1 + \frac{\lambda - h(\vec{p}, \vec{q})}{\sqrt{U_x^2 + (h - \lambda)^2}}\right) \tag{88}$$

This result was derived by Bengtsson and Schuck, 1980.

REFERENCES

Abramowitz, M., and Stegun, I. A. (1966) *Handbook of Mathematical Functions, Applied Mathematics Series 55,* (Washington: National Bureau of Standards).

Akyüz, Ö, and Winther, A. (1981) in *Proceedings of the Enrico Fermi International School of Physics, 1979,* course on "Nuclear Structure and Heavy Ion Reactions," (ed. R. A. Broglia, C. H. Dasso and R. Ricci) (Amsterdam: North Holland).

Alder, K., and Winther, A. (1960) Mat. Fys. Medd. Dan. Vid. Selsk. 32, no 8; reprinted in K. Alder and A. Winther, Coulomb Excitation (New York: Academic Press, 1966).

Alder, K., and Winther, A. (1975) *Electromagnetic Excitation,* (Amsterdam: North Holland), 277.

Anantaraman, N. (1973) *Phys. Rev. C 8,* 2245.

Anderson, P. W., and Rowell, J. M. (1963) *Phys. Rev. Lett. 10,* 230.

Arnould, M., and Howard, W. M. (1976) *Nucl. Phys. A274,* 295.

Austern, N., and Blair, J. S. (1965) *Ann. Phys. 33,* 15.

Baeza, A., Bilwes, B., Bilwes, R., Diaz, J., and Ferrero, J. L. (1984) *Nucl. Phys. A419,* 412.

Baldo, M., Rapisarda, A., Broglia, R. A., and Winther, A. (1987) *Nucl. Phys. A472,* 333.

Baltz, A. J., Bond, P. D., Garrett, J. D., and Kahana, S. (1975) *Phys. Rev. C 12,* 136.

Baltz, A. J., Kauffmann, S. K., Glendenning, N. K., and Preuss, K. (1978) *Phys. Rev. Lett. 40,* 20.

Bang, D. M., Gareev, F. G., Pinkston, W .T., and Vaagen, J. S. (1985) *Phys. Rev. 125,* 253.

Bayman, N. and Lande, A. (1966) *Nucl. Phys. 77,* 1.

Bengtsson, R., and Schuck, P. (1980) *Phys. Lett. 89B,* 321.

Berry, M. V., and Mount, K. E. (1972) *Rep. Prog. Phys. 35,* 315.

Bertrand, F. (1976) *Ann. Rev. Nucl. Sci. 26,* 457.

Bertsch, G. F., and Schaeffer, R. (1977) *Nucl. Phys. A277,* 509.

Bertsch, G. F., and Tsai, S. F. (1975) *Phys. Rep. 18C,* 125.

Bes, D. R., Broglia, R. A., and Nilsson, B. S. (1975) *Phys. Rep. 16C,* 1.

Beyer, K., and Winther, A. (1969) *Phys. Lett. 30B,* 296.

Beyer, K., Winther, A., and Smilansky, U. (1970) *Proceedings of the International Conference on Nuclear Reactions Induced by Heavy Ions, Heidelberg* (North Holland).

Bjerregaard, J. H., Hansen, O., Nathan, O., Vistisen, L., Chapman, R., and Hinds, S. (1968) *Nucl. Phys. A110*, 1.

Bjerregaard, J. H., Hansen, O., Nathan, O., Chapman, R., and Hinds, S. (1968) *Nucl. Phys. A131*, 481.

Blocki, J., Randrup, J., Swiatecki, W. H., and Tsang, C. F. (1977) *Ann. of Phys. 105*, 427.

Bohlen, H. G., and von Oertzen, W. (1971) *Phys. Lett. B37*, 451.

Bohr, A., and Mottelson, B. R. (1969) *Nuclear Structure*, Vol. I, (Reading, Mass.: Benjamin).

Bohr, A., and Mottelson, B. R. (1975) *Nuclear Structure*, Vol. II, (Reading, Mass.: Benjamin)

Bohr, A., and Mottelson, B. R. (1974) *Fysisk Tidsskrift 72*, No. 2, 58.

Bohr, A., and Mottelson, B. R. (1974a) *Physica Scripta 10A*, 13.

Bohr, A., and Mottelson, B. R. (1975) *Nuclear Structure* Vol. II (Reading, Mass.: Benjamin, Advanced Book Program).

Bonaccorso, A., Brink, D. M., and LoMonaco, L. (1987) *J. Phys. G 13*, 1407.

Bond, P. D., Barrette, J., Baktash, C., Thorn, C. E., and Kreiner, A. J. (1981) *Phys. Rev. Lett. 46*, 1565.

Bond, P. D., Garret, J. D., Hansen, O., Kahana, S., Le Vine, M. J., and Schwartzchild, A. Z. (1973) *Phys. Lett. 47B*, 231.

Bortignon, P. F., Broglia, R. A., Bes, D. R., and Liotta, R. (1977) *Phys. Rep. 30C*, 305.

Breit, G., Chun, K. W., and Wahsweiler, H. G. (1964) *Phys. Rev. 133*, B403.

Breit, G., and Ebel, M. E. (1965) *Phys. Rev. 103*, 679.

Brink, D. M., and Takigawa, N. (1977) *Nucl. Phys. A279*, 159.

Brink, D. M. (1972) *Phys. Lett. 40B*, 37.

Broglia, R. A., Hansen, O., and Riedel, C. (1973) *Adv. in Nucl. Phys., 6*, 287.

Broglia, R. A., Liotta, R., Nilsson, B. S., and Winther, A. (1977) *Phys. Rep. 29C*, 291.

Broglia, R. A., Götz, U., Ichimura, M., Kammuri, T., and Winther, A. (1973) *Phys. Lett. 45B*, 23.

Broglia, R. A., Pollarolo, G., and Winther, A. (1981) *Nucl. Phys. A361*, 307.

Broglia, R. A., and Winther, A. (1972) *Nucl. Phys. A182*, 112.

Broglia, R. A., and Winther, A. (1972) *Phys. Rep.4C*, 153.

Broglia, R. A., Landowne, S., and Winther, A. (1972) *Phys. Lett. 40B*, 293.

Broglia, R. A., Landowne, A., Malfliet, R. A., Rostokin, V., and Winther, A., (1974) *Phys. Rep. 11*, 1.

Broglia, R. A., Dasso, C. H., Pollarolo, G., and Winther, A. (1978) *Phys. Rep. 48C*, 351.

Buttle, P., J. A., and Goldfarb, L. J. B. (1966) *Nucl. Phys.78*, 409.

Buttle, P., J. A., and Goldfarb, L., J. B., *Nucl. Phys. A176* (1971) 299

Cheung, H. C., High, M. D., and Cujec, B. (1978) *Nucl. Phys. A296*, 333.

Christensen, P. R., and Winther, A., (1976) *Phys. Lett. 65B*, 19.

Cohen, S., Plasil, F., and Swiatecki, W. J. (1974) *Ann. of Phys. 82*, 557.

da Silveira, R. (1973) *Phys. Lett. 45B*, 211.

Dasso, C. H., and Pollarolo, G. (1985) *Phys. Lett. 155B*, 223.

Dasso, C. H., Landowne, S., Pollarolo, G. and Winther,A. (1986) *Nucl. Phys. A459*, 134.

Davies, K. T. R., and Nix, J. R. (1976) *Phys. Rev. C14*, 1977.

Delos, J. B. (1981) *Rev. Mod. Phys. 53*, 287.

De Vries, R. M., and Clover, M. R. (1975) *Nucl. Phys. A243*, 529.

Dietrich, K., and Hara, K. (1973) *Nucl. Phys. A211*, 349.

Dietrich, K., Hara, K., and Weller, F. (1971) *Phys. Lett. 35B*, 349.

Dobaczewski, J., Flocard, H., and Treiner, J. (1984) *Nucl. Phys. A422*, 103.

Dover, C. B., and Vary, J. P., *Proceedings of the Symposium on Heavy Ion Collisions, Heidelberg 1974* (Berlin: Springer), 1.

Døssing, T., Frauendorf, S., and Schultz, H. (1977) *Nucl. Phys. A287*, 137.

Eisen, Y., and Vager, Z. (1972) *Nucl. Phys. A187*, 219.

Erdelyi, A. (1954) *Tables of Integral Transforms*, (New York: McGraw-Hill).

Esbensen, H., Broglia, R. A., and Winther, A. (1983) *Ann. of Phys. 146*, 149.

Ferreira, L., Liotta, R., Dasso, C. H., Broglia, R. A., and Winther, A. (1983) *Nucl. Phys. A426*, 276.

Ferreira, L. S., Liotta, R. J., Winther, A., and Dasso, C. H. (1988) *Nucl. Phys. A480*, 62.

Ford, K. W. and Wheeler, J. A. (1959) *Ann. Phys. (N.Y.) 7*, 259.

Frahn, W. E. (1966) *Nucl. Phys. 75*, 577.

References

Fröman, N., and Fröman, P. O. (1965) *JWKB approximation* (Amsterdam: North Holland).

Gelbke. C. K., Bock, R., Braun-Munzinger, P., Fick, D., Hildebrand, K. D., Weiss, W., and Richter, A. (1972) *Phys. Rev. Lett. 29*, 1683.

Glendenning, N. K., and Wolschin,G. (1977) *Nucl. Phys.A281*, 486.

Goldberg, D. A., and Smith, S. M. (1972) *Phys. Rev. Lett. 29*, 500.

Goldberg, D. A., Smith, S. M., and Burdzik, G. F. (1974), *Phys. Rev. C10*, 1367.

Goldstein, H. (1951) *Classical Mechanics,* (Reading, Mass.: Addison-Wesley).

Götz, U., Ichimura, M., Broglia, R. A., and Winther, A. (1975) *Phys. Rep.16C*, 115.

Hasan, H., and Brink, D. M. (1978) *J.Phys. G4*, 1573.

Heading, J. (1962) *An Introduction to Phase Integral Methods,* (London: Methuen).

Hillis, D. L., Gross, E. E., Hensley, D. C., Rickertson, L. D., Bingham, C. R., Baker, F. T., and Scott, A., (1977), *Phys. Rev. C16*, 1467.

Hodgson, P. E. (1971) *Nuclear Reactions and Nuclear Structure,* (Oxford: Clarendon Press).

Igo, G. (1958) *Phys. Rev. Lett. 1*, 72.

Igo, G. (1959) *Phys. Rev. Lett. 3*, 308.

Josephson, B. D. (1962) *Phys. Lett.1*, 251.

Kahana, S., and Baltz, A. J. (1977) *Adv. in Nucl. Phys. 9*, 1.

Knoll, J. and Schaeffer, R. (1975) in *Extended Seminar on Nuclear Physics,* I.C.T.P. Trieste, September 1973, I.A.E.A. Vienna, 1975.

Knoll, J., and Schaeffer, R. (1976) *Ann. of Phys.97*, 307.

Knoll, J., and Schaeffer, R. (1977) *Phys. Rep. 31C*, 160.

Kovar, D. G., Henning, W., Zeidman, B., Eisen, Y., Erskine, J. R., Fortune, H. T., Ophel, T. R., Sperr, P., and Vigdor, S. E. (1978) *Phys. Rev. C17*, 83.

Krappe, H. J., and Nix, J. R. (1974) *Proceedings of Third IAEA Symposium on Physics and Chemistry of Fission, Rochester, 1973,* Vol. I, (Vienna: IAEA).

Landowne, S., Dasso, C. H., Nilsson, B. S., Broglia, R. A., and Winther, A. (1976) *Nucl. Phys. A259*, 99.

Lee, I. Y., Cline, D., Butler, P. A., Diamond, R. M., Newton, J. O., Simon, R. S., and Stephens, F. S. (1977) *Phys. Rev. Lett. 39*, 684.

Levit, S., Smilansky, U., and Pelte, D. (1974) *Phys. Lett. 53B*, 39.

Lilley, J. S. (1984) unpublished.

Lilley, J. S., Fulton, B. R., Nagarajan, M. A., Thompson, I. J., and Banes, D. W. (1985) *Phys. Lett. 151B*, 181.

LoMonaco, L., and Brink, D. M. (1985) *J. Phys. G11*, 935.

Love, W. G. (1977) *Proceedings of the Symposium on Heavy Ion Elastic Scattering, Rochester*, p. 266.

Love, W. G., Terasawa, T., and Satchler, G. R. (1977) *Phys. Rev. Lett. 39*, 6.

Lotti, P., Vitturi, A., Broglia, R. A., and Winther, A. (1990) (to be published).

Maglione, E., Pollarolo, G., Vitturi, A., Broglia, R. A., and Winther, A. (1985) *Phys. Lett. 162B*, 59.

Maglione, E., Pollarolo, G., Vitturi, A., Broglia, R. A., and Winther, A. (1987) *Phys. Lett. 191B*, 237.

Mahaux, C., Ngô, H., and Satchler, G. R. (1986) *Nucl. Phys. A449*, 354.

Malfliet, R., Landowne, S., and Rostokin, V. (1973) *Phys. Lett. 44B*, 238.

Massmann, H., and Rasmussen, J. O. (1975) *Nucl. Phys. A243*, 155.

Miller, W. H. (1974) *Advan. Chem. Phys. 25*, 69.

Moretto, L. G., and Schmitt, R. (1976) *Journal de Physique* C5, *37*, 109.

Morse, P., and Feshbach, H. (1953) *Methods of Mathematical Physics*, (New York: McGraw-Hill).

Mott, N. F., and Massey, H. S. (1949) *The Theory of Atomic Collisions*, (Oxford: Oxford University Press).

Myers, W. D. (1970) *Nucl. Phys. A145*, 387; *A204* (1973) 465.

Nagarajan, M. A., Mahaux, C., and Satchler, G. R. (1985) *Phys. Rev. Lett. 54*, 1136.

Nilsson, B. S. (1977) DIWRI (unpublished).

Pieper, S. C., Macfarlane, M. H., Gloeckner, D. H., Kovar, D.G., Becchetti, F. D., Harvey, B. G., Hendrie, D. L., Homeyer, H., Mahoney, J., Pühlhofer, F., von Oertzen, W., and Zisman, M. S. (1978) *Phys. Rev. C18*, 180.

Poling, J. E., Norbeck, E., and Carlson, R. R. (1976) *Phys. Rev. C13*, 648.

Pollarolo, G., Broglia, R. A., and Winther, A. (1983) *Nucl. Phys. A406*, 369.

Quesada, J. M., Pollarolo, G., Broglia, R. A., and Winther, A. (1985) *Nucl. Phys. A442*, 381.

Quesada, J. M. (1985) Thesis, University of Seville (unpublished).

Randrup, J. (1975) Lawrence Berkeley Laboratory report LBL-4317.

Rawitscher, G. H. (1966) *Nucl. Phys. 83*, 259.

Rehm, K. E., Körner, H. J., Richter, M., Rother, H. P., Schiffer, J. P., and Spieler, H. (1975) *Phys. Rev. C12*, 1945.

Roberts, A. (1972) *Nucl. Phys. A196*, 465.

Robson, D. (1973) *Phys. Rev. C7*, 1.

Satchler, G. R. (1983) *Direct Nuclear Reactions*, (Oxford: Clarendon).

Satchler, G. R. (1979) *Nucl. Phys. A329*, 233.

Schaeffer, R. (1978) *Nuclear Physics with Heavy Ions and Resonances, Les Houches Summer School, Session XXX*, (ed. R. Balian, M. Rho, and G. Ripka) (Amsterdam: North Holland), 69.

Schaeffer, R. (1978) Proceedings of Les Houches, École d'Éte de Physique Theorique, Session XXX, on Nuclear Physics with Heavy Ions and Mesons, Eds. R. Balian, M. Rho, and G. Ripka, North Holland, Amsterdam, Vol. I, p. 69.

Schröder, W. U., Birkelund, J. R., Huizenga, J. R., Wolf, K. L., and Viola, V. E. (1978) *Phys. Rep. 45*, 301.

Schwartzchild, A. Z., Auerbach, E. H., Fuller, R. C., and Kahana, S. (1976) *Symposium on Macroscopic Features of Heavy Ion Collisions, Argonne, 1976*, (ANL/PHY-76-2), Vol. II, p. 753.

Steadman, S. G., Belote, T. A., Grodzins, L., Cline, D., Voigt, J. A., and Videbaeck, F., (1974) *Phys. Rev. Lett. 33*, 499.

Switkowski, Z. E., Wu, Shiu-Chin, Overley, J. C., and Barnes, C. A. (1977) *Nucl. Phys. A289*, 236.

Sørensen, J. H. (1988) Thesis, University of Copenhagen (unpublished).

Sørensen, J. H., Quesada, J. M., Broglia, R. A., and Winther, A. (1988) (to be published).

Tohyama, M. (1972) *Phys. Lett. 38B*, 147.

Traber, A., Trautman, D., and Rösel, F. (1977) *Nucl. Phys. A291*, 221.

Trautman, D., and Alder, K. (1970) *Helv. Phys. Acta 43*, 363.

Videbaek, F., Christensen, P. R., Hansen, O., and Ulbak, K. (1976) *Nucl. Phys. A256*, 301.

Videbaeck, F., Goldstein, R. B., Grodzins, L., Steadman, S. G., Belote, T. A., and Garrett, J. D. (1977) *Phys. Rev. C15*, 954.

Vigezzi, E., and Winther, A., (1989) *Ann. of Phys. 192*, 432.

Voit, H., Bischof, N., Tiereth, W., Weitzenfelder, I., von Oertzen, W., and Imanishi, B. (1988) *Nucl. Phys. A476*, 491.

von Oertzen, W., Brown, R. E., Flynn, E. R., Peng, S. C., and Sunier, J. W. (1983) *Z. Physik A313*, 371.

Wilczynski, J. (1973) *Nucl. Phys. A216*, 386.

Winther, A., and Alder, K. (1979) *Nucl. Phys. A319*, 518.

INDEX

Absorption,
due to Coulomb excitation, 87–91
Absorptive potential, 120, 287,
363–364, 368, 371
dependence on impact parameters,
362
due to inelastic scattering, 364
and recoil, 365
due to transfer, 362, 364
in oscillator model, 248–258
selfconsistent equation, 256
Acceleration at turning point, 253,
275
Acceleration field,
and neutron transfer, 338
and proton transfer, 338
at distance of closest approach,
364
Action integral, 155, 163
and transfer, 349
Adiabatic cutoff, 349, 356–358, 369
and two-nucleon transfer, 427
in Coulomb excitation, 52–53
limit, 468
Adiabaticity,
cut off function, 352
parameter, 349
for inelastic scattering, 196
Adiabatic description, 470
Adjoint amplitude, 300, 304
and phase shift, 398
Alignment, 361
and absorption, 370
in Coulomb excitation, 52
Amplitude,
adjoint, 304, 309
first order amplitude, 302
of surface mode in Coulomb
excitation, 67
second order perturbation, 307
zeroth order, 303
Analogy between,
pairing and surface modes, 430
Angular distribution,
interference between Coulomb and
nuclear branch, 395–396
one-particle transfer, 378, 379,
383–385

Angular distribution (Cont'd.)
orbiting branch, 395–396
proton transfer, 394–395
two-nucleon transfer, 425, 462
Angular momentum,
and transverse recoil, 331
classical interpretation, 326
coupling in intrinsic frame
(Z-function), 376
coupling in laboratory frame
(Clebsch-Gordon), 376
coupling scheme, 321
transferred, 321
Antisymmetricized,
transfer matrix element, 301
wavefunction, 312–313
Antisymmetrization, 301, 311,
317, 454
Apex line, 176
Asymptotic expression,
for Clebsch-Gordon coefficients,
268
for spherical harmonies, 269
Atomic collisions, 468
Atomic physics, 469
Average energy loss by Coulomb
excitation, 66
Average trajectory,
and accuracy WKB approximation, 381
angular momentum, 382
complex turning point, 381

Backward rise,
elastic scattering, 367
Barrier penetration, 151
BCS,
wavefunction, 419
occupation probabilities, 419
B-coefficients,
member pairing rotational band,
421
pairing vibrations, 418
two-particle transfer amplitudes,
421
two-quasi particle excitations, 421
$B(E\lambda)$, 60

Index

WKB,
  accuracy, 381, 386
  approximation, 144–158
  radial wavefunction, 378

Yukawa folded density, 107
Yukawa folding, 106, 171–174

Z-coefficient, 376, 382
Zero point amplitude,
  pair addition, 418
  pair removal, 418
Zero point fluctuations,
  of surface-surface distance, 250
  renormalization of ion-ion poten-
    tial, 220
Zeroth-order,
  amplitude, 303

The Addison-Wesley **Advanced Book Program** would like to offer you the opportunity to learn about our new physics and scientific computing titles in advance. To be placed on our mailing list and receive pre-publication notices and special offers, just **fill out this card completely** and return to us, postage paid. Thank you.

Title and Author of this book: **Date purchased:**

_____   _____

Name _____

Title _____

School/Company _____

Department _____

Street Address _____

City _____ State _____ Zip _____

Telephone/s(   ) _____ (   ) _____

Where did you buy/obtain this book?

☐ Bookstore   ☐ Mail Order   ☐ School (Required for Class)
☐ Campus Bookstore   ☐ Toll Free # to Publisher   ☐ Professional Meeting
☐ Other _____ ☐ Publisher's Representative

What professional scientific and engineering associations are you an active member of?

☐ AAPT (Amer Assoc of Physics Teachers)   ☐ APS (Amer Physical Society)   ☐ SPS (Society of Physics Students)
☐ AIP (Amer Institute of Physics)   ☐ Sigma Pi Sigma   ☐ AAAS (Amer Assoc for the Advancement of Science)
☐ Other _____

Check your areas of interest.

(10)   ✔**Physics**

11 ☐ Quantum Mechanics	18 ☐ Materials Science	25 ☐ Geophysics
12 ☐ Particle/Astro Physics	19 ☐ Biological Physics	26 ☐ Medical Physics
13 ☐ Condensed Matter	20 ☐ High Polymer Physics	27 ☐ Optics
14 ☐ Mathematical Physics	21 ☐ Chemical Physics	28 ☐ Vacuum Physics
15 ☐ Nuclear Physics	22 ☐ Fluid Dynamics	
16 ☐ Electron/Atomic Physics	23 ☐ History of Physics	
17 ☐ Plasma/Fusion Physics	24 ☐ Statistical Physics	

29 ☐ Other _____

Are you more interested in: ☐ theory ☐ experimentation?

Are you currently writing, or planning to write a textbook, research monograph, reference work, or create software in any of the above areas?
      ☐ Yes   ☐ No
      Area: _____

(If Yes) **Are you interested in discussing your project with us?**
      ☐ Yes   ☐ No

Physics

fold and staple

BUSINESS REPLY MAIL
FIRST CLASS PERMIT NO. 828   REDWOOD CITY, CA   94065

Postage will be paid by Addressee:

**ADDISON-WESLEY
PUBLISHING COMPANY, INC.**®

Advanced Book Program
350 Bridge Parkway, Suite 209
Redwood City, CA 94065-1522